civics+ citizenship

history

geography

economics+ business

contents

Good Humanities 7

Victorian Curriculum

1st edition

Ben Lawless

Danielle O'Leary

Peter van Noorden

Publisher: Catherine Charles-Brown
Publishing director: Olive McRae
Copy editors: Kate McGregor, Philip Bryan
Project editors: Kate McGregor, Lin Li Ng
Aboriginal and Torres Strait Island content consultant: Aleryk Fricker
Proofreaders: Kelly Robinson, Carly Slater
Indexer: Max McMaster
Text and cover designer: Jo Groud
Typesetters: Paul Ryan, Kerry Cooke
Illustrator: QBS Learning
Cartography: MAPgraphics
Map reviewer: Katrina Spencer
Production controllers: Nicole Ackland, Sarah Blake
Digital production: Erin Dowling
Permissions researchers: Samantha Russell-Tulip, Hannah Tatton
Cover images photographed by:
Geography: Alessandra Meniconzi
History: James Henry
History cover model: Lockeah Wapau

First published in 2021 by Matilda Education Australia, an imprint of Meanwhile Education Pty Ltd
Melbourne, Australia
T: 1300 277 235
E: customersupport@matildaed.com.au
www.matildaeducation.com.au

National Library of Australia Cataloguing-in-Publication data
Author: Ben Lawless, Danielle O'Leary, Peter van Noorden
Title: Good Humanities 7 Victorian Curriculum
ISBN: 9781420247169

A catalogue record for this book is available from the National Library of Australia at www.nla.gov.au

Warning: It is recommended that Aboriginal and Torres Strait Islander peoples exercise caution when viewing this publication as it may contain names or images of deceased persons.

Printed in Malaysia by Vivar Printing
Jun-2025

Introduction to Geography

G0

WHY DOES GEOGRAPHY MATTER?

page 2

geography skills

page 4

HOW DO I STUDY GEOGRAPHY?

learning ladder

page 8

HOW DO I USE THIS BOOK?

Why does Geography matter?

Geography is the study of our world and its characteristics, patterns and changes over time. In Geography, we focus on two main ideas: human activity and natural processes. We are interested in how humans influence the *space* in which they live and, in return, how the natural *environment* influences how we as humans live. We use maps, images, graphs and other sources of data to explore patterns, and conduct fieldwork to answer questions. Geography is key to understanding the world around us, because *without Geography, you are nowhere!*

Thinking like a geographer

As a Geography student, you are a trainee geographer! Geography is not just about reading and memorising; geographers *learn by doing*, using their geography skills to help them along the way. You have most likely already been practising many geographic skills in your day-to-day life.

If you like travelling or the outdoors, Geography helps you find your way and explain processes. If you like sport, you may already know a range of countries because of the team names. You may have experience using a map to find a friend's house or using social media to locate the best restaurant near your house.

Source 1

As a Geography student, you are a trainee geographer!

Three reasons why Geography matters

Exploring Geography will help you develop powerful skills

Geography is an active and vibrant study where you will *learn by doing*. To help you complete your learning, you will develop a new set of skills. These skills will remain with you and will help you in all aspects of your life. Some of these valuable new skills are:

- reading maps
- interpreting data
- creating visual representations
- navigating
- understanding how other people experience the places they live.

Exploring Geography lets you see the world (sometimes without leaving your desk!)

Do you want to visit the Amazon jungle? Are you concerned about coral bleaching on the Great Barrier Reef? Do you wonder where you would end up if you travelled 17 525 kilometres north of your exact location right now?

The study of Geography allows you to access every corner of the world: either on a fieldtrip, or at your desk using geographic information systems such as Esri or Google Earth. Through this process you will come to understand the *interconnected* nature of our world and the role we all play as custodians of our planet.

Exploring Geography makes us informed global citizens

Studying Geography helps us to understand ourselves and our planet. Geography has two main focuses:

- physical or natural geography
- human or built geography.

Physical geography is the study of the Earth's landscapes, landforms and natural processes, some which have been occurring for hundreds of thousands of years. Human geography is the study of how we interact with this natural world.

You might investigate how we change environments; how we use the environment; how different cultures, languages or religions are spread around the world; or even how we break up the world into land masses called continents and countries with political borders. Most critically, the study of Geography will empower you to be an active global citizen who makes informed decisions.

Learning Ladder G0.1

1. Why are skills so important in Geography?
2. What do you already know about Geography? Brainstorm a word or phrase that relates to Geography for every letter of the alphabet. (Hint: Your 'x' word could be *x*-axis on a graph!)
3. Geography is an 'umbrella' study that brings in many other subjects, such as Maths, Science, History, Civics and citizenship, and Economics and business. List five things that you enjoy learning about and explain how they link to Geography.
4. Look at the image of Huangguoshu Waterfall National Park on pages 4–5 of this chapter. Identify one example of human activity and one natural process that you can see in this image.
5. As a class, discuss how we all use Geography in everyday life. Think about apps, books or even plans you might make.

How do I study Geography?

In Year 7 Geography, you will focus on five geographical skills:

1 spatial characteristics
2 interconnections
3 geographical challenge
4 collect, record and display data
5 analyse data.

You will also practise the important skills of research and fieldwork. On the following pages you will be introduced to geographic concepts that help explain spatial characteristics and different types of maps that can be used to display data. The Geo How-To section on pages 151–167 will support you step by step when you begin to use the other geography skills in your work.

Geographic concepts

Geographic concepts will help you to think like a geographer. The acronym **SPICESS** can be used as a prompt to help you remember the seven geographic concepts:

- **S**pace
- **P**lace
- **I**nterconnection
- **C**hange
- **E**nvironment
- **S**cale
- **S**ustainability.

S Space

In a geographical context, *space* does not refer to *outer space*. Instead, it refers to the way that we use, *change* and distribute things on the Earth's surface. For example, this *space* is classified as a National Park. While tourists are allowed to visit this *space*, they must be respectful of the *environment* and remain on the designated pathways.

P Place

The concept of *place* allows humans to identify the location or position of something within a *space*. We can identify place through describing the relative or absolute location of that area. This *place* is called Huangguoshu Waterfall and it is located in China.

I Interconnection

Interconnection is the idea that two things or phenomena are related, interact or are linked in some way. For example, there is an *interconnection* between the abundance of trees and the movement of water through this environment. Without this water source, the trees would not be able to survive in such quantities.

C Change

Change refers to how a *place* is altered due to shifts in the environment or to meet the needs of humans. Change can be positive or negative, and can occur over short or long term periods of time. In this *place*, humans have *changed* the *environment* by adding boardwalks, handrails and access points so that tourists can view the waterfall.

E Environment

An *environment* can be defined by its **geographic characteristics**. Some environments are largely natural and are untouched by humans, such as coastlines, islands and forests. Other environments have undergone significant change and are largely unnatural, such as cities and other urban areas. Within environments we can observe processes, interconnections between **phenomena** and change over time. This image is an example of tourists visiting a natural environment.

S Scale

Scale usually refers to the size of something. Scale can be literal, such as a scale on a map using data to show you how big something is in real life. It can also be used as a word when describing a region. For example, patterns can exist on a local, regional, national or global *scale*. This image is an example of tourism on a local scale.

S Sustainability

Sustainability is the concept of maintaining and preserving resources and environments for future generations. This could be through the use of sustainable, **renewable energy** such as hydropower or wind-generated electricity. In this image, the boardwalk has been added so tourists do not damage the environment, so it can be sustained for the pleasure of future visitors.

Source 1

Huangguoshu Waterfall National Park in Guizhou province, China. This is one of the largest waterfalls in China and East Asia.

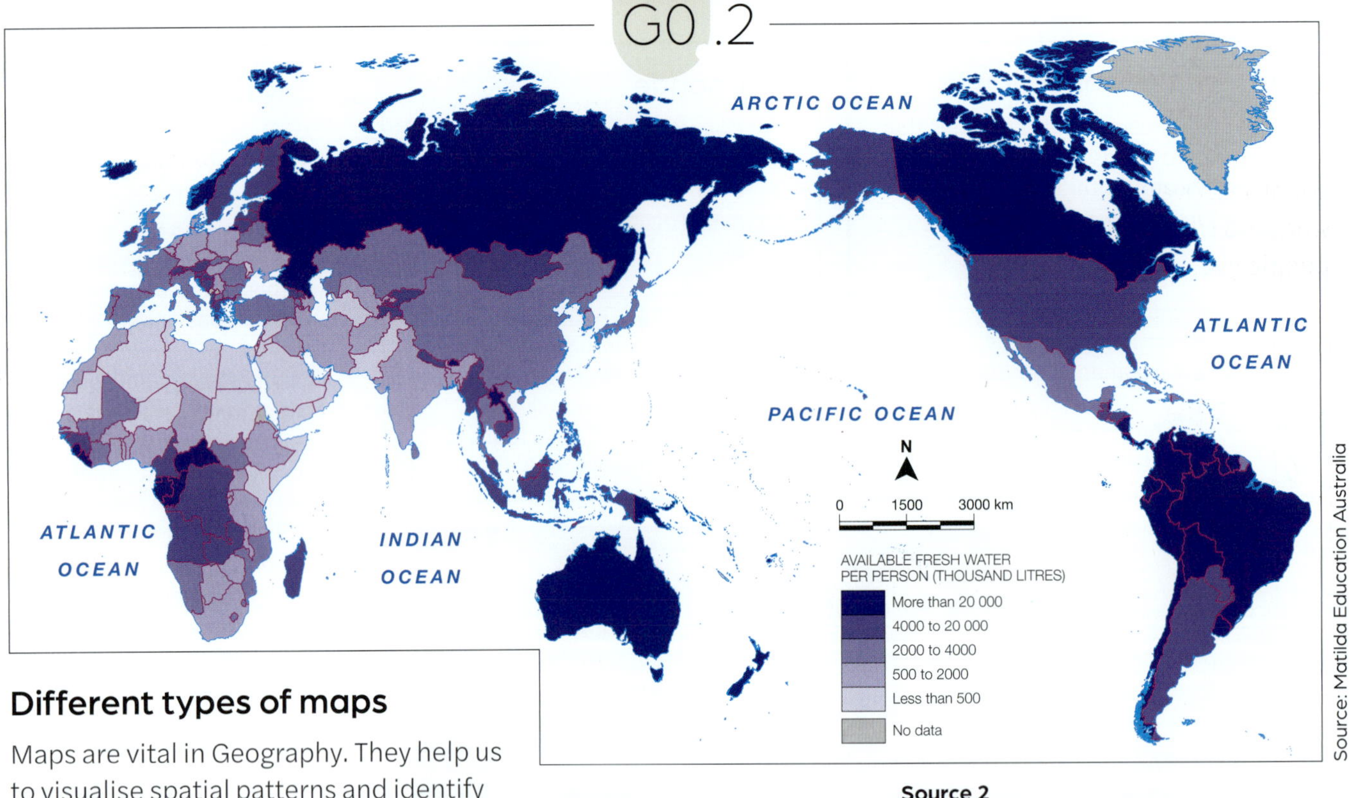

Source 2

A choropleth map showing predicted water availability per capita on a global scale in 2025

Different types of maps

Maps are vital in Geography. They help us to visualise spatial patterns and identify **interconnections**. There are many different types of maps, which all have slightly different uses. The Geo How-To section on pages 151–167 will be a handy reference as you begin to work with maps. Make sure you refer to it frequently.

Choropleth maps

Choropleth maps are among the most common maps used to learn Geography.

Choropleth maps use dark and light shades of colour so that the reader can instantly see a pattern. The choropleth map in Source 2 uses darker shades to represent the areas with the highest levels of available water and lighter shades to represent the areas with the lowest levels of available water. It is best to use the same colour in different shades, to clearly show the pattern, rather than different colours. Can you identify how much water will be available to people in Australia in 2025?

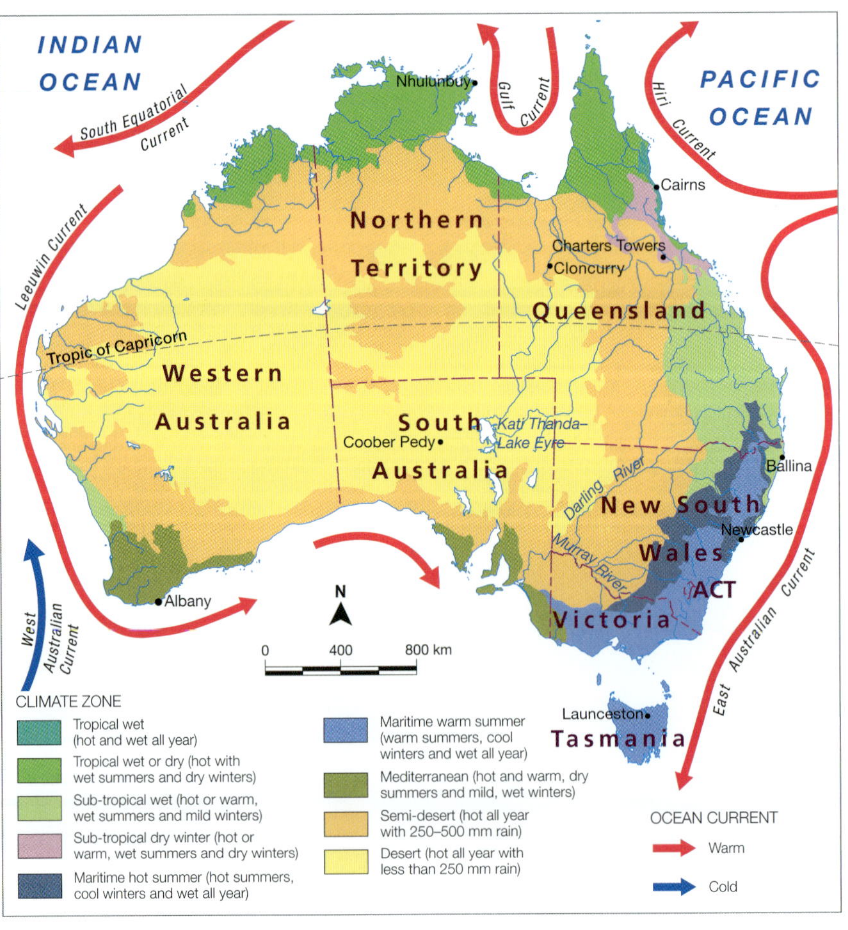

Source 3

A map of the Australian climate

Climate map

In Geography, **climate** is important for determining land cover and how liveable a place is. While weather is what we observe day to day, climate refers to the average conditions (usually temperature and **precipitation**) within a region over a period of time. Climate is important when studying water, as it gives us an indication of expected rainfall.

Political maps

Political maps show the boundaries of different countries. Over time, these boundaries change as wars and treaties have determined where different people live.

Source 4

An Asia political map showing country borders

Topographic maps

Topographic maps show how mountainous a region is. Topographic maps have a series of wiggly lines with numbers along them. These are known as **contour lines**. The closer together the contour lines, the steeper the slope is. The numbers indicate how far above sea level that particular point is. On this topographic map, we can see steep slopes on either side of the rivers. This indicates that the rivers are located in valleys.

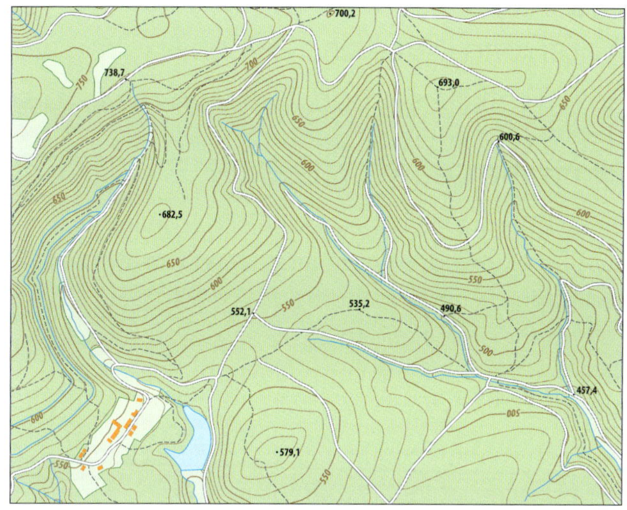

Source 5

An extract of a topographic map showing a river valleys

Learning Ladder G0.2

1 How are maps used in Geography?

2 Why do geographers present data on different kinds of maps?

3 Download an outline map of Australia that shows the states and territories. Survey your classmates to see how many people have travelled to each Australian state or territory. As a class, review your results and create a suitable legend. It may look something like this:

	Number of people in my class who have travelled to a state or territory
	0–2 people have travelled here
	3–5 people have travelled here
	6–8 people have travelled here
	9–11 people have travelled here
	12 or more people have travelled here

Using coloured pencils, illustrate your findings as a choropleth map.

4 Research online to find a few examples of topographic maps. In pairs, annotate the following features on one of the maps:
- the steepest terrain shown in the map
- the location with the highest altitude
- geographical features such as lakes, valleys, volcanoes or ridges.

5 Look at Source 3 – the Australian climate map.
   a In small groups, discuss how this map may be useful in determining which regions in Australia would be most 'liveable' (refer to page 86).
   b What other information could we gain from these types of maps? Make a list to share with the rest of your class.

Mapping with BOLTSS, page 152
Mapping skills, page 154

How do I use this book?

Good Humanities has been built to help you thrive as you move through the Level 7 Humanities curriculum and to enable you to demonstrate your progress in every single lesson. The Geography section of this book explores two geographical topics: Water, and Place and Liveability. You will also find a Fieldwork section and a Geo How-To skills section. The Geo How-To section is vital and you should refer to it often.

Climb the Learning Ladder

Geography is a skills-based subject, so learning content alone does not mean that you have geographical understanding. In order to be a geographer, you need to be able to interpret spatial data and conduct and communicate your own research.

Each chapter begins with a Learning Ladder. The Learning Ladder is your 'plan of attack' for the skills you will practise in each chapter. It lists the five geographical skills you will be learning, and has five levels of progression for each of those skills.

Each skill described in the Learning Ladder is of a higher difficulty than the one below it. To be able to achieve the higher-level skills, you need to be able to master the lower ones. Practising activities at the earlier levels will help you develop the skills to complete more involved tasks, such as evaluating. This approach is called 'developmental learning' and it puts you in charge of your own learning progression!

Read the ladder from the bottom to the top. As you progress through the chapter, you will climb up the Learning Ladder.

learning ladder

Source 1 The Learning Ladder helps you to take charge of your own learning!

Step	Spatial characteristics	Interconnections	Geographical challenge	Record, collect and display data	Analyse data
step 5	I can analyse the impact of change on places	I can explore spatial association and interconnections	I can plan action to tackle a geographical challenge	I can evaluate data	I can draw conclusions by analysing collected data
step 4	I can predict changes in the characteristics of places	I can evaluate the implications of significant interconnections	I can evaluate alternatives for a geographical challenge	I can use data to support claims	I can analyse relationships between different data
step 3	I can explain processes influencing places	I can identify and explain the implications of interconnections	I can compare strategies for a geographical challenge	I can choose, collect and display appropriate data	I can explain the reasons behind a trend or spatial distribution
step 2	I can explain spatial characteristics	I can explain interconnections	I can compare responses to a geographical challenge	I can recognise and use different types of data	I can describe patterns and trends
step 1	I can identify and describe spatial characteristics	I can identify and describe interconnections	I can identify responses to a geographical challenge	I can record, collect and display data in simple forms	I can use geographic terminology to interpret data

Check your progress

Each chapter is divided into about 12 sections. Each section is designed to cover one lesson, but sometimes your teacher might decide to spend more time (or less time) on a particular section. A section is usually two pages long – but some are four pages long.

At the end of every section, you will find a block of questions called 'Learning Ladder', which has two different types of questions or activities:

1. **Show what you know:**

   These questions ask you to look back at the content you have read and viewed, and to show your understanding of it by listing, describing and explaining.

2. **Learning Ladder:**

   These activities are linked to the Learning Ladder. You can complete one of the questions or all of them. In each chapter you will complete at least one activity for each level of the Learning Ladder. This will sharpen your geographical skills.

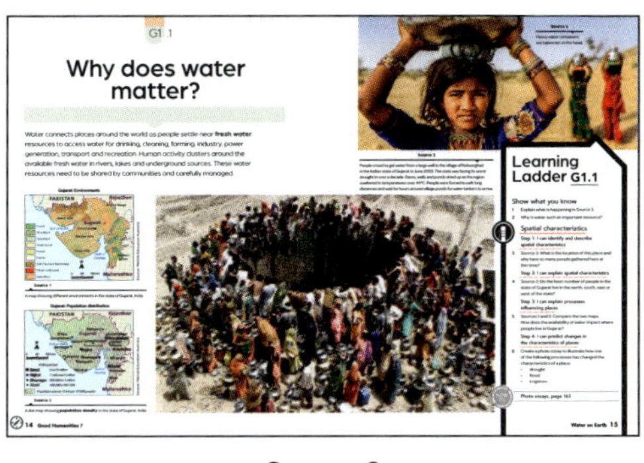

G1.1

Why does water matter?

Learning Ladder G1.1

Source 2

Check your progress regularly. You can attempt one or all of the Learning Ladder questions.

civics+ citizenship

economics+ business

The study of Humanities is more than just History and Geography – it is also the study of Civics and citizenship, and Economics and business. In every chapter of this book you will discover either a Civics and citizenship lesson or an Economics and business lesson. School is busy and you have a lot to cover, so designing a textbook where the important Civics and citizenship and Economics and business content is placed meaningfully next to relevant History or Geography lessons makes good sense, and will help you to connect your learning.

As you work through the Civics and citizenship and Economics and business sections in this book, you will also be working your way up a Learning Ladder for these subjects too!

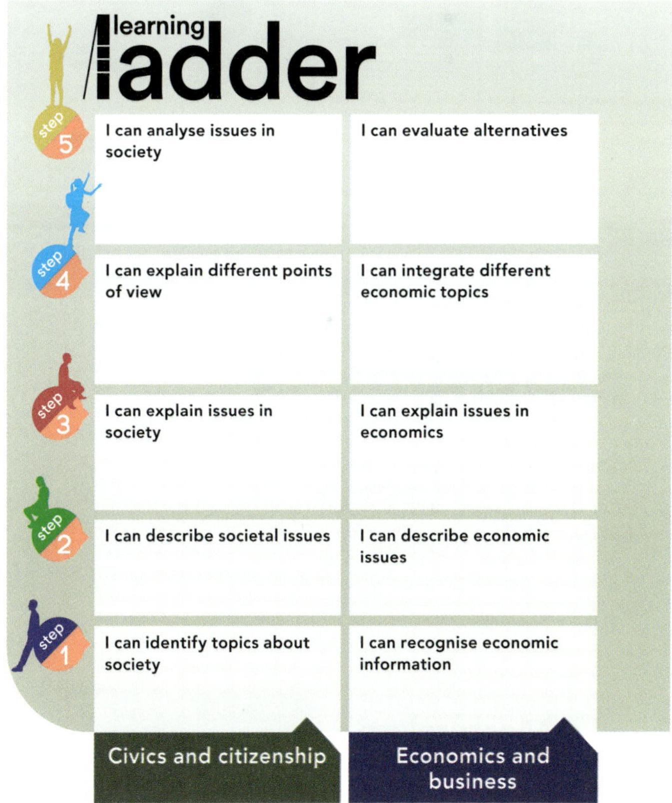

Case studies

Throughout every chapter, you will discover a variety of case studies that explore local, national and global places, issues and events. A case study is an in-depth exploration of a single subject (the case) and is usually based on data or research. The local case studies are focused on Victorian contexts. The national case studies explore other parts of Australia and the global case studies describe places and events in other countries.

G2.7

thinking globally

Why does rice need so much water to grow?

Learning Ladder G2.7

Source 3

Case studies are an important part of your Geography course.

Geo How-To

In the middle of the book, you will find a skills section called 'Geo How-To'. In this section, there are explanations about how to perform each skill. There are *lots* of examples. Refer to it often, especially when answering the Learning Ladder questions.

The Fieldwork section of your textbook explains all the skills you need for hands-on research, as well giving you several suggested tasks.

G6

Mapping skills

Direction

Grid referencing

Scale

154 Good Humanities 7

Geo How-To 155

Source 4

The Geo How-To section is your key to success – refer to it often!

Masterclass

Masterclass is a review section at the end of each chapter. The questions here are organised by the steps on the Learning Ladder. You can complete all the questions, or you might be directed by your teacher to complete just some of them.

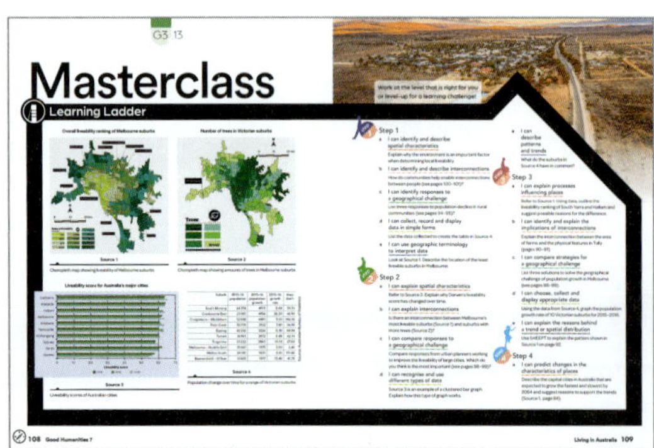
G3 13

Masterclass

Learning Ladder

108 Good Humanities 7

Living in Australia 109

Source 5

You did it! We knew you could! The Masterclass is your opportunity to show your progress. Take charge of your own learning and see if you can extend yourself.

Capstone

After you complete a chapter, it's time to put your new knowledge and understanding together for the capstone project to show what you *know* and what you *think*. In the world of building, a capstone is an element that finishes off an arch, or tops off a building or wall. That is what the capstone project will offer you, too: a chance to top off and bring together your learning in interesting and creative ways. It will ask you to think critically, to use key concepts and to answer 'big picture' questions. The capstone project is accessible online; scan the QR code to find it quickly.

G4 12

Masterclass

Capstone

Source 6

The capstone project brings together the learning and understanding of each chapter. It provides an opportunity to engage in creative and critical thinking.

Learning Ladder G0.3

1. How can you use the Learning Ladder to monitor your progress in Year 7 Geography?
2. As a class, discuss the idea of 'monitoring your own progress'. Why is this important?
3. Read through the steps of the Geography Learning Ladder and consider where you might already be up to for each skill, based on your prior learning.

G1
Water
on Earth
WHERE IS THE
WORLD'S WATER? page 18
interconnection
page 20
HOW DOES WATER MOVE?
thinking globally
page 24
WHY IS MAWSYNRAM THE WORLD'S WETTEST PLACE?
economics + business
page 40
WHAT ARE THE COSTS OF FLOODS?

How can I understand water on Earth?

Without water, no living thing can survive. Two-thirds of Earth is covered in water, but just 2.5 per cent of this water is fresh water. Most of Earth's fresh water is locked up as ice. Fresh water is one of our most precious environmental resources, but increases in population are putting the quality and availability of water at risk.

Step	Spatial characteristics	Interconnections	Geographical challenge
step 5	**I can analyse the impact of change on places** I can analyse and evaluate the implications of changing climatic conditions at different scales and calculate its impact on people and environments.	**I can explore spatial association and interconnections** I can compare distribution patterns and the interconnections between them; e.g. land uses such as farms and sporting fields on floodplains.	**I can plan action to tackle a geographical challenge** I can frame questions, evaluate findings, plan actions and predict outcomes to tackle a water-based geographical challenge.
step 4	**I can predict changes in the characteristics of places** I can predict changes in the characteristics of places over time due to variations in climate.	**I can evaluate the implications of significant interconnections** I can identify, analyse and explain key water-based interconnections within and between places, and evaluate their implications over time and at different scales.	**I can evaluate alternatives for a geographical challenge** I can weigh up alternative views and strategies on a water-based geographical challenge using environmental, social and economic criteria.
step 3	**I can explain processes influencing places** I can explain the series of actions leading to change in a place, such as the development of rainfall.	**I can identify and explain the implications of interconnections** I can identify, analyse and explain water-based interconnections and explain their implications.	**I can compare strategies for a geographical challenge** I can compare strategies for a geographical challenge, taking into account a range of factors and predicting the likely outcomes.
step 2	**I can explain spatial characteristics** I can identify concepts of Space, Place, Interconnection, Change, Environment, Sustainability and Scale (SPICESS) when I read about water on Earth.	**I can explain interconnections** I can describe and explain interconnections and their effects, such as the formation and use of aquifers.	**I can compare responses to a geographical challenge** I can identify and compare responses to a geographical challenge and describe its impact on different groups.
step 1	**I can identify and describe spatial characteristics** I can talk about spatial characteristics at a range of scales; e.g. drainage basins of the world.	**I can identify and describe interconnections** I can identify and explain simple interconnections involved in phenomena such as the water cycle and flooding.	**I can identify responses to a geographical challenge** I can find responses to a geographical challenge such as flooding and understand the expected effects.

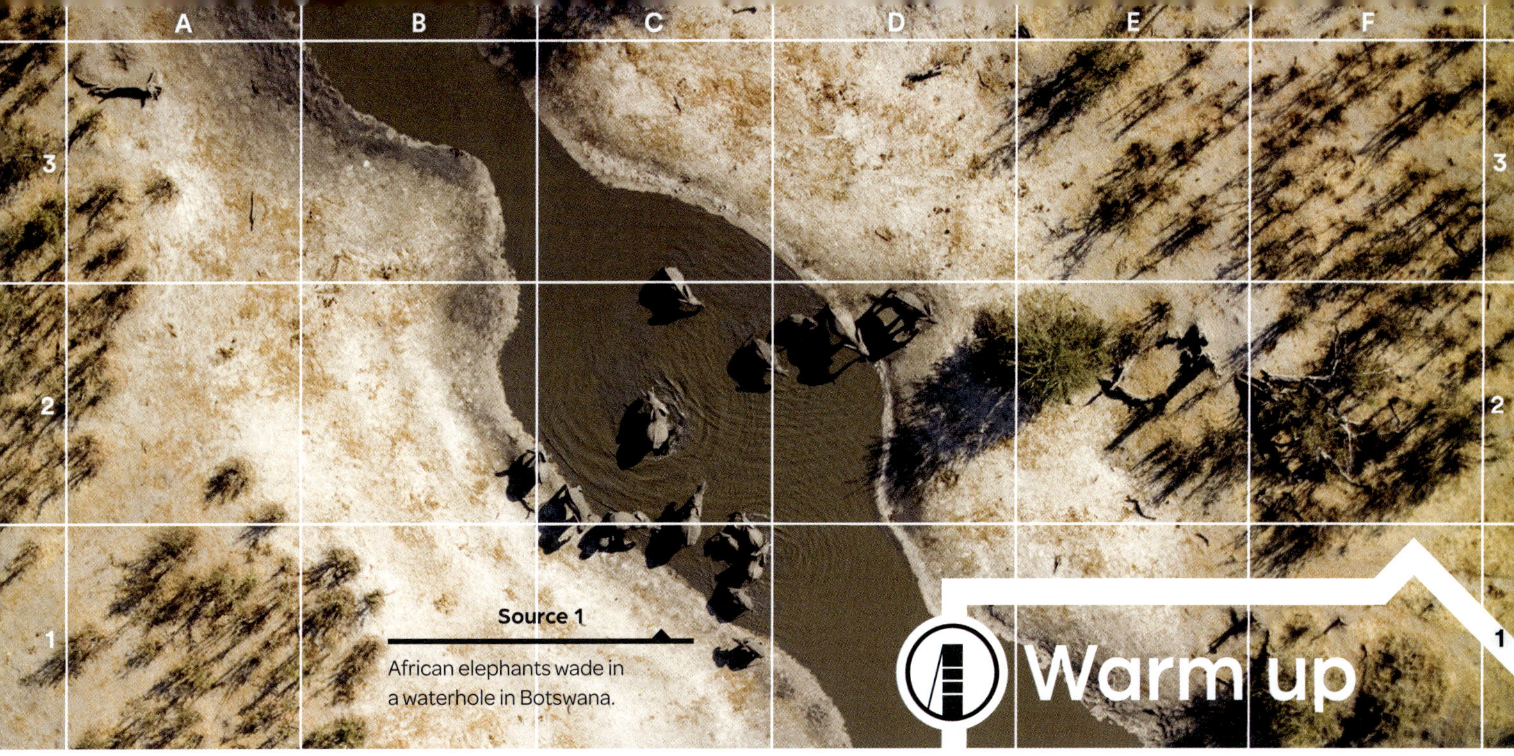

Source 1

African elephants wade in a waterhole in Botswana.

Warm up

Spatial characteristics

1 What place is shown in Source 1?

2 Is this place natural or built?

Interconnections

3 *Interconnection* is the idea that two things or phenomena are related, interact or are linked in some way. Why might there be an interconnection between herds of animals and waterholes?

Geographical challenge

4 Illegal poaching is a huge threat to elephants. Why might there also be an interconnection between poachers and waterholes?

Collect, record and display data

5 a How many elephants are located at this waterhole?

b Is this quantitative or qualitative data? (Quantitative data is a measurement; qualitative data is a description.)

6 What survey method might you use to count smaller animals, such as frogs?

Analyse data

7 Use the grid references in Source 1 to describe the location of the elephants. Refer to page 152 to help you do this.

8 Why are the elephants located here?

I can evaluate data
I can determine whether data presented about water on Earth is reliable and assess whether the methods I used in the field or classroom were helpful in answering a water-based research question.

I can use data to support claims
I can select or collect the most appropriate data and create specialist maps and information using ICT to support investigations into water and its use on Earth.

I can choose, collect and display appropriate data
I can select useful sources of water data and represent them to conform with geographic conventions.

I can recognise and use different types of data
I can define the terms primary, secondary, qualitative and quantitative data and represent data in more complex forms.

I can collect, record and display data in simple forms
I can identify that maps and graphs use symbols, colours and other graphics to represent data.

Collect, record and display data

I can draw conclusions from analysing collected data
I can summarise findings and use collected data to support key patterns and trends I have identified for water-based research.

I can analyse relationships between different data
I can use multiple data sources, overlays and GIS to find links and relationships that exist in patterns of water use on Earth.

I can explain the reasons behind a trend or spatial distribution
I can identify Social, Historical, Economic, Environmental, Political and Technological (SHEEPT) factors to help me explain patterns in data.

I can describe patterns and trends
I can identify Patterns, Quantify them and point out Exceptions (PQE) to describe the patterns I see.

I can use geographic terminology to interpret data
I can identify increases, decreases or other key trends on a map, graph or chart about water.

Analyse data

Why does water matter?

Water connects places around the world as people settle near **fresh water** resources to access water for drinking, cleaning, farming, industry, power generation, transport and recreation. Human activity clusters around the available fresh water in rivers, lakes and underground sources. These water resources need to be shared by communities and carefully managed.

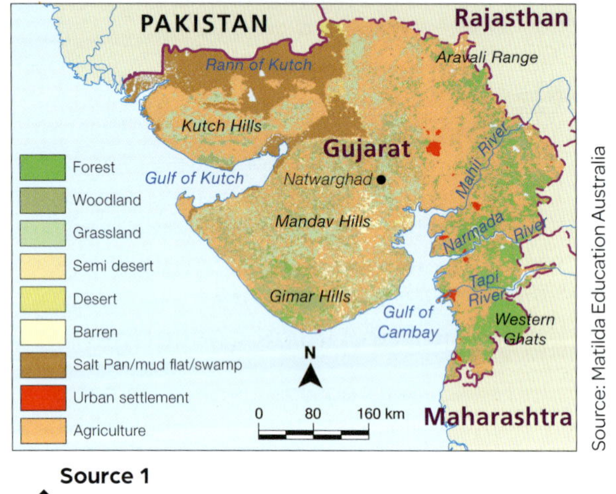

Source 1

A map showing different environments in the state of Gujarat, India

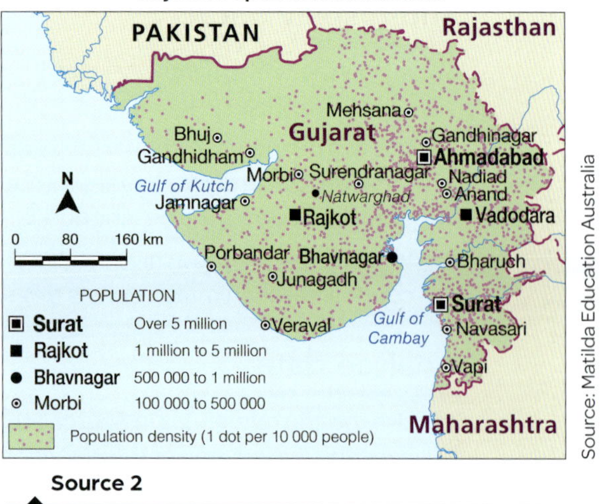

Source 2

A dot map showing **population density** in the state of Gujarat, India

Source 4

Heavy water containers are balanced on the head.

Source 3

People crowd to get water from a large well in the village of Natwarghad in the Indian state of Gujarat in June 2003. The state was facing its worst drought in over a decade. Dams, wells and ponds dried up as the region sweltered in temperatures over 44°C. People were forced to walk long distances and wait for hours around village ponds for water tankers to arrive.

Learning Ladder G1.1

Show what you know

1 Explain what is happening in Source 3.

2 Why is water such an important resource?

Spatial characteristics

Step 1: I can identify and describe spatial characteristics

3 Source 3: What is the location of this place and why have so many people gathered here at this time?

Step 2: I can explain spatial characteristics

4 Source 2: Do the least number of people in the state of Gujarat live in the north, south, east or west of the state?

Step 3: I can explain processes influencing places

5 Sources 1 and 2: Compare the two maps. How does the availability of water impact where people live in Gujarat?

Step 4: I can predict changes in the characteristics of places

6 Create a photo essay to illustrate how one of the following processes has changed the characteristics of a place:

- drought
- flood
- irrigation.

HOW TO

Photo essays, page 163

Will we always have water?

Humans depend on the Earth's **environmental resources**: water from rainfall; food from the land and oceans; and energy from the atmosphere, land and water. Some environmental resources will last forever (they are perpetual) and others are renewable. But some resources are not renewable – if we overuse them, they will run out.

Types of environmental resources

1 Renewable resources

Renewable resources will naturally renew themselves if we do not use them too quickly. We need to manage renewable resources to ensure they have a chance to recover (or regenerate).

Fresh water is considered to be a renewable resource, but this does not mean that it is an unlimited resource. Water pollution and population increases can put our fresh water resources at risk.

2 Perpetual resources

Perpetual resources are renewable resources that don't run out, regardless of how much they are used. Examples of perpetual resources include:

- wind energy
- solar energy
- waves and tides
- **geothermal energy**
- **hydroelectricity**.

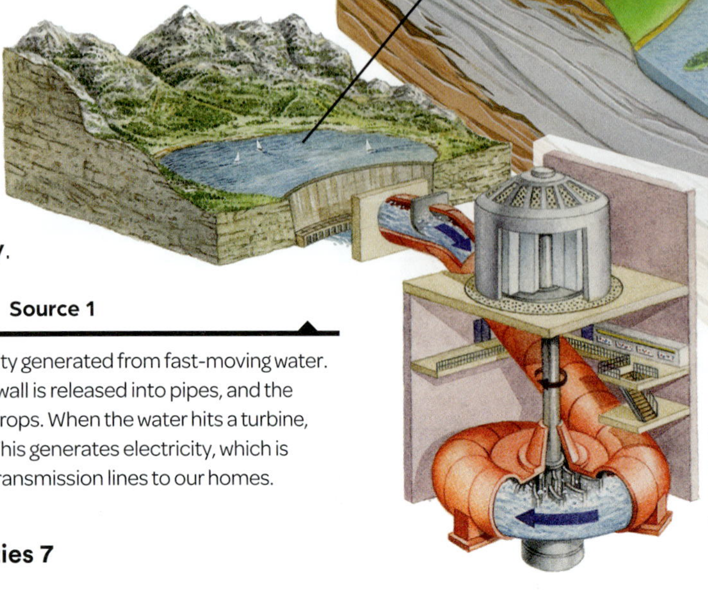

Source 1

Hydroelectricity is electricity generated from fast-moving water. Water held back by a dam wall is released into pipes, and the water gathers speed as it drops. When the water hits a turbine, it makes the turbine spin. This generates electricity, which is then transferred through transmission lines to our homes.

Bores are dug to tap into underground water storages.

3 Non-renewable resources

Non-renewable resources are resources that are only available in limited amounts. If we overuse them, they will run out. These resources are found inside Earth and they took millions of years to form. Non-renewable resources include the **fossil fuels**: oil, natural gas and coal. Today, 84 per cent of the total amount of energy used around the world comes from fossil fuels. As demand for energy rises, fossil fuels are running out. Scientists are exploring the potential of renewable energy sources for the future, such as solar and wind power.

Source 2

Flowing water connects places and is used and stored by humans as a resource for drinking, growing food, electricity generation, waste removal and transport.

Source 3

Floating platforms are used to extract fossil fuels such as oil from beneath the surface of Earth. Massive drills up to 2.5 kilometres long are used. Oil is drawn up to the surface, then pumped into tanker ships that take it to be processed.

The natural flow of a river is sometimes interrupted by a **dam**. A dam provides water storage for use in towns, industry and farms, as well as for power generation.

Water is redirected over long distances using pipes to move the water to where it is required.

Trees are a good example of a renewable resource. They can be cut down for timber, but they can also regenerate through seeds. Trees protect the soil from erosion and improve water quality.

Water treatment plants filter waste water from cities to produce clear, fresh water. The recycled water can be used by humans or returned into rivers.

Irrigated farms draw water from a river or lake to water crops and pastures.

A raised area called a **levee** protects areas from flooding.

Learning Ladder G1.2

Show what you know

1 Is fresh water an unlimited resource?

2 Study Source 2. Give three examples of how humans change the river for their own use.

Collect, record and display data

Step 1: I can collect, record and display data in simple forms

3 Draw a table to record examples of renewable and non-renewable resources.

Step 2: I can recognise and use different types of data

4 Sources 1-3 are block diagrams. Why are block diagrams useful when showing water resources? (see also page 162).

Step 3: I can choose, collect and display appropriate data

5 Conduct a field sketch or use the internet to find a photo of an environment in your local community.

a Label the key elements in this environment that are linked with water.

b Highlight any renewable or non-renewable resources in the environment.

Step 4: I can use data to support claims

6 Research the top 10 renewable energy-using nations in the world and present this in a visual form.

Sketches and annotating, page 160
Block diagrams, 162

Where is the world's water?

Earth is referred to as the 'blue planet' because 70 per cent of its surface is covered in water. Just 2.5 per cent of all water is fresh water – a key resource we need to survive. Most of that fresh water is frozen, and locked up in glaciers and ice caps.

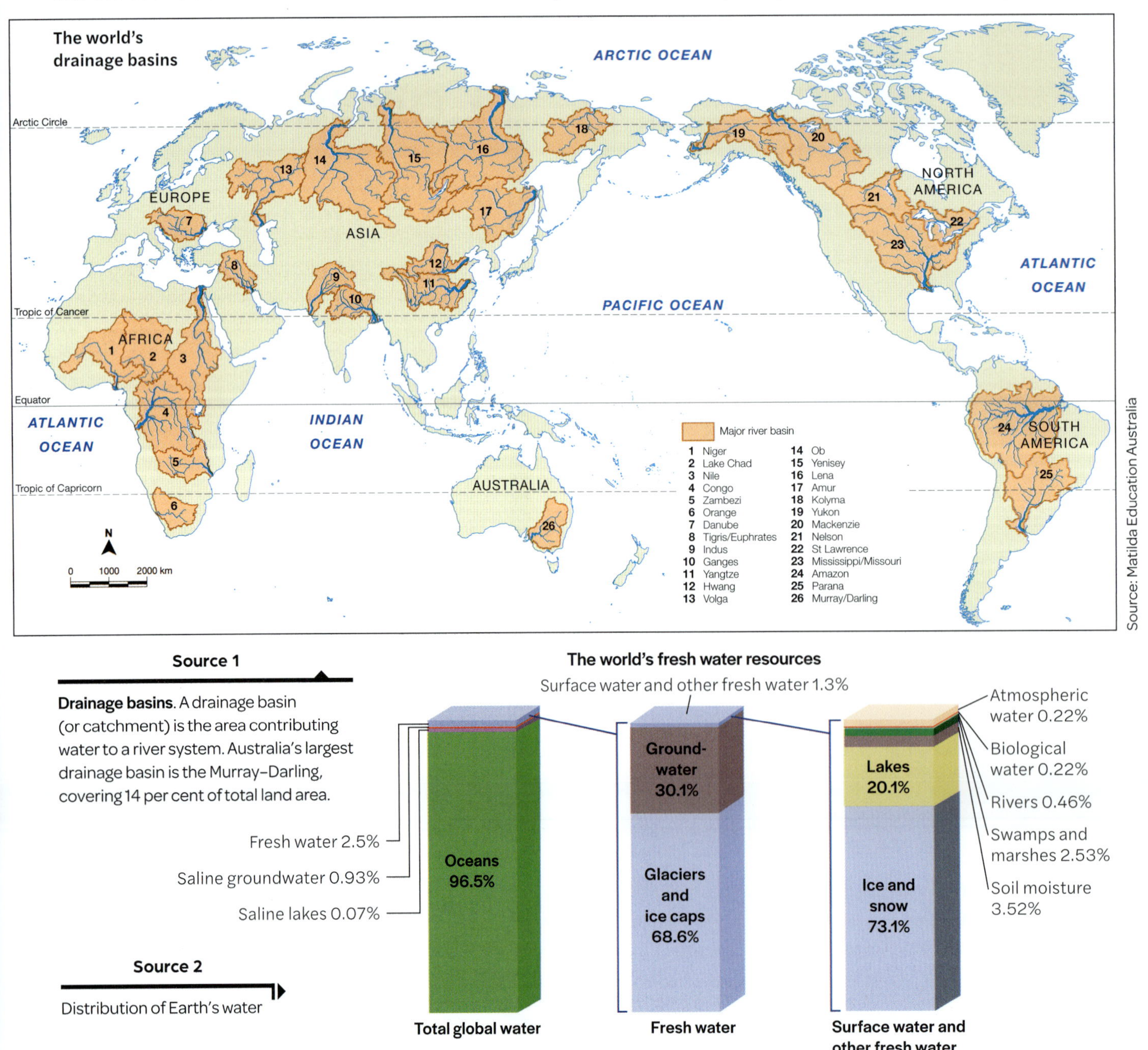

Source 1

Drainage basins. A drainage basin (or catchment) is the area contributing water to a river system. Australia's largest drainage basin is the Murray–Darling, covering 14 per cent of total land area.

Source 2

Distribution of Earth's water

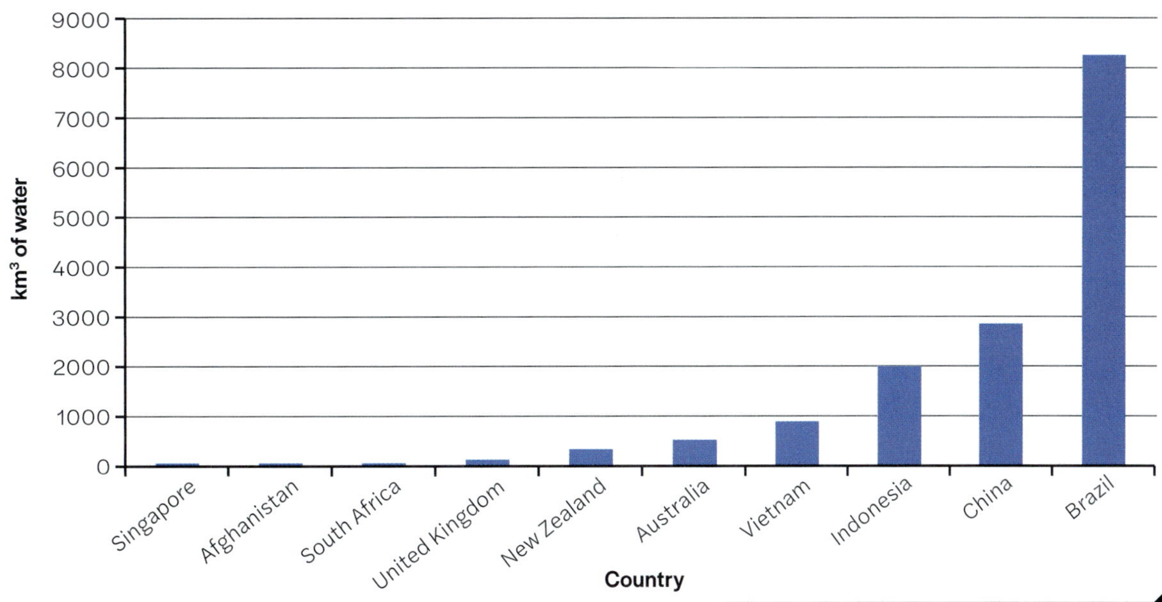

Source 3

Fresh water distribution by country. Fresh water is not evenly distributed across the planet, as some countries have more water than others. These countries are **water rich**. Water-rich countries are those with high rainfall, such as Indonesia, or with large river systems, such as the Amazon River in Brazil. Dry countries, such as Australia and Afghanistan, are **water poor**.

Water on Earth

Water is an essential resource that we need to survive. It is also a renewable resource that occurs naturally on Earth. The same amount of water that was available in the time of the dinosaurs is still available today.

Water constantly changes state and moves through oceans, rivers and the atmosphere in a system known as the water cycle (see pages 20 and 21). To say water 'changes state' means that it changes from being a solid (ice), to a liquid (water), to a gas (vapour).

Water is a renewable resource because of the water cycle, which continually returns water to rivers, lakes and underground storage for humans to use.

Over 70 per cent of the Earth's surface is covered in water, but 96.5 per cent of that is salt water in our oceans.

The fresh water that all living things need to survive makes up only 2.5 per cent of our total water resource. Only a tiny amount of all fresh water is easily accessible in rivers, lakes, ice and snow – just 1.3 per cent. The rest of the fresh water is locked away underground or frozen in **glaciers** and **ice caps**.

Learning Ladder G1.3

Show what you know

1 Why can't we easily access all of Earth's fresh water?

2 Define what is meant by 'water poor' and 'water rich'.

Collect, record and display data

Step 1: I can collect, record and display data in simple forms

3 What data does Source 1 display?

Step 2: I can recognise and use different types of data

4 Source 2: Is this a primary or secondary piece of data? Explain your answer.

Step 3: I can choose, collect and display appropriate data

5 Source 3: As a class, list how the data for this source may have been collected.

Step 4: I can use data to support claims

6 Source 2: How do each of these percentages help explain the amount of fresh water on Earth:
- fresh water 2.5%
- surface water and other fresh water 1.3%.

PQE, page 156
Sketches and annotating, page 160

How does water move?

In the water cycle, water changes from liquid to gas, and then back to liquid or solid again. The water cycle is powered by the Sun, which 'evaporates' water and turns it into a gas called vapour, and causes winds that move the water vapour.

Evaporation, condensation and precipitation

The **water cycle** is charged by the sun, which heats up water so it **evaporates** and changes it into a gas called **water vapour**. The water vapour rises into the atmosphere until it cools. Cold air doesn't hold as much moisture as warm air, so the water vapour changes back into a liquid and forms small droplets of water. This process is called **condensation**. We see these drops of water as clouds.

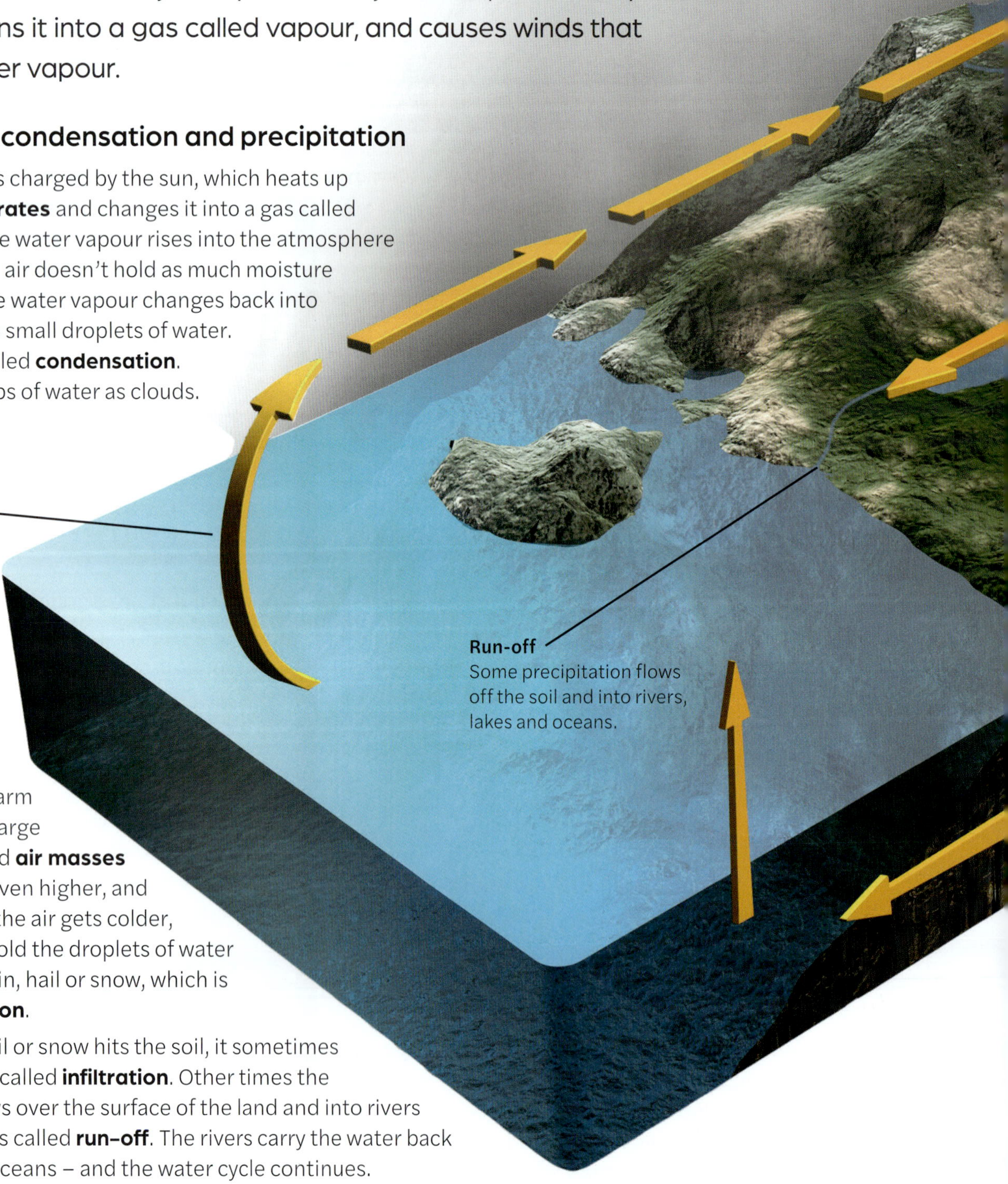

Evaporation
The Sun heats water on the surface of Earth and turns it into a gas called water vapour, which rises into the air.

Run-off
Some precipitation flows off the soil and into rivers, lakes and oceans.

Mountains, warm temperatures or large bodies of air called **air masses** can push the air even higher, and it cools again. As the air gets colder, it can no longer hold the droplets of water and they fall as rain, hail or snow, which is called **precipitation**.

When rain, hail or snow hits the soil, it sometimes soaks in, which is called **infiltration**. Other times the precipitation flows over the surface of the land and into rivers and lakes, which is called **run-off**. The rivers carry the water back to the lakes and oceans – and the water cycle continues.

Transpiration
The evaporation of water from plants, mainly from the leaves.

Condensation
As water vapour rises into the air, it cools and condenses into small drops of water. The drops are so light and small that they can float in the air. We see the water vapour as clouds.

Precipitation
Moist air cools as it is pushed higher into the colder parts of the atmosphere, and it can no longer hold moisture. The tiny water droplets in the clouds join up and become heavier. They fall to the ground as rain, hail or snow.

Infiltration
Some water that falls to the ground soaks into the soil. Water that filters beneath the surface of Earth is called groundwater.

Source 1
The water cycle

Water balance

The water cycle has a **water balance**. If there is too much precipitation on land, it is returned to the oceans by run-off. If there is too much **evaporation**, it is returned to the land by winds that blow the moist air from the sea onto the land.

Learning Ladder G1.4

Show what you know

1 How does the Sun power the water cycle?

2 What causes air to rise?

Interconnections

Step 1: I can identify and describe interconnections

3 Outline the process of water turning from a liquid to a gas and to a solid as a series of steps.

Step 2: I can explain interconnections

4 Construct a flow diagram to show how water moves through the environment. Use these labels: evaporation, condensation, precipitation, transpiration, run-off.

Step 3: I can identify and explain the implications of interconnections

5 Source 1: Look carefully at the dam in the diagram.
 a Why has the flow of the river been interrupted?
 b Suggest two ways that water can enter the dam.
 c Suggest two ways that water can leave the dam.

Step 4: I can evaluate the implications of significant interconnections

6 Give an example to explain how the water cycle has a water balance.

HOW TO
Flow diagrams, page 161
Geographic concepts, page 4

Where does rain occur?

Rain occurs when warm moist air cools and condenses into water droplets. Warm air can hold more water than cool air, so when warmer air rises and cools, it eventually reaches a point where it cannot hold any more water. This is called the dew point.

When warm air cools above its **dew point**, the water vapour it is holding condenses into a liquid and it rains. There are three different ways of forcing air to rise and cool and condense into rain:

- **frontal rainfall**: when cold air meets warm air
- **orographic rainfall**: when warm air is lifted over a mountain
- **convectional rainfall**: when land heats the air above it.

1 Frontal rainfall

Frontal rain occurs when a warm air mass meets a cold air mass. The lighter warm air is less dense, and is pushed up over the heavier cold air. This creates a 'front'. As the warm air mass rises, it cools. Then the water vapour in the air mass condenses into water and falls as raindrops along the front.

2 Orographic rainfall

Orographic rainfall is when moist air meets high land and is forced to rise. As it rises, the air cools and condenses into raindrops on the side of the mountain the wind is blowing from (the windward side). As the air descends from the high land on the other side of the range (the leeward side), it warms and creates dry regions. The **deserts** of central Australia are located on the leeward side of the Great Dividing Range.

3 Convectional rainfall

Convectional rainfall is common in **tropical** areas where the ground is heated by the hot sun. The sun heats the land and, as the moisture from the ground evaporates, the heat from the ground heats the air above it. The warm air rises rapidly, then condenses to form clouds as it starts to cool. Convection can produce towering **cumulonimbus clouds**, which produce heavy rainfall, thunderstorms and lightning.

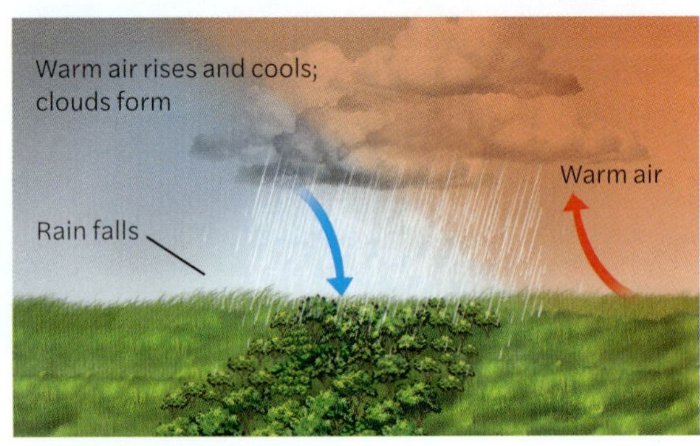

Source 1

How frontal rainfall occurs

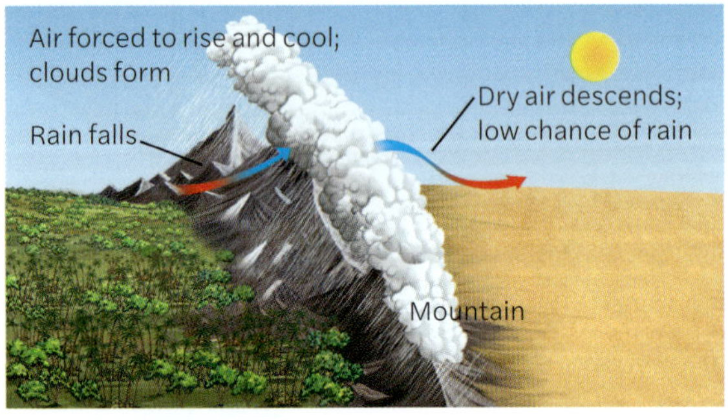

Source 2

How orographic rainfall occurs

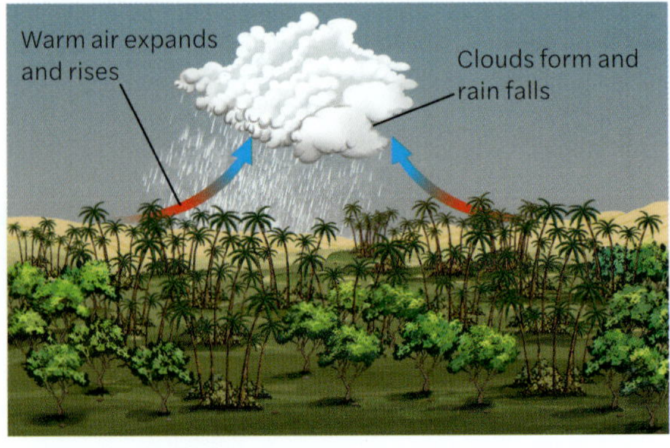

Source 3

How convectional rainfall occurs

World average annual rainfall

ARCTIC OCEAN

Arctic Circle

Tropic of Cancer

Equator

Tropic of Capricorn

PACIFIC OCEAN

ATLANTIC OCEAN

INDIAN OCEAN

N

0 2500 5000 km

AVERAGE ANNUAL RAINFALL, mm

Over 2000	1000 to 1500	250 to 500
1500 to 2000	500 to 1000	Under 250

Source: Matilda Education Australia

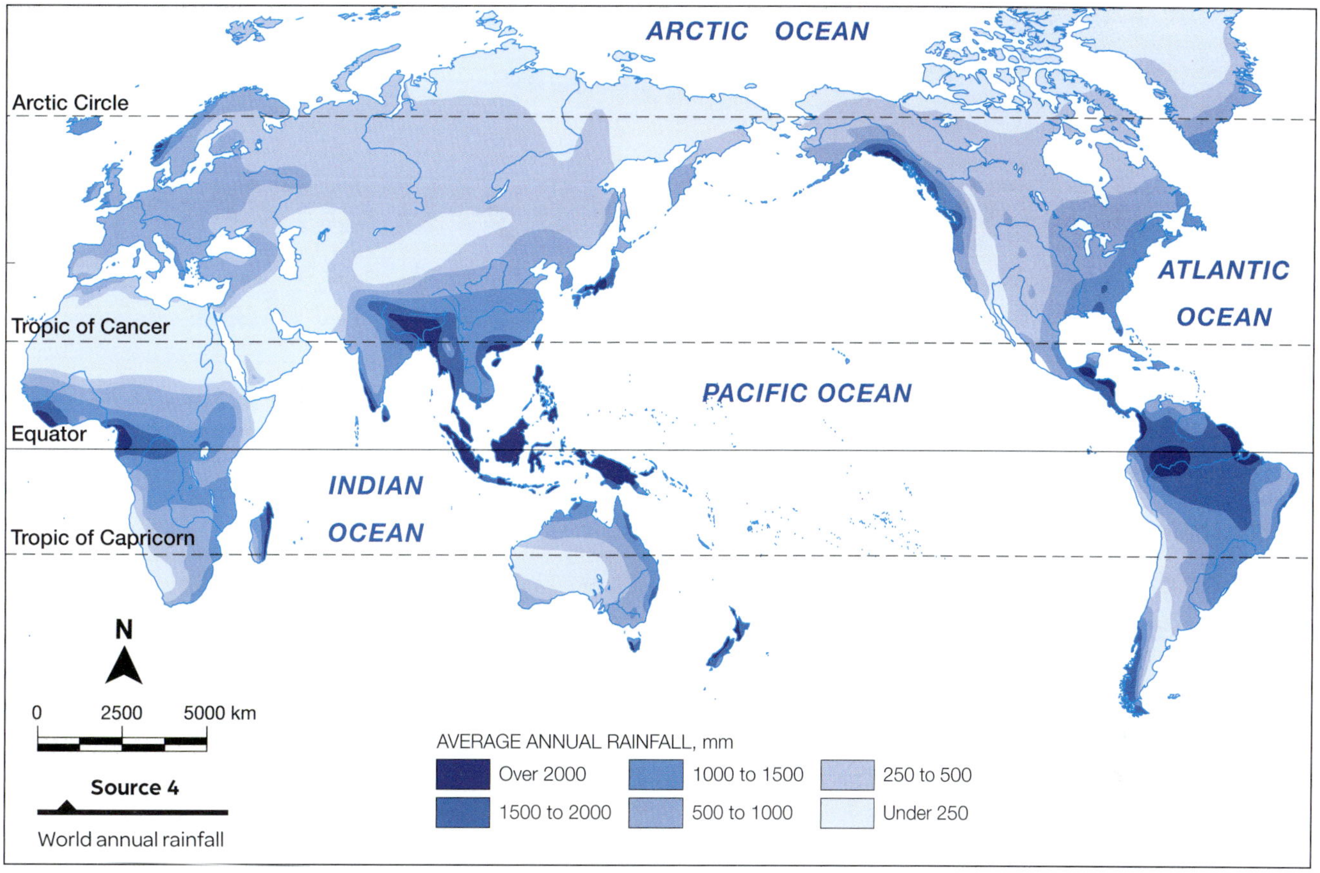

Source 4

World annual rainfall

Learning Ladder G1.5

Show what you know

1 When does rainfall occur?

2 Outline the key similarities and differences between frontal and orographic rainfall.

3 Source 5: Create a diagram or download an image from the internet of a supercell thunderstorm. Add labels to show how it causes rainfall.

Source 5

A supercell thunderstorm in Kansas. Supercell thunderstorms are the biggest and most severe type of thunderstorm. They form when warm, humid air is trapped under a layer of cold air. The rising warm air and sinking cold air create powerful spiralling convection currents that produce hail, downpours and sometimes tornadoes.

Spatial characteristics

Step 1: I can identify and describe spatial characteristics

4 Source 4: Use the PQE method to describe the pattern of world rainfall.

Step 2: I can explain spatial characteristics

5 Explain why Australia's deserts are located where they are.

Step 3: I can explain processes influencing places

6 Source 2: How does orographic rainfall cause high rainfall areas on the windward side of mountains but deserts on the leeward side?

Step 4: I can predict changes in the characteristics of places

7 Each day we see a weather map showing the state of the weather for a region. The maps show air masses that meteorologists use to predict the weather over the next few days. Research weather maps and provide an example of a weather map with labels to show how it works.

HOW TO

PQE, page 156

thinking globally

Why is Mawsynram the world's wettest place?

Mawsynram in India has an annual rainfall of 11 871 mm, and is the wettest inhabited place in the world. Cherrapunji, just 15 kilometres from Mawsynram, is the second-wettest place in the world. Mawsynram and Cherrapunji are both located in north-eastern India, about 1400 metres above **sea level**.

During the rainy season, or **monsoon**, much of India receives heavy rainfall. The monsoon months from June to September produce the heaviest rainfall in the north-eastern part of India, on the foothills of the Himalayan Mountains. Winds coming from the Indian Ocean absorb moist air and are forced to rise as they meet the mountains. The air cools and condenses into raindrops, in a process known as orographic rainfall (see page 22). On the other side of the Himalayas it is dry and warm.

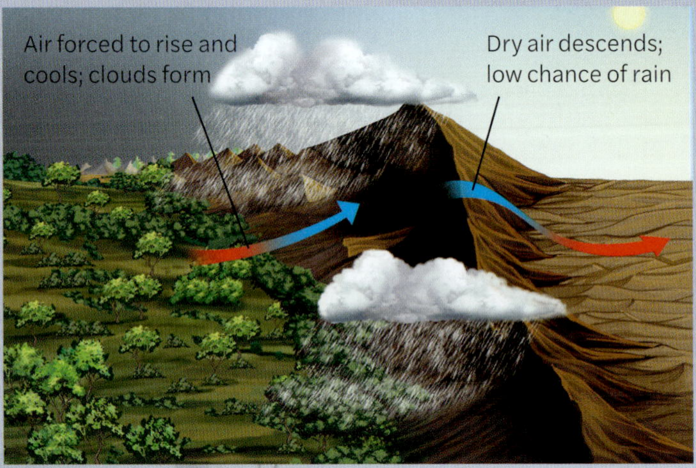

Source 1

The monsoon season in Mawsynram produces heavy rain over an extended period. People plan for its arrival by stocking up on food. Women make body umbrellas, called *knups*, from bamboo, grass and plastic.

Air forced to rise and cools; clouds form

Dry air descends; low chance of rain

Source 2

Orographic rainfall brings heavy rainfall to Mawsynram.

MAWSYNRAM VILLAGE

Source 3

Mawsynram in India is the world's wettest place.

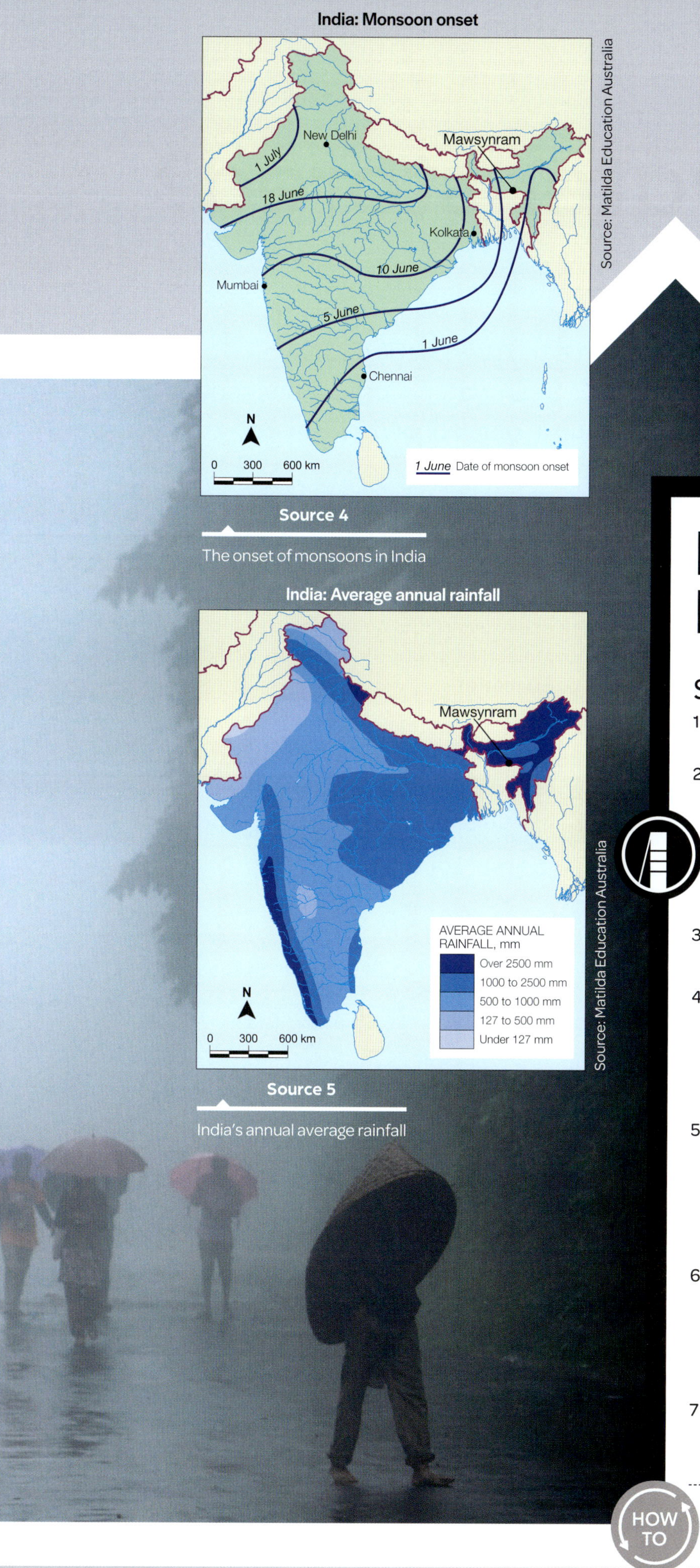

Source 4

The onset of monsoons in India

Source 5

India's annual average rainfall

Mawsynram climate graph Source: Climate-Data.org, 2019

Maximum temperature — Rainfall

°C — mm

Jan Feb Mar Apr May Jun Jul Aug Sep Oct Nov Dec

Source 6

Rainfall and temperature in Mawsynram, by month

Learning Ladder G1.6

Show what you know

1 How does orographic rainfall deliver so much rain to Mawsynram?

2 Give two examples of how climate affects the lives of people living in Mawsynram.

Analyse data

Step 1: I can use geographic terminology to interpret data

3 Define the term 'monsoon'.

Step 2: I can describe patterns and trends

4 Source 5: Look at the map of India's average rainfall. Use compass points to describe the driest and wettest regions in India.

Step 3: I can explain the reasons behind a trend or spatial distribution

5 Source 6: Look at the Mawsynram climate graph.
   a Compare the similarities and differences in the patterns of rainfall and temperature over the year.
   b What is the connection between the highest rainfall months and the monsoon (Source 4)?

6 Source 1: Using SHEEPT, explain the factors that people in Mawsynram would use to plan for the monsoon.

Step 4: I can analyse relationships between different data

7 Compare Source 4 and Source 6: Why do you think the north-westerly winter winds don't bring much rain to Mawsynram?

HOW TO
SHEEPT, page 158
Block diagrams, page 162
Climate graphs, page 165

How is water stored?

Fresh water is an essential renewable resource that we need to survive. Only a tiny percentage of all water on Earth is fresh water – just 2.5 per cent. Most of the fresh water is frozen and locked up in glaciers and ice sheets, or stored beneath the surface as groundwater.

Frozen water

Nearly 70 per cent of all fresh water on Earth is stored in **ice sheets**, **ice shelves**, **glaciers** and **icebergs**. In Antarctica, a huge ice sheet with an average thickness of 2.5 kilometres covers the solid rock below.

The freezing Antarctic temperatures prevent most of the snow from melting and the ice sheet it has formed is up to 4.7 kilometres deep at its deepest level. If the Antarctic ice sheet were to melt completely, the world's sea levels would rise by up to 60 metres.

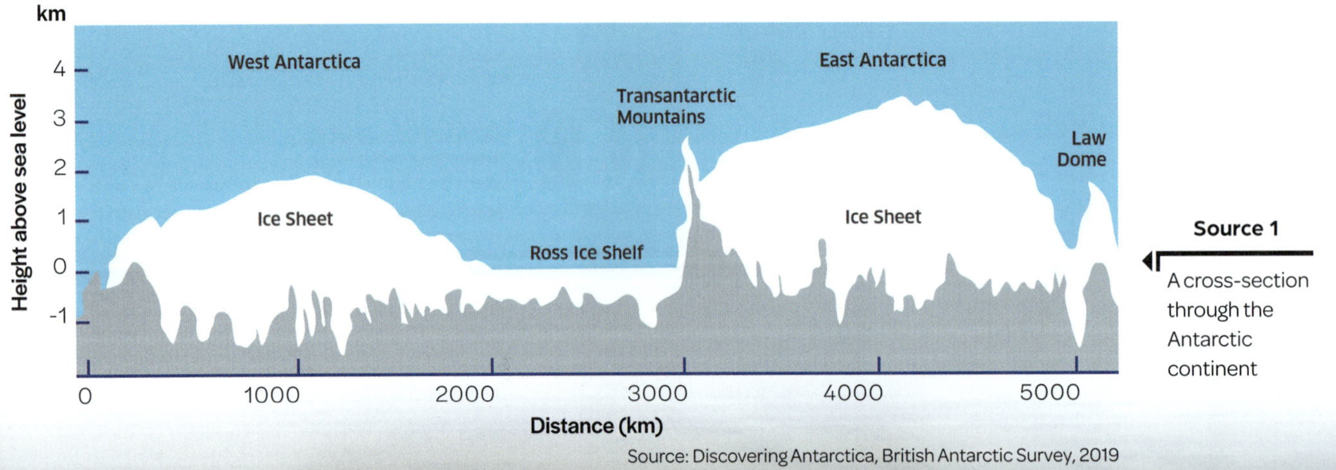

Source: Discovering Antarctica, British Antarctic Survey, 2019

Source 1

A cross-section through the Antarctic continent

Source 2

Most of the world's fresh water is stored as ice. Huge ice sheets move and break off into icebergs when they reach the sea in a process known as calving. These Adelie penguins are on an iceberg that calved from Antarctica's Cape Adare peninsula.

Groundwater

Groundwater is water that is located under Earth's surface. It is fed by surface water from rainfall and rivers. Groundwater can appear on the surface at low points as a **spring** or in dry areas as an **oasis**.

Groundwater fills spaces between grains of soil or rock. In an area way below the soil, called the **saturated zone**, all the spaces between soil and rock particles are filled with water. The top of this zone is called the **water table**.

Australia is the driest inhabited continent on Earth, and surface water resources are limited. Groundwater is a very important renewable resource, and it provides a third of Australia's total water consumption. About 90 per cent of the Northern Territory's water is sourced from groundwater.

Farmers sink bores to pump water to the surface from underground rocks that hold water, called **aquifers**. Windmills use wind energy to pump water to the surface.

Building dams

Humans create their own water storages, called **dams** or **reservoirs**, which collect and retain water. Large dams on rivers provide a reliable water supply for **irrigation**, hydroelectricity or for urban use.

To build a dam, the flow of a river is stopped by the dam wall (see page 29). Water builds up behind the wall and floods the valley. The release of water in front of the dam wall is then regulated to make sure there is enough river water for people and for the river ecosystem.

Learning Ladder G1.7

Show what you know

1 Why can't we use 70 per cent of the fresh water on Earth?

2 What geographic challenge would the world have if the Antarctic ice sheet were to melt?

Geographical challenge

Step 1: I can identify responses to a geographical challenge

3 How do humans make their own water storages?

Step 2: I can compare responses to a geographical challenge

4 Where is most of the usable fresh water on Earth stored and how is it accessed?

Step 3: I can compare strategies for a geographical challenge

5 Look at Source 3 and discuss:
- a How is an aquifer like a river?
- b How do aquifers help connect human settlements in deserts?

Step 4: I can evaluate alternatives for a geographical challenge

6 Create an experiment to show what happens when a dam wall blocks river flow. What happens to the environments on the land behind the wall? What alternative could give humans a reliable water supply?

HOW TO Block diagrams, page 162

Source 3

A desert oasis

thinking nationally

What is the purpose of the Gordon Dam?

Lake Gordon in south-west Tasmania contains up to 12 359 040 megalitres of water (a megalitre is one million litres), making it Australia's largest reservoir. The combination of Lake Gordon and Lake Pedder is by far Australia's largest water storage.

How was Australia's largest water storage formed?

Lake Pedder used to be much smaller. It was flooded in 1972 to make a larger lake as part of the Gordon **hydroelectric** power scheme. The water in Lake Pedder flows to Lake Gordon via the McPartlan Pass Canal, providing 40 per cent of the water used in the Gordon Power Station.

The Gordon Power Station is located on the Gordon River. The power station and the dam were constructed and are now operated and maintained by Hydro Tasmania. Water from the combined water reservoir falls 173 metres through a vertical shaft into the power station. The force of the moving water turns **turbines** and a shaft connected to a rotor and a stator where electrons are stimulated, resulting in the generation of electricity. The Gordon Power Station provides 13 per cent of Tasmania's electricity or 450 megawatts from three turbines.

Source 1

The Gordon River hydroelectric power scheme

Source: Hydro Tasmania – Gordon Power Scheme

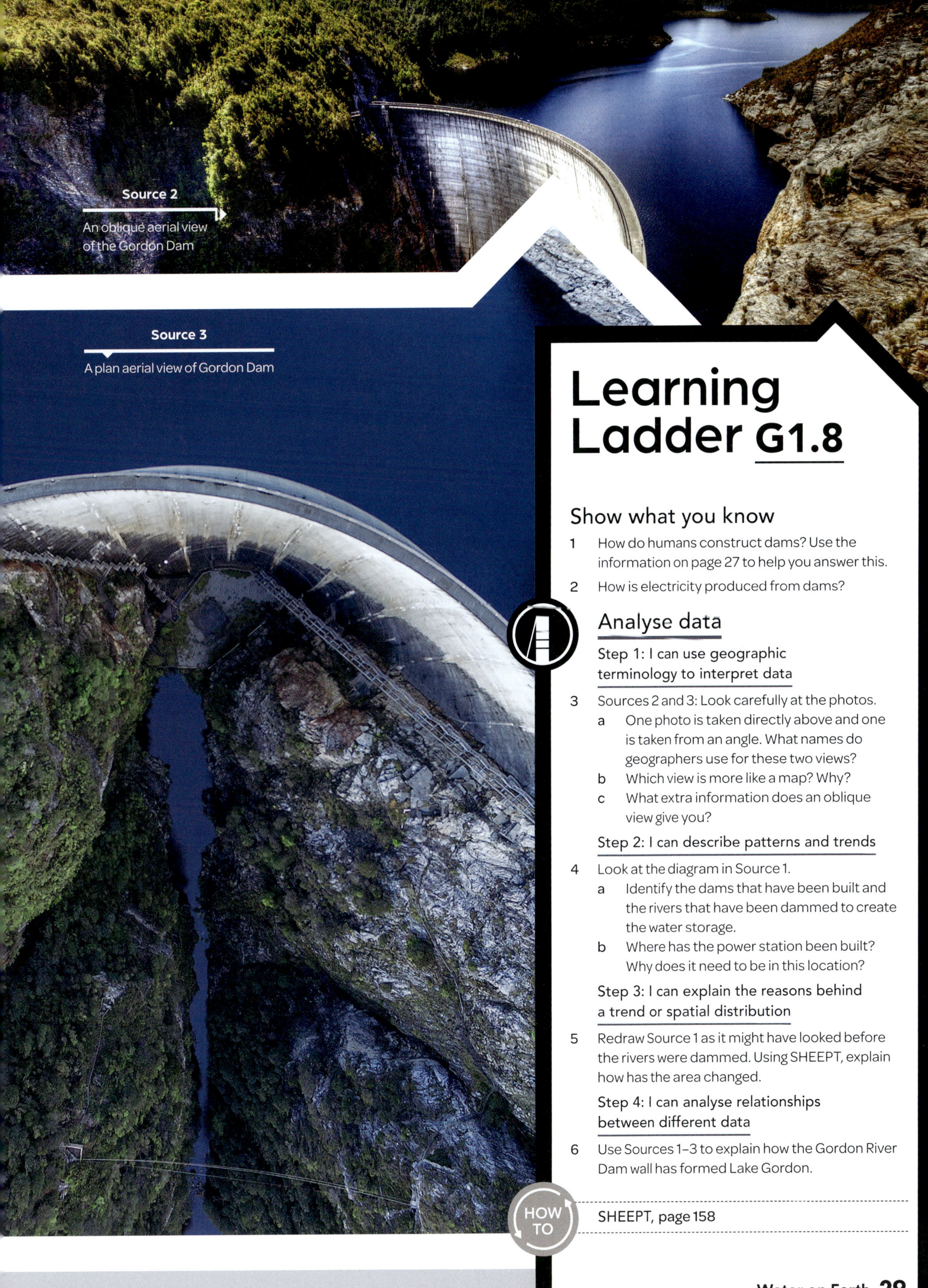

Source 2

An oblique aerial view of the Gordon Dam

Source 3

A plan aerial view of Gordon Dam

Learning Ladder G1.8

Show what you know

1 How do humans construct dams? Use the information on page 27 to help you answer this.

2 How is electricity produced from dams?

Analyse data

Step 1: I can use geographic terminology to interpret data

3 Sources 2 and 3: Look carefully at the photos.

a One photo is taken directly above and one is taken from an angle. What names do geographers use for these two views?

b Which view is more like a map? Why?

c What extra information does an oblique view give you?

Step 2: I can describe patterns and trends

4 Look at the diagram in Source 1.

a Identify the dams that have been built and the rivers that have been dammed to create the water storage.

b Where has the power station been built? Why does it need to be in this location?

Step 3: I can explain the reasons behind a trend or spatial distribution

5 Redraw Source 1 as it might have looked before the rivers were dammed. Using SHEEPT, explain how has the area changed.

Step 4: I can analyse relationships between different data

6 Use Sources 1–3 to explain how the Gordon River Dam wall has formed Lake Gordon.

HOW TO

SHEEPT, page 158

How do businesses set goals?

Successful businesses set clear goals to ensure all the staff have a clear idea of what they are striving to achieve. The Thankyou project had the goal of selling fresh water to customers to fund the provision of clean water to countries where millions die because of unsafe water. The key objective of most businesses is to make a profit, but there are other types of goals that businesses aim for. These include financial goals and organisational goals.

Financial goals

Financial goals are directly related to money, such as sales, costs and profits. These goals might be to increase sales, improve efficiency or increase profits by reducing costs.

Organisational goals

Organisational goals are goals that will help the business to become (or remain) successful, but are not measured in dollars. For example, training staff, having satisfied customers or improving safety in the work environment.

Businesses set strategies (or plans) to help them achieve goals. They review the success of these plans at regular intervals.

Thankyou

Thankyou is an Australian **social enterprise** with very clear business goals. Its original financial goals were to tap into the global billion-dollar bottled water business. To compete with large multinational corporations such as Coca-Cola, Thankyou promised consumers that 100 per cent of the profits from the business would go to helping the 700 million people around the world without access to safe water.

Setting the goals

One of the co-founders of Thankyou is Daniel Flynn. Flynn had a very clear organisational goal for the business. 'What if we could turn bottled water into getting clean water, and make it convenient for people who actually need it?' Flynn says.

'We wanted to tackle a huge global problem with a huge global product, and the product to get us there was bottled water. We felt the link was very important for customers: buying water and funding water, and buying food and funding food, and so on.'

Setting the strategy

Turning the Thankyou idea into a business – and competing with large corporations – required a very specific strategy. Thankyou used social media to show retailers there was a demand for a product that would help people in need around the world.

Before an important meeting with the 7-Eleven chain of convenience stores, Thankyou asked its Facebook followers to jump on to

Source 1

Thankyou co-founders Jarryd Burns (left), Daniel Flynn (right) and his partner Justine Flynn (centre). Thankyou had a very clear business goal. Profit from the sale of every bottle of Thankyou water is used to fund water wells, filters, safe toilets and hygiene education in countries that need help. Thanks to a following of 300 000 people across social media, Thankyou is stocked in retail stores such as Coles and 7-Eleven, and it has raised more than $6.2 million to help over 857000 people to access safe water, sanitation and maternal and child health services in 20 countries.

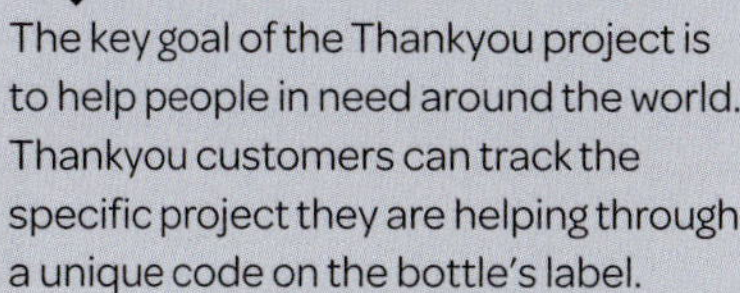

Source 2

The key goal of the Thankyou project is to help people in need around the world. Thankyou customers can track the specific project they are helping through a unique code on the bottle's label.

7-Eleven's Facebook wall and tell them that if they stocked Thankyou water, the followers would buy it. Within a day, 7-Eleven was inundated with Facebook posts – and within weeks Thankyou water was on their shelves.

Learning Ladder G1.9

Economics and business

Step 1: I can recognise economic information

1 What financial goals does Thankyou have?

Step 2: I can describe economic issues

2 Why is it important for businesses to have clear goals?

Step 3: I can explain issues in economics

3 How did Thankyou compete with large distributors of bottled water such as Coca-Cola?

Step 4: I can integrate different economic topics

4 Interview someone you know who works in a business to help answer these questions.
   a What financial goals does your business have?
   b What organisational goals does your business have?
   c What strategies do you have to help you reach your business goals?

Step 5: I can evaluate alternatives

5 Thankyou used the market for bottled water to help those without access to safe water. Develop a plan for a social enterprise of your own. Outline your financial goals, organisational goals and strategy.

How do rivers work?

Rivers are moving bodies of water that flow quickly downhill, eroding the land to form a valley. Rivers slow and spread out as they reach flatter land and then finally flow into the ocean, lake or larger river.

Source 1

The movement of water from mountain to sea

Waterfall
A waterfall occurs where a river flows over a steep drop into in a plunge pool below. The river erodes the soft rock, leaving only hard rock, like granite, to form cliffs and ledges for the water to spill over.

Rivers in mountains

When a river is near its source, the force of water moving down a steep slope erodes the ground vertically to form v-shaped **valleys**. The moving water picks up rocks and soil (called sediment). The sediment erodes the riverbed further, in a process called **abrasion**.

Waterfalls often form in the upper stages of a river, where rivers meet harder rock that is not easily worn away, or eroded. Instead, water spills over the harder rock and erodes the soft rock below, forming a large pool of water called a **plunge pool**. Eventually, continued erosion will undermine the harder rock, and it will crack and fall into the plunge pool. Then a new waterfall forms further upstream. As the process continues, it forms a steep-sided valley, called a **gorge**.

Rivers on flatter land

As the river reaches lower, flatter land, the eroded stones, sand and soil carried in the river begin to wear away the floor and sides of the valley. On the flatter land (called a floodplain, because it floods when there is too much water), rivers slow down and erode horizontally, creating snake-like **meanders**. The water speed is greatest on the outside of a bend, which is where erosion takes place. Water speed is slowest on the inside of a bend, and that's where eroded rock and soil is dropped, in a process called **deposition**.

Source 2

Cross-section of a waterfall

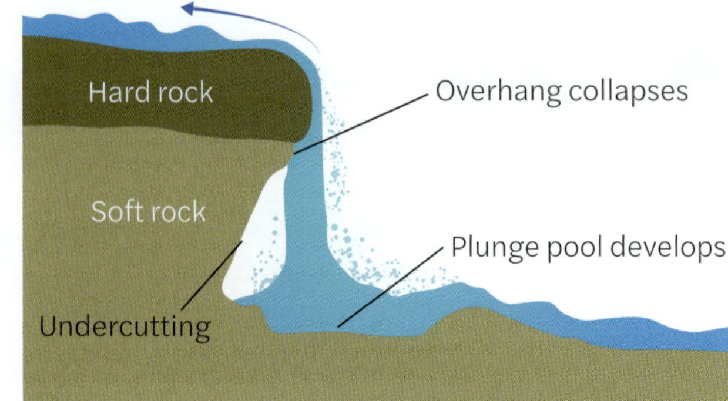

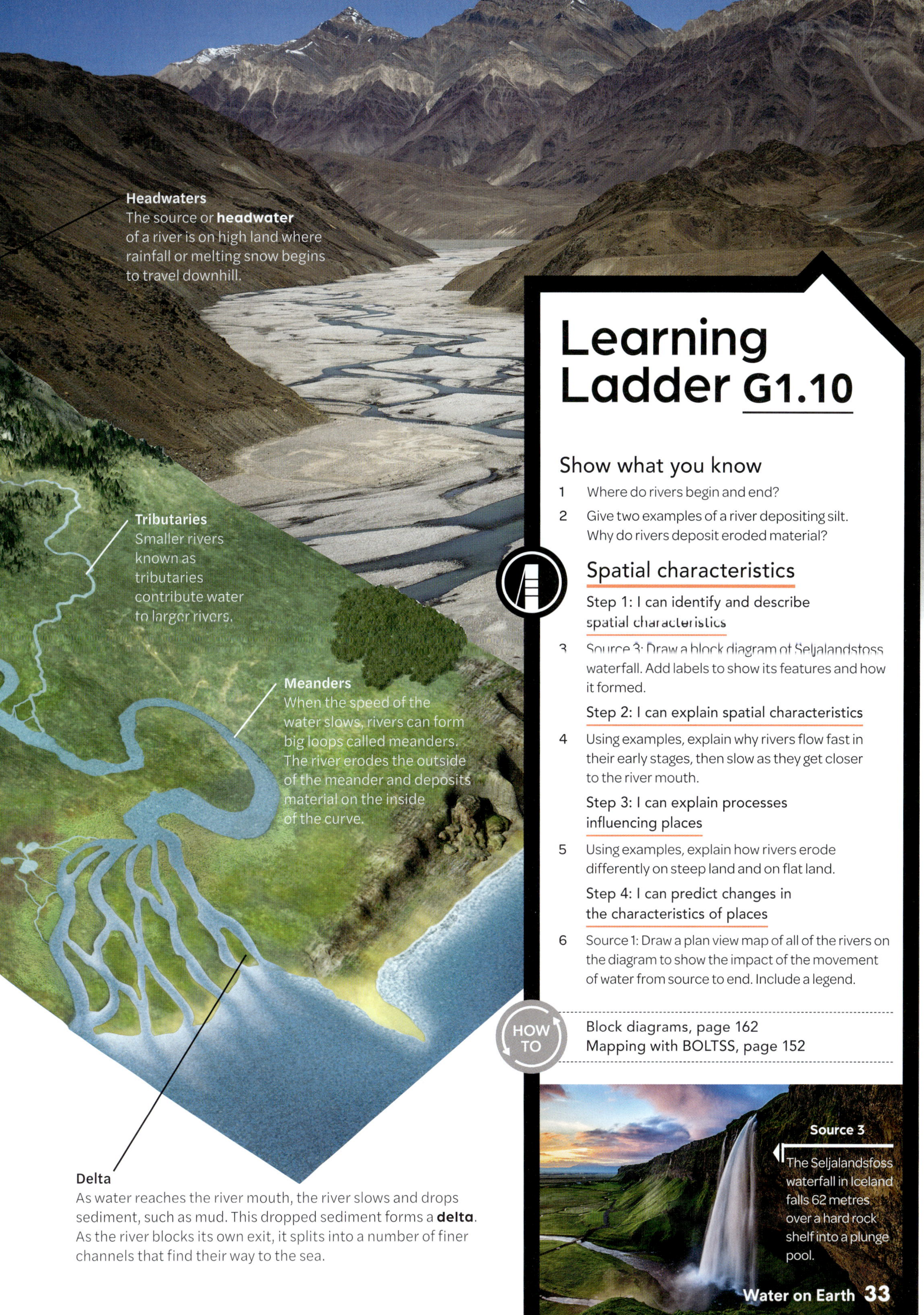

Learning Ladder G1.10

Show what you know

1 Where do rivers begin and end?

2 Give two examples of a river depositing silt. Why do rivers deposit eroded material?

Spatial characteristics

Step 1: I can identify and describe spatial characteristics

3 Source 3: Draw a block diagram of Seljalandsfoss waterfall. Add labels to show its features and how it formed.

Step 2: I can explain spatial characteristics

4 Using examples, explain why rivers flow fast in their early stages, then slow as they get closer to the river mouth.

Step 3: I can explain processes influencing places

5 Using examples, explain how rivers erode differently on steep land and on flat land.

Step 4: I can predict changes in the characteristics of places

6 Source 1: Draw a plan view map of all of the rivers on the diagram to show the impact of the movement of water from source to end. Include a legend.

HOW TO

Block diagrams, page 162
Mapping with BOLTSS, page 152

Source 3

The Seljalandsfoss waterfall in Iceland falls 62 metres over a hard rock shelf into a plunge pool.

How do we use the Murray River?

At 2508 kilometres, the Murray River is Australia's longest single river. It connects places on its banks that rely on water for farming, industry and town water. In the 20th century, a large number of dams, **locks** and **weirs** were constructed along the river to control its flow. These changes made more water available for transport, and for farmers to **irrigate** their crops. This made the Murray River valley Australia's most productive agricultural region – but it also seriously damaged **ecosystems** along the river.

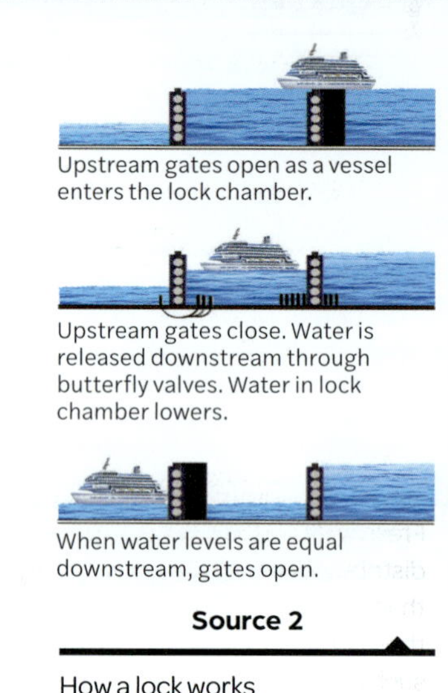

Upstream gates open as a vessel enters the lock chamber.

Upstream gates close. Water is released downstream through butterfly valves. Water in lock chamber lowers.

When water levels are equal downstream, gates open.

Source 2

How a lock works

Source 1

Murray River locks, weirs, dams and barrages

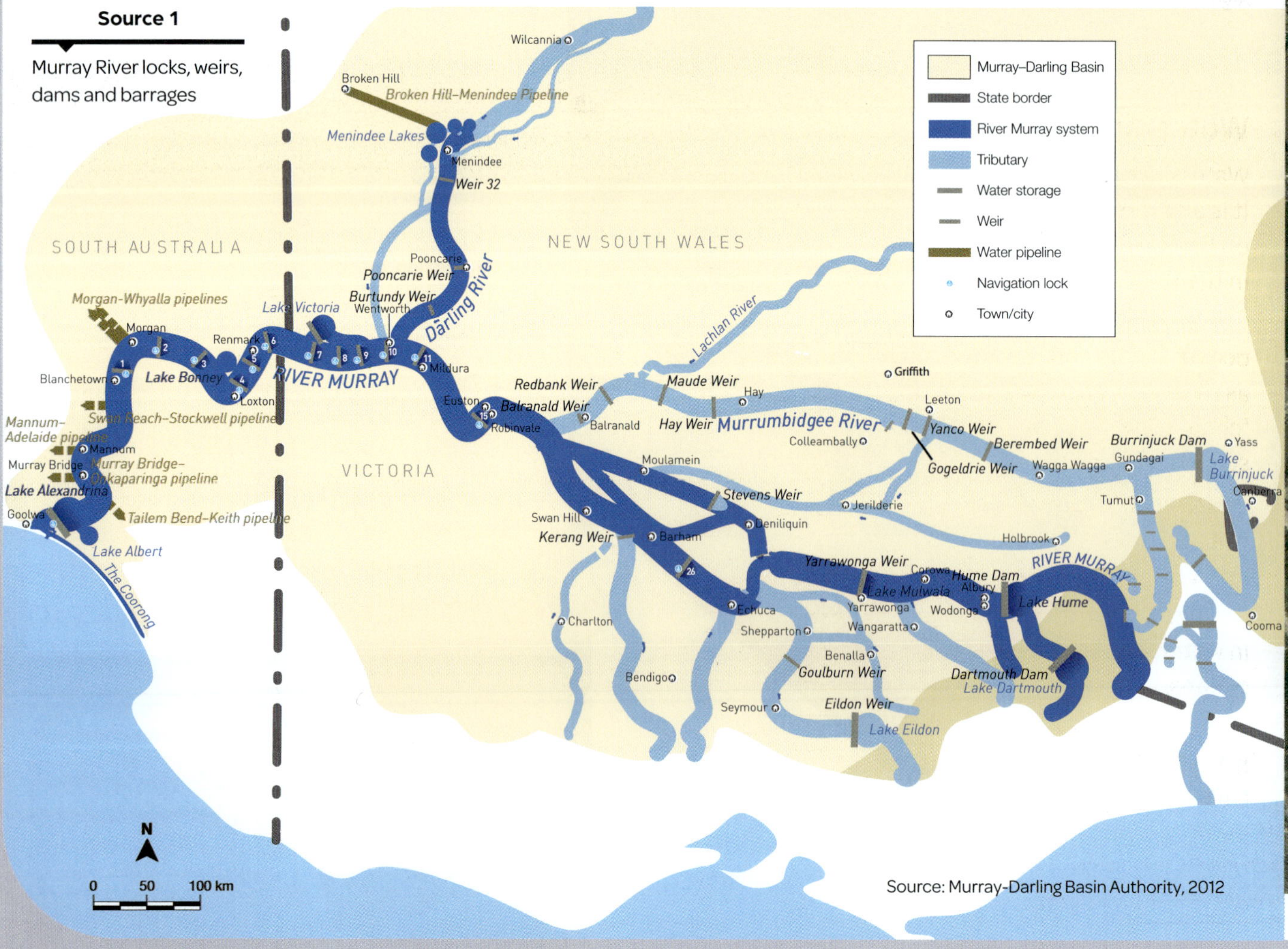

Source: Murray-Darling Basin Authority, 2012

Source 3

The first river boat (or paddle steamer) travelled up the Murray River in 1853. After that, the Murray River became a major inland 'highway' for farmers, settlers and travellers. Fleets of river boats carried produce from stations and farms to towns along the river. Many settlements developed along the Murray River because of the trade brought by the river boats.

Source 4

A satellite image of the Murray River at Renmark

Learning Ladder G1.11

Show what you know

1 Where does the Murray River flow?

2 Why did people build weirs and dams along the Murray River?

Interconnections

Step 1. I can identify and describe interconnections

3 Source 4: Draw a map of the area shown in this satellite image using BOLTSS. Then create a key to show activities that are interconnected with the Murray River – irrigated farmland and towns.

4 On the map you prepared in question 3, label the following:
- the large town of Renmark on the north bank of the river
- Paringa on the south bank
- the weir that crosses the Murray River to the south-west of Renmark

Step 2: I can explain interconnections

5 Source 4: Places and their people are interconnected with other places. How does the satellite image show that rivers connect communities?

Step 3: I can identify and explain the implications of interconnections

6 Source 2: How do locks help improve interconnection along the Murray River?

Step 4: I can evaluate the implications of interconnections

7 Source 1: How has the building of weirs helped and hindered interconnection along the Murray River?

HOW TO Mapping with BOLTSS, page 152

Why do floods occur?

Floods are caused by heavy rains that force rivers to carry more water than their channels can hold. The water spills over the river banks and onto the flat land next to rivers, known as floodplains.

Why rivers flood

Each river can hold only a certain amount of water – this is known as its **carrying capacity**. Heavy rain or a collapsed dam wall or levee can lead to a river exceeding its carrying capacity. When this happens, the river bursts its banks and **floods** the surrounding land.

People, cattle and wildlife can become stranded.

Floodplains

The flat lands next to river channels are called **floodplains**. Land on floodplains is usually reserved for parks, golf courses and farmland, with few permanent buildings.

Floodplains are some of the richest and most productive land in the world. The soil carried by floodwaters is spread over the surrounding land. After each flood, more and more soil is **deposited** on the floodplain.

Billions of people around the world live on floodplains because their fertile soil and direct access to fresh water make it easy to live there and grow food.

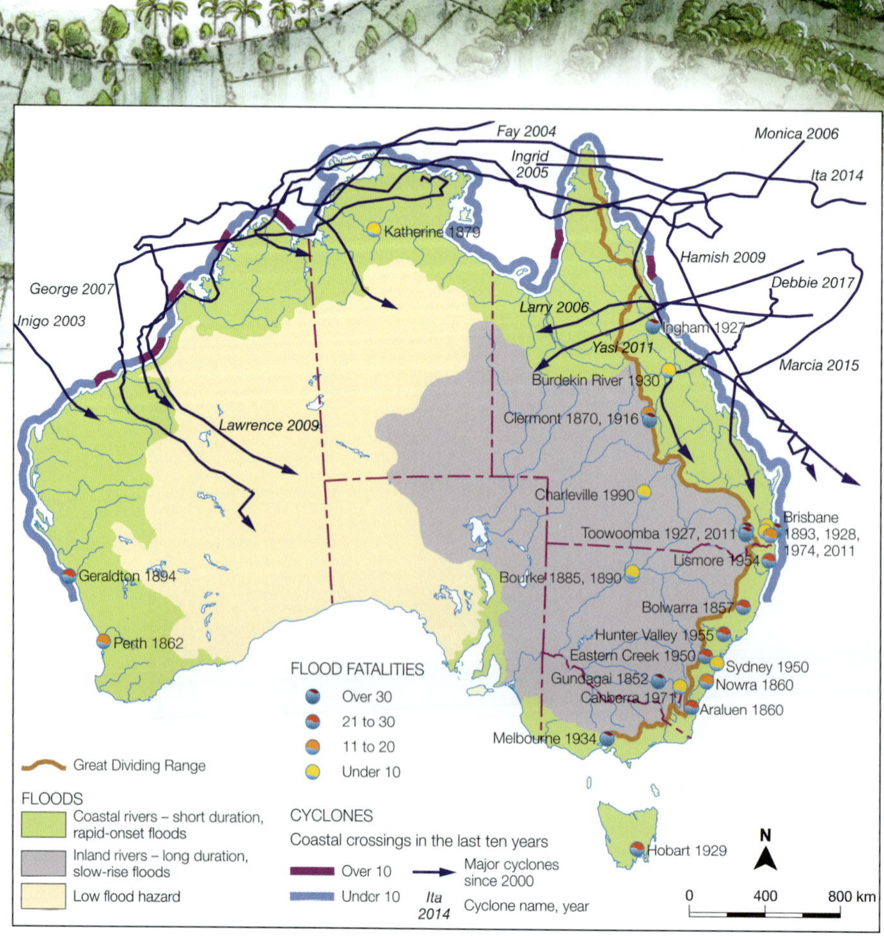

Source: Matilda Education Australia

Source 1

Map of Australian flood and cyclone areas

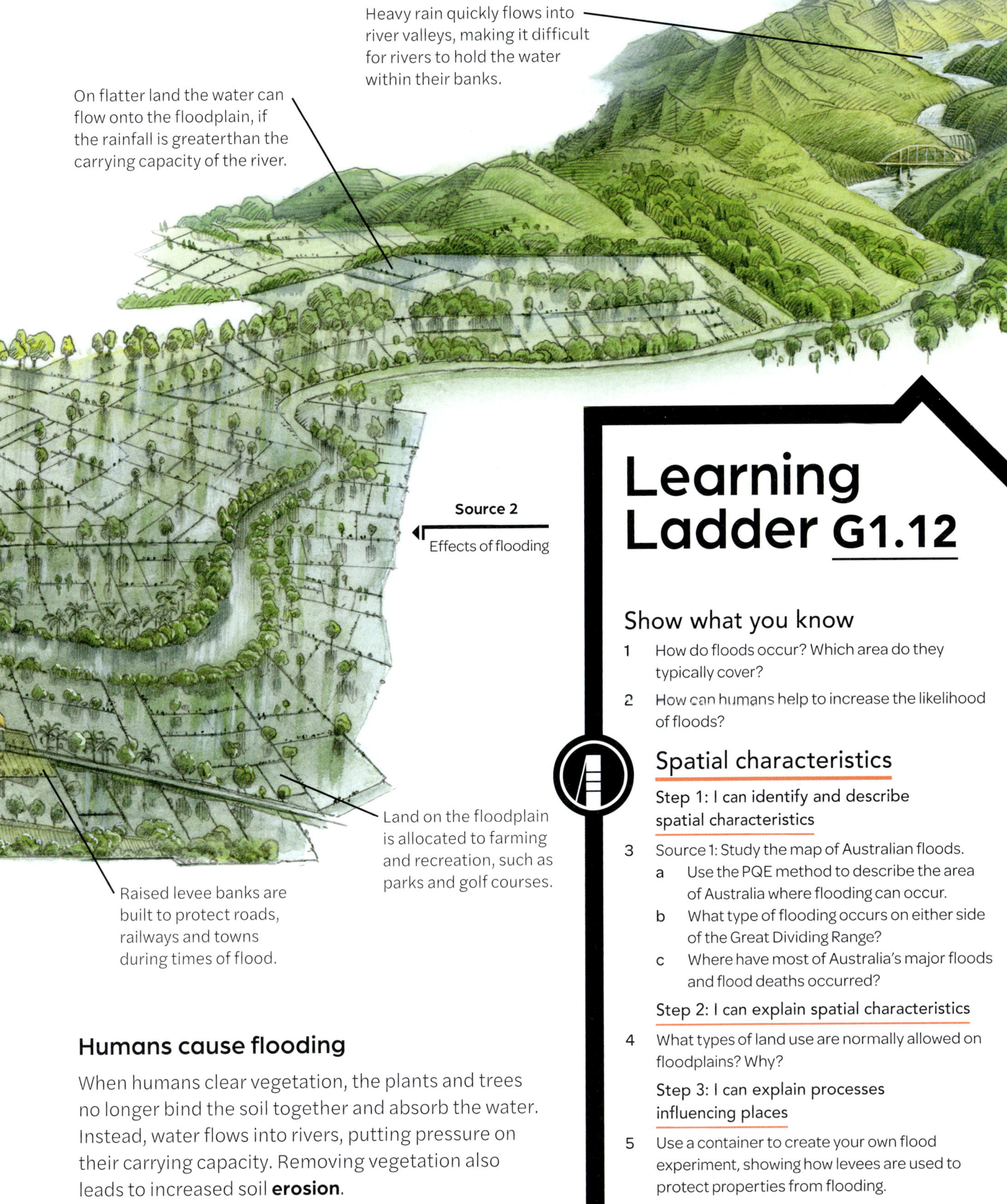

Source 2 Effects of flooding

Humans cause flooding

When humans clear vegetation, the plants and trees no longer bind the soil together and absorb the water. Instead, water flows into rivers, putting pressure on their carrying capacity. Removing vegetation also leads to increased soil **erosion**.

When we build towns and cities, we replace natural environments with hard surfaces – such as concrete and roads – that prevent water from soaking into the ground. When we build levee banks along rivers, we stop floodwaters, but we also increase the carrying capacity of the river. With levees, the river runs faster and deeper, and carries more water.

Learning Ladder G1.12

Show what you know

1 How do floods occur? Which area do they typically cover?

2 How can humans help to increase the likelihood of floods?

Spatial characteristics

Step 1: I can identify and describe spatial characteristics

3 Source 1: Study the map of Australian floods.
   a Use the PQE method to describe the area of Australia where flooding can occur.
   b What type of flooding occurs on either side of the Great Dividing Range?
   c Where have most of Australia's major floods and flood deaths occurred?

Step 2: I can explain spatial characteristics

4 What types of land use are normally allowed on floodplains? Why?

Step 3: I can explain processes influencing places

5 Use a container to create your own flood experiment, showing how levees are used to protect properties from flooding.

Step 4: I can predict changes in the characteristics of places

6 Source 1: Use the legend to locate the cyclone prone areas on the map. Is there an interconnection between these areas and floods in Australia?

HOW TO

PQE page 156

How can we plan for floods?

Flooding costs Australians an average of $400 million each year in reconstruction, loss of business and compensation to flood victims. We can help reduce the impact of flooding by predicting when and where a flood will hit, building protective structures and practising emergency drills.

Preparing for floods

Flooding occurs naturally for many rivers, so it is almost impossible to stop floods altogether. However, there are three steps we can take to reduce the impact of floods.

1 Predicting floods

Flood forecasters predict when and where floods will occur, giving people time to prepare and evacuate. Forecasters use a network of **rain gauges** that monitor rainfall along the river **catchment**, and use **stream gauging stations** along the rivers to measure river heights.

Forecasters use computer models to estimate how much rainfall will run off the catchment, and how long the run-off will take to reach the river and travel downstream.

2 Preparing communities

Flood engineers study past flood records to design flood protection strategies. Sometimes large raised banks, called levees, are built in flood-prone areas to deflect the flow of water and make it run in a certain direction. During large floods, levees sometimes help the river to run faster and deeper, which transfers the flood problem downstream.

Local councils map flood-prone areas, and make and enforce rules about building in these zones. Sporting fields and farming are allowed on flood-prone land, but housing and industry are not, and are usually zoned for development on higher land or in areas protected by levees.

3 Responding effectively

Planning for a flood emergency means that we can respond quickly to save lives and make the flooded area safe. Critical facilities, such as emergency hospitals and evacuation shelters, are located in areas that will not flood. Evacuation plans, emergency drills and education programs are conducted to help communities get ready for a flood emergency.

During a flood emergency, warnings are sent to mobile phones, and spread via social media networks such as Facebook and Twitter.

Using spatial technology

In Geography, we rely on maps and other visual representations of our landscape to make decisions and to understand patterns. These maps and images are made using digital software and hardware, which is referred to as **spatial technology**.

Maps and satellite images can be used during emergencies, such as floods, to help plan, prepare for and manage the situation. For example, emergency vehicles can use digital maps to know which roads have been flooded and need to be avoided. Global positioning systems (**GPS**) are also used to locate missing people or to pinpoint rescue operations. Spatial technology is very helpful for geographers – it can be constantly updated, and data is usually free for users.

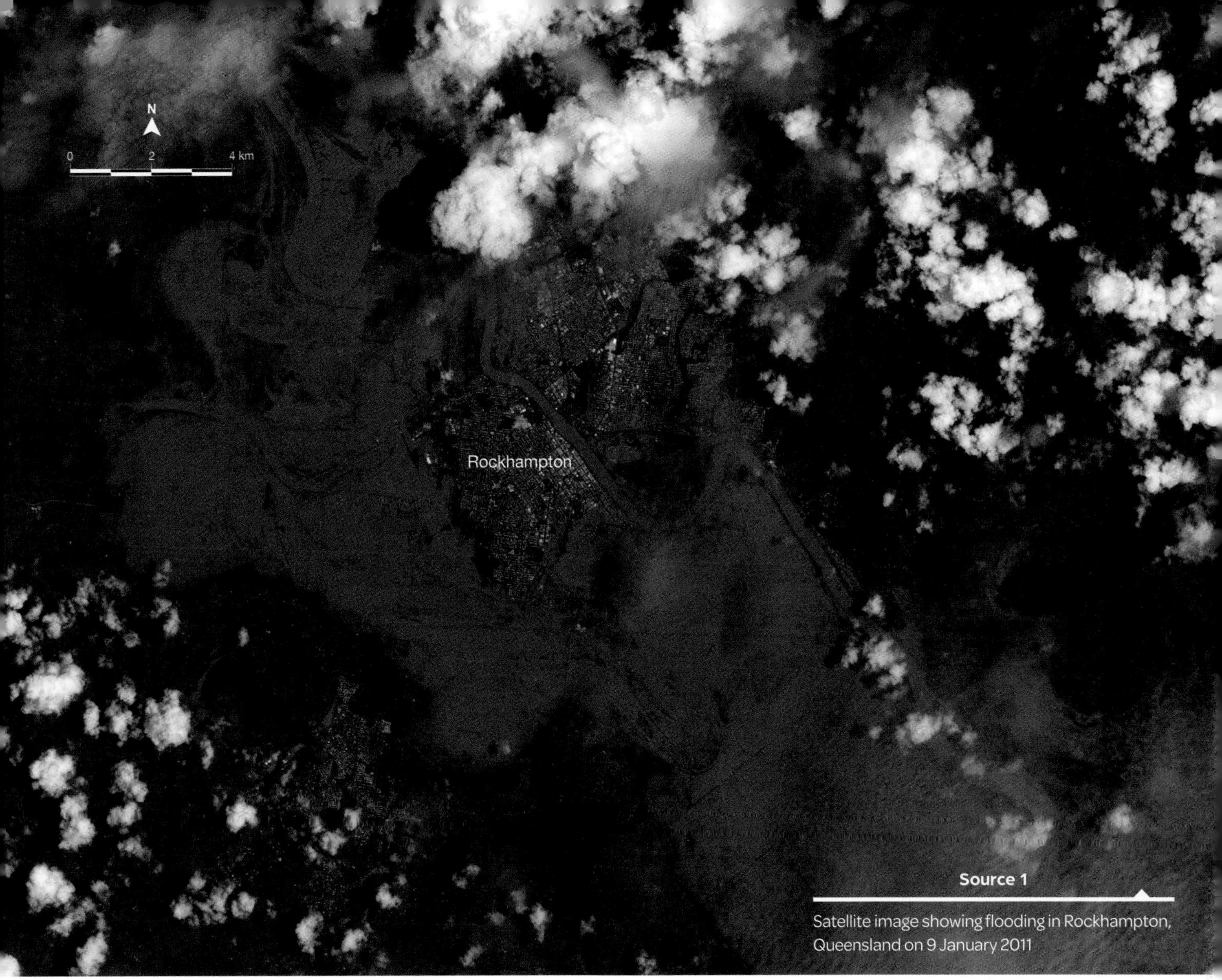

Rockhampton

Source 1

Satellite image showing flooding in Rockhampton, Queensland on 9 January 2011

Learning Ladder G1.13

Show what you know

1 What data do we collect to help predict floods?

2 Define the term 'spatial technology'.

Geographical challenge

Step 1: I can identify responses to a geographical challenge

3 Why are floods a geographic challenge for humans?

Step 2: I can compare responses to a geographical challenge

4 What rules do local councils enforce about building in flood-prone areas? Why?

Step 3: I can compare strategies for a geographical challenge

5 Look carefully at Source 1. Use Google Maps to help you prepare a sketch map of this area. Mark in the flooded area using the information shown on the satellite image. Explain why satellite images are useful in managing and predicting flooding events.

Step 4: I can evaluate alternatives for a geographical challenge

6 Make a list of the types of data collection methods useful in planning for floods and predicting them. Put an asterisk next to each of the methods that involve spatial technologies.

Satellite images, page 167

What are the costs of floods?

Floods have social, economic and environmental consequences. As an economic issue, floods are responsible for increasing costs around the world. This includes costs directly related to the flooding, such as stock losses and damage to property, and the cost to the economy from having business disrupted.

Economics of flooding

A flood is an **economic issue** that brings both **benefits** and **costs**. The reason that large numbers of people live on floodplains is because of the rich soil deposited on them during floods. The fertile land on floodplains supports the healthy growth of pasture and crops.

However, because so many people are attracted to living on or near floodplains, the economic cost of floods is increasing every year. Flooding affects around 250 million people in the world each year.

Floods are Australia's most expensive natural disaster, with the economic cost estimated at $400 million dollars each year. The costs can be divided into direct costs and indirect costs.

- *Direct costs* are caused by damage to buildings and contents, vehicles, livestock and crops crops, as well as damage to **infrastructure** such as roads, bridges and powerlines.
- *Indirect costs* include disruption to transport and business, loss of income and legal costs.

As a result of these costs, the **flow-on effect** for consumers is higher prices caused by a shortage of supply. For example, in March 2017, the price of bananas soared after banana plantations in north-east Queensland were destroyed by Cyclone Debbie and flood damage.

Source 1

Former Prime Minister Malcolm Turnbull surveying the flooded areas in Bowen, Queensland, after Cyclone Debbie dumped record rains in March 2017. The damage bill was estimated at $2 billion.

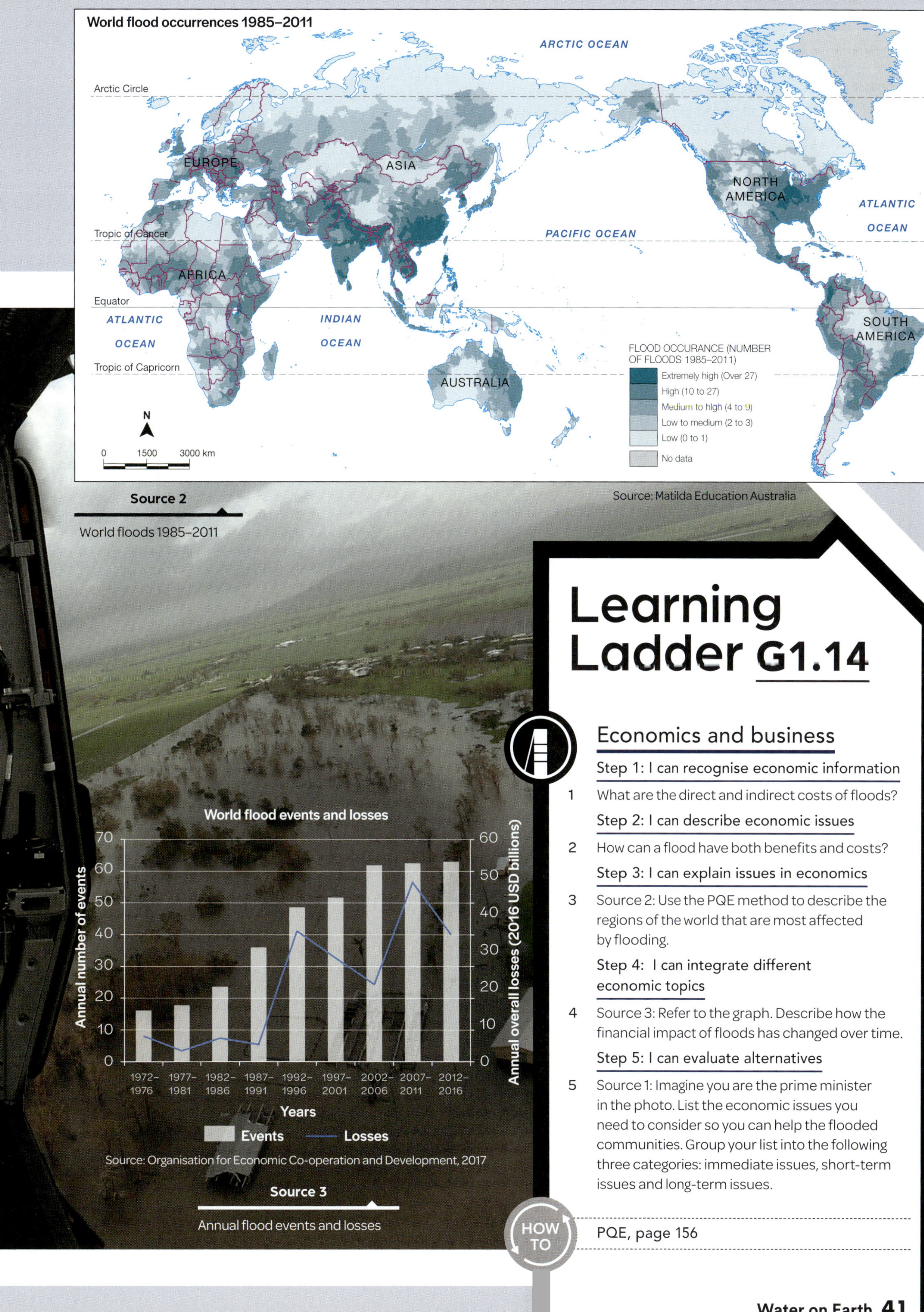

Source 2

World floods 1985–2011

Source 3

Annual flood events and losses

Learning Ladder G1.14

Economics and business

Step 1: I can recognise economic information

1 What are the direct and indirect costs of floods?

Step 2: I can describe economic issues

2 How can a flood have both benefits and costs?

Step 3: I can explain issues in economics

3 Source 2: Use the PQE method to describe the regions of the world that are most affected by flooding.

Step 4: I can integrate different economic topics

4 Source 3: Refer to the graph. Describe how the financial impact of floods has changed over time.

Step 5: I can evaluate alternatives

5 Source 1: Imagine you are the prime minister in the photo. List the economic issues you need to consider so you can help the flooded communities. Group your list into the following three categories: immediate issues, short-term issues and long-term issues.

HOW TO

PQE, page 156

Masterclass

Learning Ladder

Work at the level that is right for you or level-up for a learning challenge!

Melbourne flood elevations

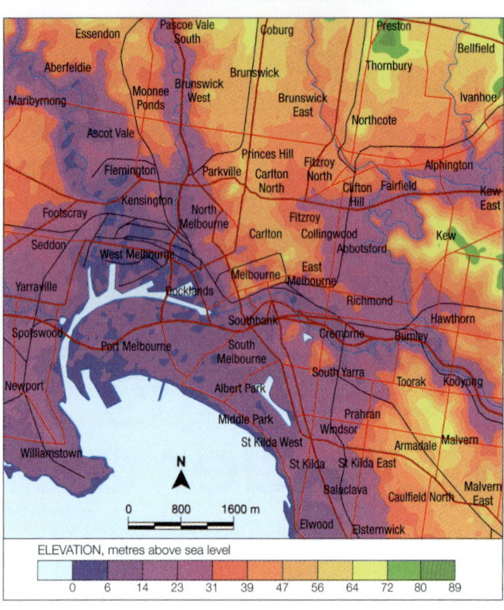

Source 1

Melbourne climate

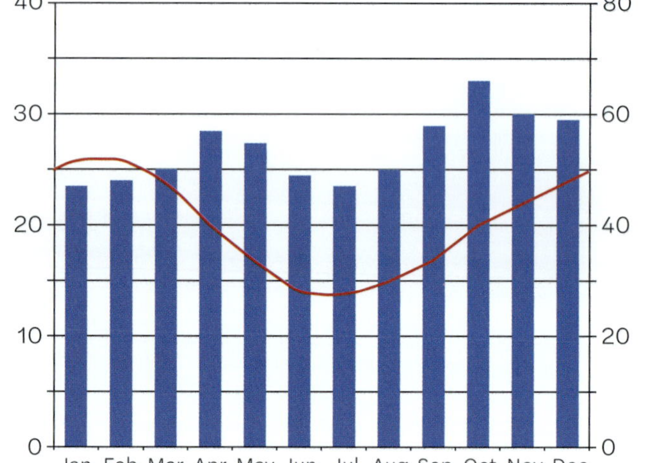

Source 2

Step 1

a **I can identify and describe spatial characteristics**

Look at Source 3. Where have floods caused the most deaths in Australia?

b **I can identify and describe interconnections**

Using Source 1, identify whether these local areas are at risk of flooding. What are their elevations?

a Spotswood b Brunswick c Kew

c **I can identify responses to a geographical challenge**

Study Source 1. Can you recognise five key impacts of this flood *environment*? How can we prepare for floods to reduce impact on *places*?

d **I can collect, record and display data in simple forms**

Using Source 2, identify what the bars and lines on the graph represent.

e **I can use geographic terminology to interpret data**

Look at Source 1. Use compass points to describe the areas with the highest elevation in Melbourne.

Step 2

a **I can explain spatial characteristics**

Source 3: Which areas of Australia are most affected by cyclones?

b **I can explain interconnections**

How is the term *interconnection* used in Geography?

c **I can compare responses to a geographical challenge**

Refer to Source 2. At which times are people in Melbourne most at risk of floods? At which times are they most at risk of bushfires?

Source 4

Floods in Kerala, India

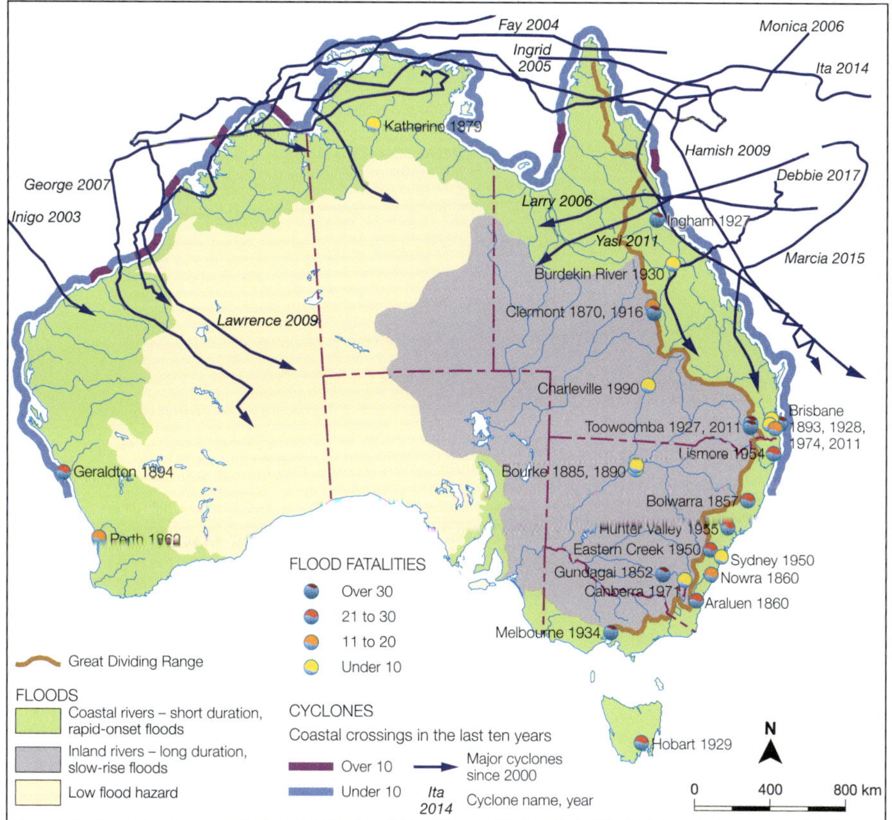

Source 3

d **I can recognise and use different types of data**

Create a table with the column headings 'Primary data collection methods' and 'Secondary data collection methods'. List three methods for each.

e **I can describe patterns and trends**

Refer to Source 1. Using your knowledge of elevation and water movement, suggest which regions in Melbourne are most at risk of flooding.

Step 3

a **I can explain processes influencing places**

What processes led to the change in the place in Source 4?

b **I can identify and explain the implications of interconnections**

What are the implications of the interconnection between cyclone paths and coastal communities in northern Australia?

c **I can compare strategies for a geographical challenge**

How do desert communities tackle the geographical challenge of providing enough water to live?

d **I can choose, collect and display appropriate data**

Create a step-by-step guide explaining how to successfully carry out a field sketch.

Masterclass

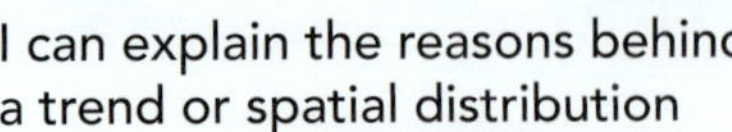

e I can explain the reasons behind a trend or spatial distribution

Source 2: Which seasons in Melbourne have the lowest and highest rainfall?

Step 4

a I can predict changes in the characteristics of places

Predict what would happen to the place in Source 4 if the floodwaters were to rise by another two metres.

b I can evaluate the implications of significant interconnections

How has the building of weirs on the Murray River helped and hindered interconnection along the Murray River?

c I can evaluate alternatives for a geographical challenge

During construction, what happens to the land behind a dam wall? Is there an alternative to give humans a reliable water supply?

d I can use data to support claims

Design a fieldwork project based on this question: 'How does water connect people and places?' Outline and justify one primary and one secondary method to help answer this question.

e I can analyse relationships between different data

Source 4: Compare the relationship between coastal river flooding and flood deaths.

Step 5

a I can analyse the impact of change on places

Source 1: Climate change models suggest that by 2070 rainfall intensity will increase by 6%, making flooding more common. What impact could this have on Melbourne's largest new growth area around Port Melbourne?

b I can explore spatial association and interconnections

Find a distribution map of world population from a reliable source. Compare the map to Source 2 on page 41. Consider the spatial association between population density and the number of floods and describe a region of the world where deaths from flooding is likely to be higher.

c I can plan action to tackle a geographical challenge

Make a list of the types of data collection methods useful in planning for floods and predicting them.

d I can evaluate data

Which form of data is most useful when investigating spatial patterns and distributions: quantitative or qualitative? Justify your answer.

e I can draw conclusions from analysing collected data

Climate change models suggest that by 2070 Melbourne's rainfall will decrease by 11% and extreme heat days will increase from 9 days to 26 days per year. What challenges will this pose for planners?

Capstone

How can I understand water on Earth?

In this chapter, you have learnt a lot about water on Earth. Now you can put your new knowledge and understanding together for the capstone project to show what you know and what you think.

In the world of building, a capstone is an element that finishes off an arch or tops off a building or wall. That is what the capstone project will offer you, too: a chance to top off and bring together your learning in interesting, critical and creative ways. You can complete this project yourself, or your teacher can make it a class task or a homework task.

Scan this QR code to find the capstone project online.

mea.digital/GHV7_G1

G2

Water for life

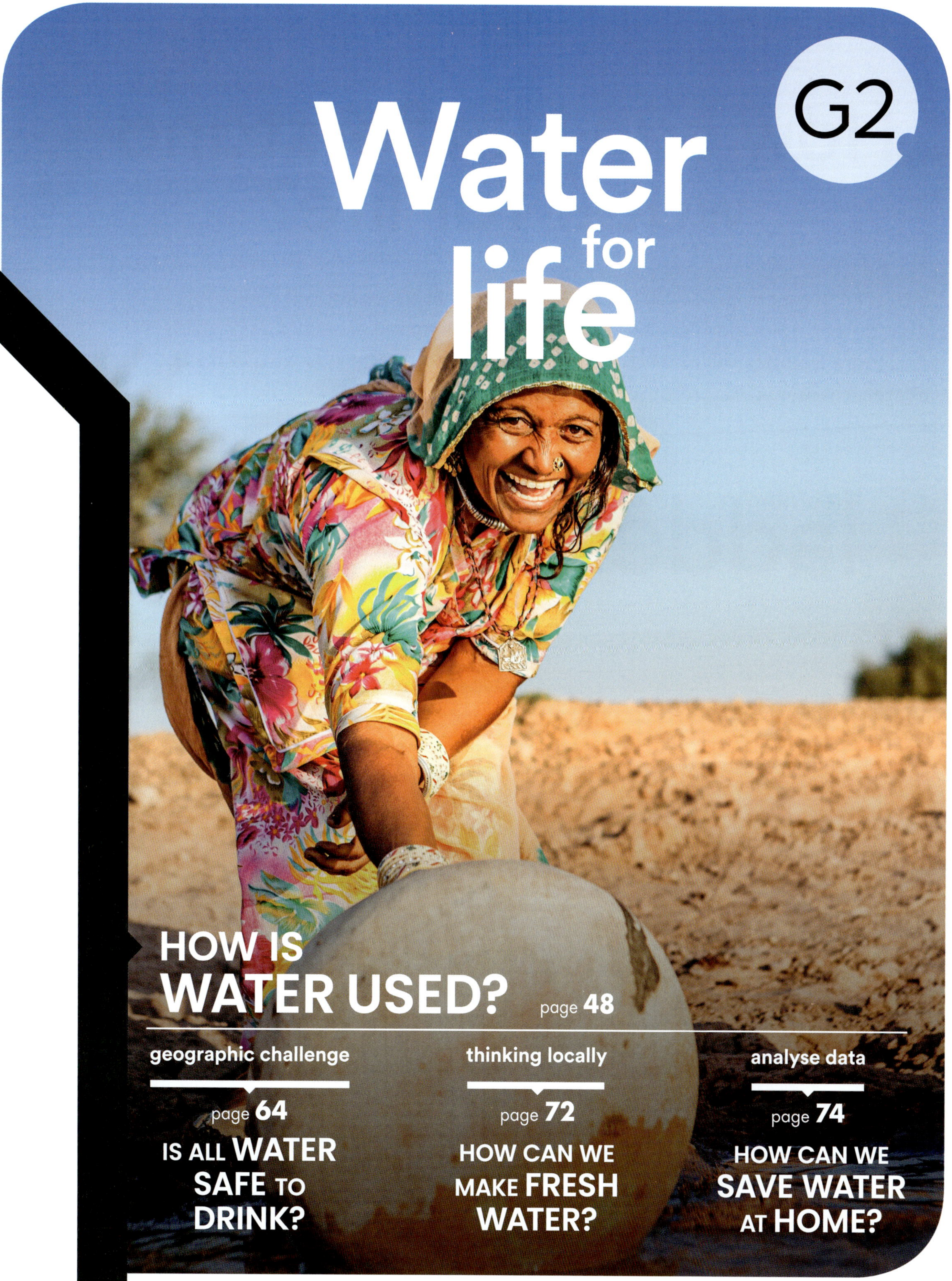

HOW IS WATER USED?

page 48

geographic challenge

page 64

IS ALL WATER SAFE TO DRINK?

thinking locally

page 72

HOW CAN WE MAKE FRESH WATER?

analyse data

page 74

HOW CAN WE SAVE WATER AT HOME?

How can I understand water for life?

Water is an essential part of life. Without water, life as we know it could not exist on Earth. Furthermore, we would not be able trade products using ships, and there would be no marine or fresh water habitats. The way we use and preserve water is important to ensure a sustainable future.

learning ladder

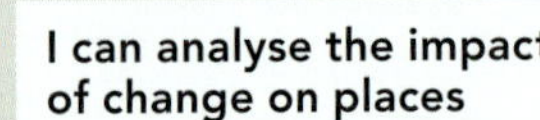

Step	Spatial characteristics	Interconnections	Geographical challenge
step 5	**I can analyse the impact of change on places** I can analyse and evaluate the implications of changing water use over time and at different scales and calculate its impact on people and environments.	**I can explore spatial association and interconnections** I can compare distribution patterns and the interconnections between them; e.g. the distribution of high rainfall or irrigation and rice production.	**I can plan action to tackle a geographical challenge** I can frame questions, evaluate findings, plan actions and predict outcomes to tackle a water-based geographical challenge.
step 4	**I can predict changes in the characteristics of places** I can predict changes in the characteristics of places over time due to variations in water supply and use.	**I can evaluate the implications of significant interconnections** I can identify, analyse and explain key water-based interconnections within and between places, and evaluate their implications over time and at different scales.	**I can evaluate alternatives for a geographical challenge** I can weigh up alternative views and strategies on a water-based geographical challenge using environmental, social and economic criteria.
step 3	**I can explain processes influencing places** I can explain the series of actions leading to change in a place, such as drought or overuse of water.	**I can identify and explain the implications of interconnections** I can identify, analyse and explain water-based interconnections and explain their implications.	**I can compare strategies for a geographical challenge** I can compare strategies for a geographical challenge, taking into account a range of factors and predicting the likely outcomes.
step 2	**I can explain spatial characteristics** I can identify concepts of Space, Place, Interconnection, Change, Environment, Sustainability and Scale (SPICESS) when I read about water for life.	**I can explain interconnections** I can describe and explain interconnections and their effects, such as the multiple uses of a river from source to mouth.	**I can compare responses to a geographical challenge** I can identify and compare responses to a geographical challenge and describe its impact on different groups.
step 1	**I can identify and describe spatial characteristics** I can talk about spatial characteristics at a range of scales; e.g. rainfall distribution patterns for Australia and the world.	**I can identify and describe interconnections** I can identify and explain simple interconnections involved in the phenomena such as the water cycle and flooding.	**I can identify responses to a geographical challenge** I can find responses to a geographical challenge such as using safe water or flooding and understand the expected effects.

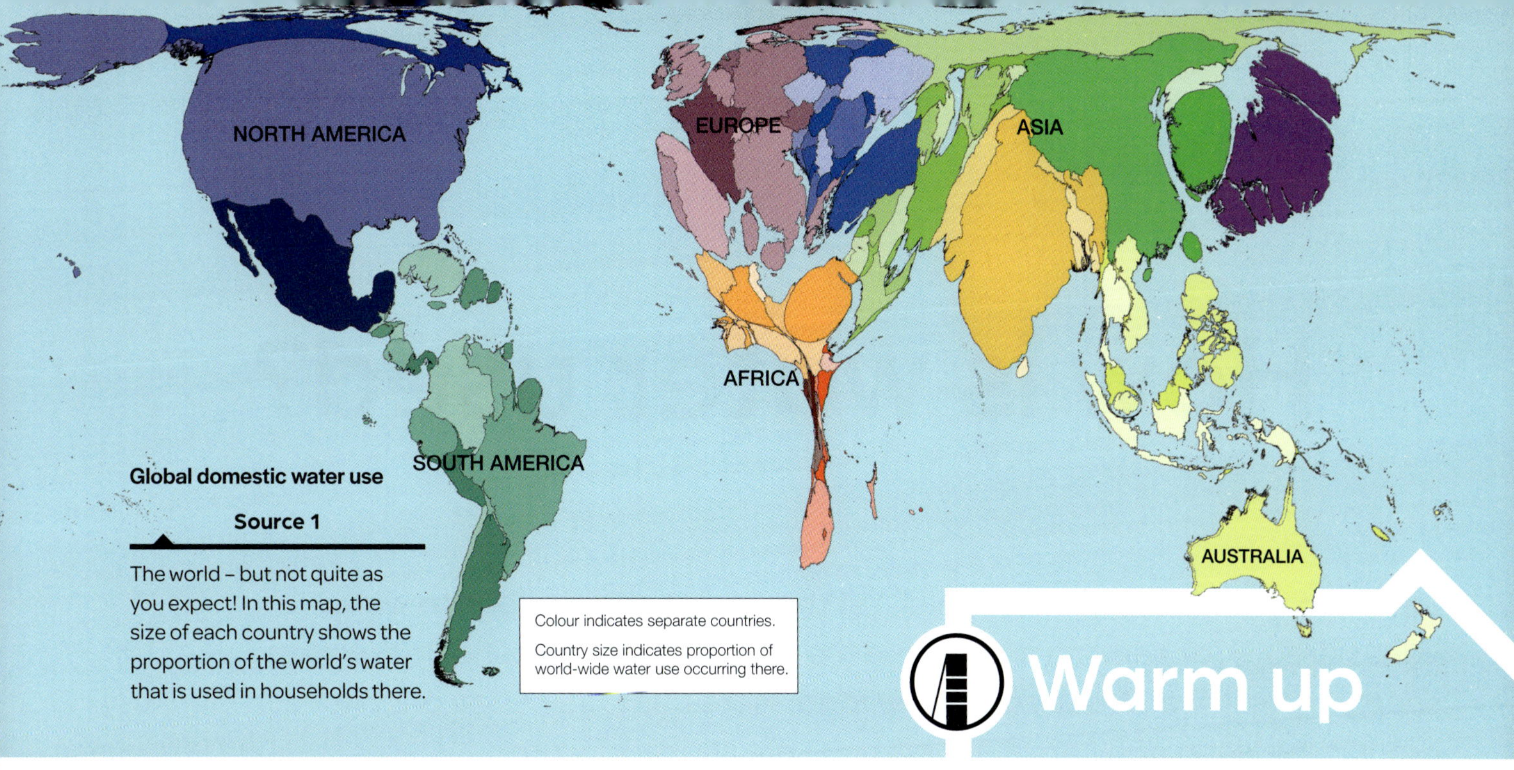

Global domestic water use

Source 1 The world – but not quite as you expect! In this map, the size of each country shows the proportion of the world's water that is used in households there.

Warm up

Spatial characteristics

1 What does the geographic concept of *space* look at (see page 4)?

2 Why are the actual sizes of countries in Source 1 changed?

Interconnections

3 *Interconnection* is the idea that two things or phenomena are related, interact or are linked in some way. Suggest what other thing might be interconnected with India and China's high domestic water use.

Geographical challenge

4 What responses have Australian governments made to the geographical challenge of providing water for the population in a dry country like Australia?

Collect, record and display data

5 What raw data is shown as a map in Source 1?

6 What are the advantages and disadvantages of using maps like Source 1?

Analyse data

7 Looking at Source 1:

a Why do you think that household water use in North America is so great?

b Why do you think that the household water use in Africa is so small?

Collect, record and display data

I can evaluate data
I can determine whether data presented about water on Earth is reliable and assess whether the methods I used in the field or classroom were helpful in answering a water-based research question.

I can use data to support claims
I can select or collect the most appropriate data and create specialist maps and information using ICT to support investigations into water and its use on Earth.

I can choose, collect and display appropriate data
I can select useful sources of water data and represent them to conform with geographic conventions.

I can recognise and use different types of data
I can define the terms primary, secondary, qualitative and quantitative data and represent data in more complex forms.

I can collect, record and display data in simple forms
I can identify that maps and graphs use symbols, colours and other graphics to represent data.

Analyse data

I can draw conclusions from analysing collected data
I can summarise findings and use collected data to support key patterns and trends I have identified for water-based research.

I can analyse relationships between different data
I can use multiple data sources, overlays and GIS to find relationships that exist in patterns of water use on Earth.

I can explain the reasons behind a trend or spatial distribution
I can identify Social, Historical, Economic, Environmental, Political and Technological (SHEEPT) factors to help me explain patterns in data.

I can describe patterns and trends
I can identify Patterns, Quantify them and point out Exceptions (PQE) to describe the patterns I see.

I can use geographic terminology to interpret data
I can identify increases, decreases or other key trends on a map, graph or chart about water.

How is water used?

Clean water is essential for survival. Yet the access people have to water varies greatly, depending on where in the world they live. Most of the world's supply of fresh water is used to grow food. More than 785 million people around the world don't have access to a safe supply of water.

Supplying the world with water

In the 1900s, the world's population tripled and water usage increased by six times. Agriculture accounts for 70 per cent of all water consumed, compared to 20 per cent for industry and 10 per cent for home use.

Supplying the world with enough water is now a huge issue. In Australia in 2016, we used 16 000 gigalitres of water – that's enough water to completely fill the Melbourne Cricket Ground 10 000 times!

Throughout the world there are places that are **water rich** and places that are **water poor**. Australia is water poor. It is the world's driest inhabited **continent**, and the quantity of water resources varies greatly across the continent.

However, Australia is a wealthy country (referred to in Geography as a More Economically Developed Country (MEDC)), so nearly everyone can access **safe water** to drink. Many other countries in the world are not so lucky. One in nine people in the world don't have access to safe water, and millions of people get sick or die from drinking dirty water.

Per person, Australians are one of the biggest users of water in the world. We need to manage our valuable water resources carefully, so that we balance the needs of different users in homes and on farms, as well as the needs of industry. By making a few simple changes, we can all help save precious water resources.

Source 1

Children drinking from a tap at a primary school in a displaced people's camp in Nigeria. The school and the fresh water pump were supplied by UNICEF – the United Nations Children's Fund aid agency.

Human water needs	Water footprint	Access to drinking water
2–4 L The daily drinking water requirement per person is 2–4 litres. **2000+ L** It takes 2000 to 5000 litres of water to produce one person's daily food. 	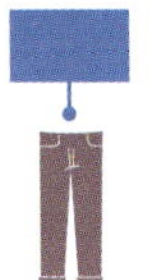    500 g of rice 1700 L; 1 pair of jeans 3800 L; 1 pair of shoes 8000 L; 1 kilo of beef 13 000–15 000 L; 1 car 30 000 L	 **785 000 000** **people don't have access to safe water.**

Source 2

This infographic displays some important facts about water on Earth, including human water needs, the water 'footprint' of various items and access to drinking water.

Learning Ladder G2.1

Show what you know

1 What is most fresh water used for?

2 Source 1: Look at the photo to answer these questions.
   a What are the children queuing up to do?
   b Why is it important that the children drink safe water?
   c Who supplied this safe water?

Collect, record and display data

Step 1: I can collect, record and display data in simple forms

3 Brainstorm a list of all the ways that water is:
- used in Australia
- transported to homes and farms
- stored for our use.

Step 2: I can recognise and use different types of data

4 Create a summary table listing key quantitative statistics about water use in Australia.

Step 3: I can choose, collect and display appropriate data

5 Prepare a graph that clearly shows the different amounts of water required to produce the following products.
- Toast: 650 litres
- Apple: 70 litres
- Cheese: 2500 litres
- Egg: 200 litres

Step 4: I can use data to support claims

6 Research the top 10 thirstiest products to produce and prepare an infographic like Source 2 to display the data.

HOW TO

Simple graphs, page 162

Where is water in Australia?

Australia is almost the driest continent on Earth – the only continent with less rain is Antarctica. Most of Australia's rain falls in thin strips along the coast, with desert occupying much of the centre of the continent. Only 12 per cent of the rain that falls is collected in rivers, so 30 per cent of Australians need to use groundwater as their main water supply.

Rainfall in Australia

Australia is the driest populated continent on Earth. Two-thirds of the continent is classified as **desert** or **semi-desert**. This classification depends on the amount of rainfall.

- Desert: less than 250 mm rainfall per year.
- Semi-desert: 250–500 mm rainfall per year.

Most of Australia's rain falls on its northern, eastern and south-western coasts. Australia's north coast receives most of its tropical rainfall in summer with the arrival of the **monsoon** (see also page 24).

The Great Dividing Range runs the entire length of Australia's east coast and forces winds to drop their moisture as rain along the coastal strip. Then, having released moisture to the east of the range, the winds sink as dry air to the desert and semi-desert regions to the west of the Great Dividing Range. This process is known as **orographic rainfall** (see page 22).

The wettest place in Australia is Bellenden Ker in north Queensland. Bellenden Ker averages 8312 mm of rainfall per year. It receives so much rain because of its tropical location on the eastern slopes of the Great Dividing Range.

The driest place in Australia is Kati Thanda (Lake Eyre) in South Australia. It averages only 125 mm of rainfall per year. **Air masses** have usually dropped their rain on the south-west corner of Western Australia, and are dry by the time they reach so far inland.

Australian rivers

Because Australia is such a dry continent, it doesn't have many large rivers or large permanent lakes. In fact, Australia has the lowest volume of water in rivers of any inhabited continent in the world. Just 12 per cent of Australia's rainfall is collected in rivers.

Australia: Average annual rainfall

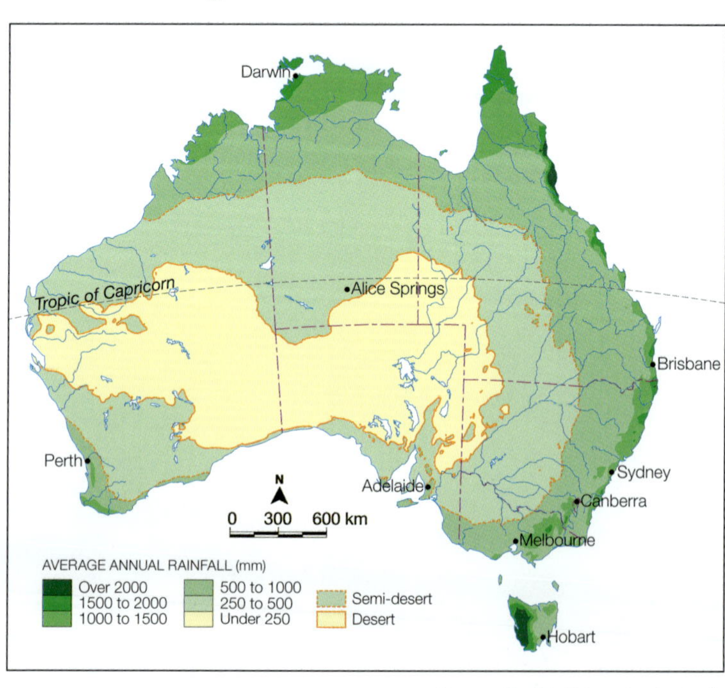

Source: Matilda Education Australia

Source 1

Australian average rainfall per year

Australia's inland rivers are different to all other rivers in the world. The volume of water in most rivers increases the further the river courses, but Australia's inland rivers tend to lose water. The Darling River, which is Australia's longest river, loses enough water through **evaporation** each year to fill Sydney Harbour four times!

Many Australian rivers flow mainly – or only – when floods come down from far upstream. Heavy tropical rain falling in north Queensland is sometimes channelled through rivers to South Australia, where it floods the Kati Thanda **salt pan**.

Australian groundwater

One-third of Australia's residents rely on **groundwater** as their main source of water. Groundwater is water that is located in **aquifers** beneath Earth's surface. The aquifers are fed by surface water from rainfall and rivers. Communities access underground aquifers by drilling a well or bore and pumping water to the surface (see page 27).

The Great Artesian Basin in eastern Australia is the world's largest aquifer as it covers more than 1.7 million square kilometres – or 22 per cent of Australia's land mass. Water in an aquifer is trapped underground in a layer of sandstone covered by **sedimentary** rock, and can be more than one million years old.

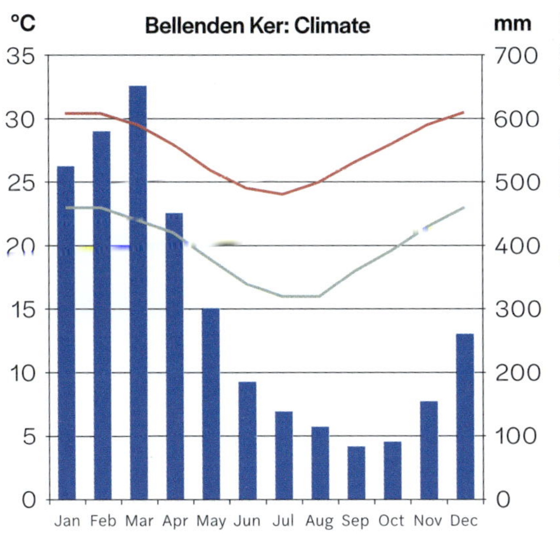

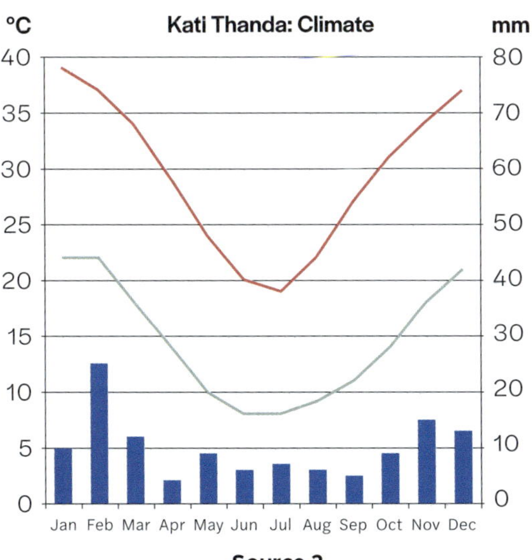

Source 2

Climate graphs of the wettest and driest places in Australia

Learning Ladder G2.2

Show what you know

1 How much of Australia is desert or semi-desert?

2 How do Australian rivers differ to most other rivers in the world? Why?

3 Why are aquifers important to Australians?

4 Look carefully at Source 2:

  a How much rainfall does Bellenden Ker receive in February?

  b How much rainfall does Kati Thanda receive for the entire year?

Collect, record and display data

Step 1: I can collect, record and display data in simple forms

5 Research and list four different weather recording instruments and what they measure.

Step 2: I can recognise and use different types of data

6 Is the information in Source 1 primary or secondary data to you? Why?

Step 3: I can choose, collect and display appropriate data

7 What primary and secondary data collection methods could you use to compare temperature and rainfall over a one-week period at your school?

Step 4: I can use data to support claims

8 What source on these pages would you select to show to an international corporation that is wanting to establish rice farms in Australia that require large amounts of water. List a further three pieces of data that the corporation would find useful in making a decision.

HOW TO

Climate graphs, page 165

How do droughts and bushfires occur?

Record-breaking temperatures and severe drought led to massive bushfires across Australia in 2019 and 2020. Much of Australia had been in drought conditions for years, providing the dry conditions that make it easier for fires to begin and to spread. Scientists have warned that global warming is resulting in a hotter and drier climate in Australia, which will continue to drive more frequent and severe bushfires.

Australia in drought 2020

A **drought** is a prolonged period of water shortages caused by below-average rainfall, increased evaporation from higher temperatures or a reduction in surface water or groundwater supplies. A drought can last for months or years and is driven by not only lack of rainfall and high temperatures but also by overuse from growing populations.

The 2020 drought has been driven by a positive Indian Ocean Dipole; an event where temperatures on the sea surface are warmer in the western Indian Ocean, and cooler in the east. The difference between the two temperatures is the largest in 60 years, leading to high rainfall and floods in eastern Africa and droughts in Australia and south-east Asia.

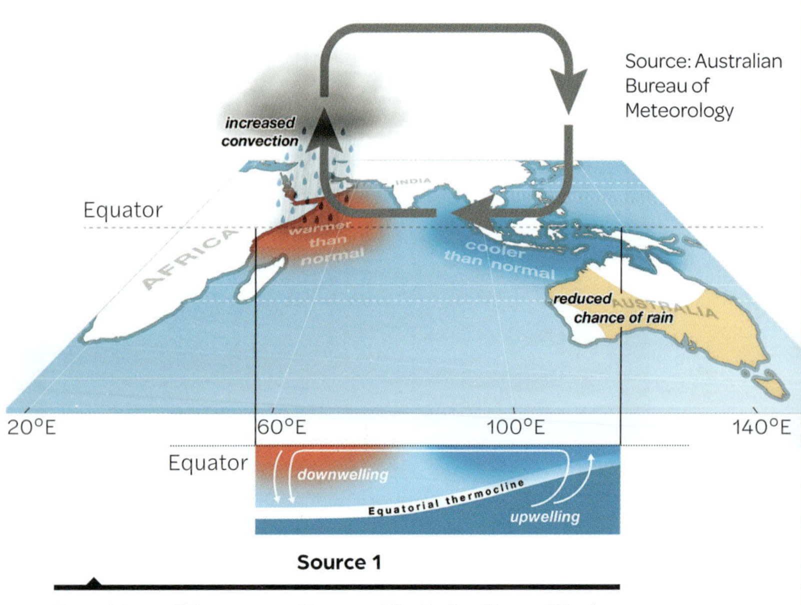

Source 1

Drought conditions caused by a positive Indian Ocean Dipole

Rainfall shortages

Lower than average rainfall has plunged most of New South Wales and Queensland into drought since early 2017. Rainfall shortages have been most extreme in the northern half of New South Wales and southern Queensland, where rainfall totals have been the lowest ever recorded.

In 2019 Australia recorded its driest year ever, with the lowest annual rainfall the country has ever seen.

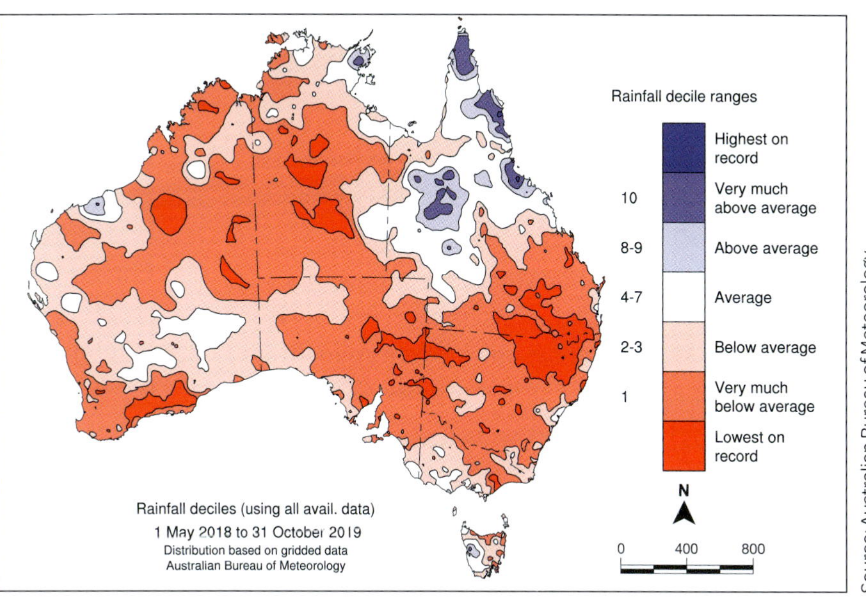

Source 2

Australian rainfall shortages in 2018 and 2019

Source 3

Farmer Gary Mooring feeds sheep on his drought-ravaged farm at Louth in northern New South Wales in February 2019. In a three-year period very little rain fell over most of New South Wales and Queensland. Families who have lived on the land for generations were forced to sell their properties, starving livestock died and some rural towns ran out of water.

Australia is getting warmer

Since 1910 Australia's climate has warmed by more than one degree Celsius. **Climate change** is largely driven by increased carbon dioxide and other human-made emissions into the atmosphere. These gases trap heat just like the glass roof of a greenhouse. The greenhouse effect is expected to deliver even hotter and drier conditions to Australia.

Climate change is already leading to more severe heatwave conditions in Australia. Nine of the 10 warmest years on record have occurred since 2005. In 2019 Australia's annual mean temperature was 1.52 degrees Celsius above average – making 2019 the hottest year ever recorded in Australia.

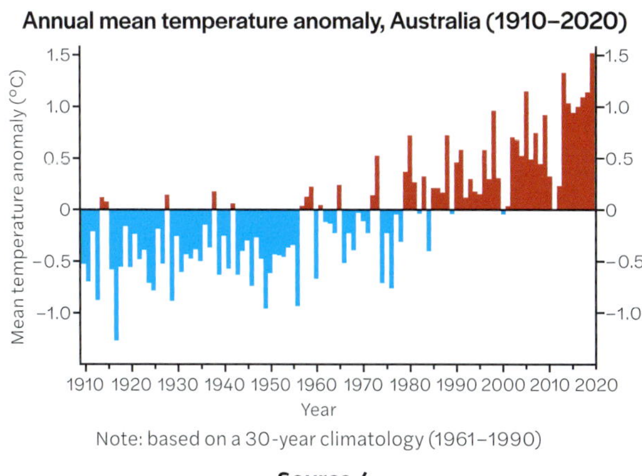

Source 4

Variations to the average annual temperature in Australia

Australian bushfires 2019–20

The combination of drought over much of Australia and heatwaves in New South Wales and Queensland combined to cause some of the worst fires in Australian history in the 2019-20 spring and summer period.

Bushfires occur when weather conditions are hot, dry and windy and there is enough dry fuel to burn. In these conditions it only takes a spark from a machine or a lightning strike for a fire to begin.

Once fires have started, wind fuels the fire with oxygen and moves the fire forward. Wind also carries burning leaves, twigs and bark, known as **embers**, ahead of the fire. Embers cause new **spot fires** to ignite. The intense heat from the 2019-20 bushfires caused rare **pyrocumulonimbus clouds** to form above them. These clouds can produce thunderstorms and lightning that can spark other bushfires or downpours to put them out.

Record-breaking temperatures and months of severe drought created perfect conditions for bushfires to spread rapidly and difficult conditions for the thousands of firefighters called into action.

The 2019-20 bushfires affected all states of Australia except the Australian Capital Territory, although smoke from the fires blanketed the nation's capital. More than 10 million hectares of land burned, an area almost the size of England. At least 24 humans died, along with 100 000 head of cattle and 480 000 000 native animals and birds. More than 1600 homes were destroyed, forcing thousands of people to seek shelter elsewhere.

A state of emergency for all of New South Wales and affected areas of Victoria came into force. Parks, camping grounds and roads were closed and holidaymakers in the peak holiday season were told to leave coastal towns in danger along a 260-kilometre stretch of NSW coast.

In the holiday town of Mallacoota in far eastern Victoria, 4000 residents and tourists fled to the beach on new year's eve in 2019. Only a change in wind direction stopped the fire from reaching them on the foreshore. The stranded people were eventually evacuated by the Australian navy.

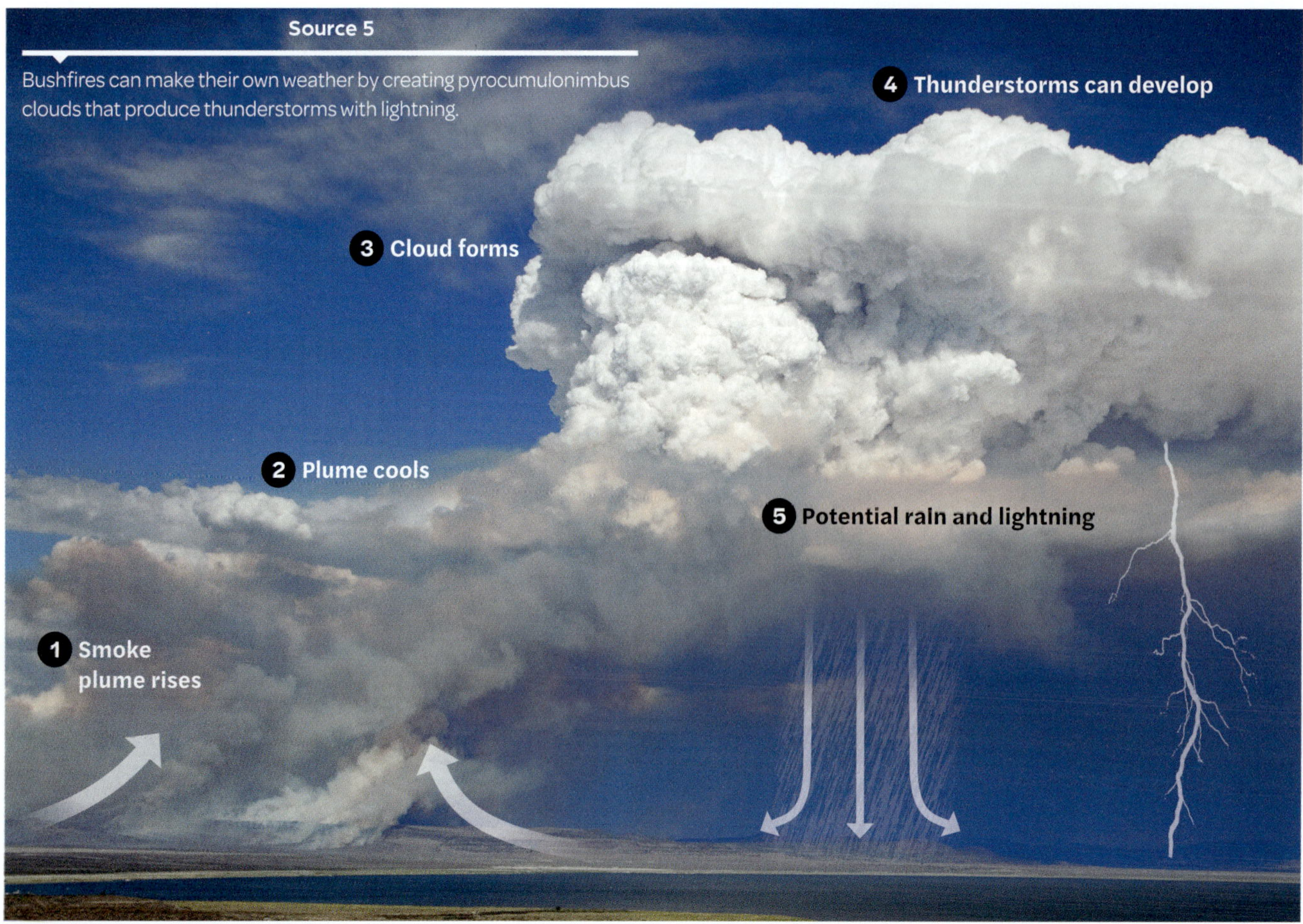

Source 5

Bushfires can make their own weather by creating pyrocumulonimbus clouds that produce thunderstorms with lightning.

Source 6

The remains of burnt out buildings in the main street in the New South Wales town of Cobargo on December 31, 2019. Thousands of holidaymakers and locals were forced to flee to coastal towns on Australia's south-east coast as bushfires moved closer.

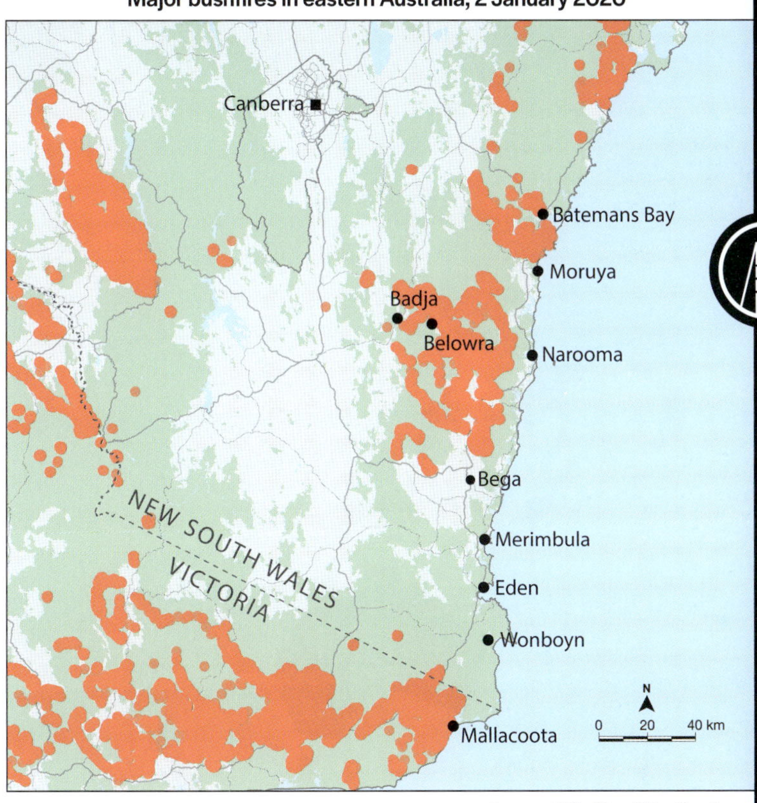

Source 7

Bushfires in NSW and eastern Victoria, early January 2020

Learning Ladder G2.3

Show what you know

1. How do bushfires occur?
2. Source 3: Describe what is happening in this aerial photograph and why it is taking place.
3. Source 2: Which areas of New South Wales had the lowest rainfall on record in 2018-19?
4. Source 4: Describe the trend in variations to Australia's average annual temperature.
5. Source 7: Which areas of New South Wales and Victoria were burned and what was the impact of bushfires across Australia?

Interconnections

Step 1: I can identify and describe interconnections

6. Suggest two ways that pyrocumulonimbus clouds are interconnected to bushfires?

Step 2: I can explain interconnections

7. Source 1: Explain how the Indian Ocean Dipole is interconnected with drought conditions in Australia.

Step 3: I can identify and explain the implications of interconnections

8. Explain how drought and bushfires are interconnected.

Step 4: I can evaluate the implications of significant interconnections

9. Evaluate the implications of the interconnection between drought and hot, dry and windy conditions. What action can be taken to reduce the likelihood of this interconnection occurring?

Who uses the most water?

In the last 50 years, the human population of Earth has grown to 7.6 billion people – and in that time, water use has doubled. India, China and the USA consume a third of all the water used globally. In total water use, Australia is a small water user, but per head of population we are one of the thirstiest nations on the planet. One in four people live in countries without the pipes, pumps and equipment to extract water.

Source 1

World water use per person per year

World water use

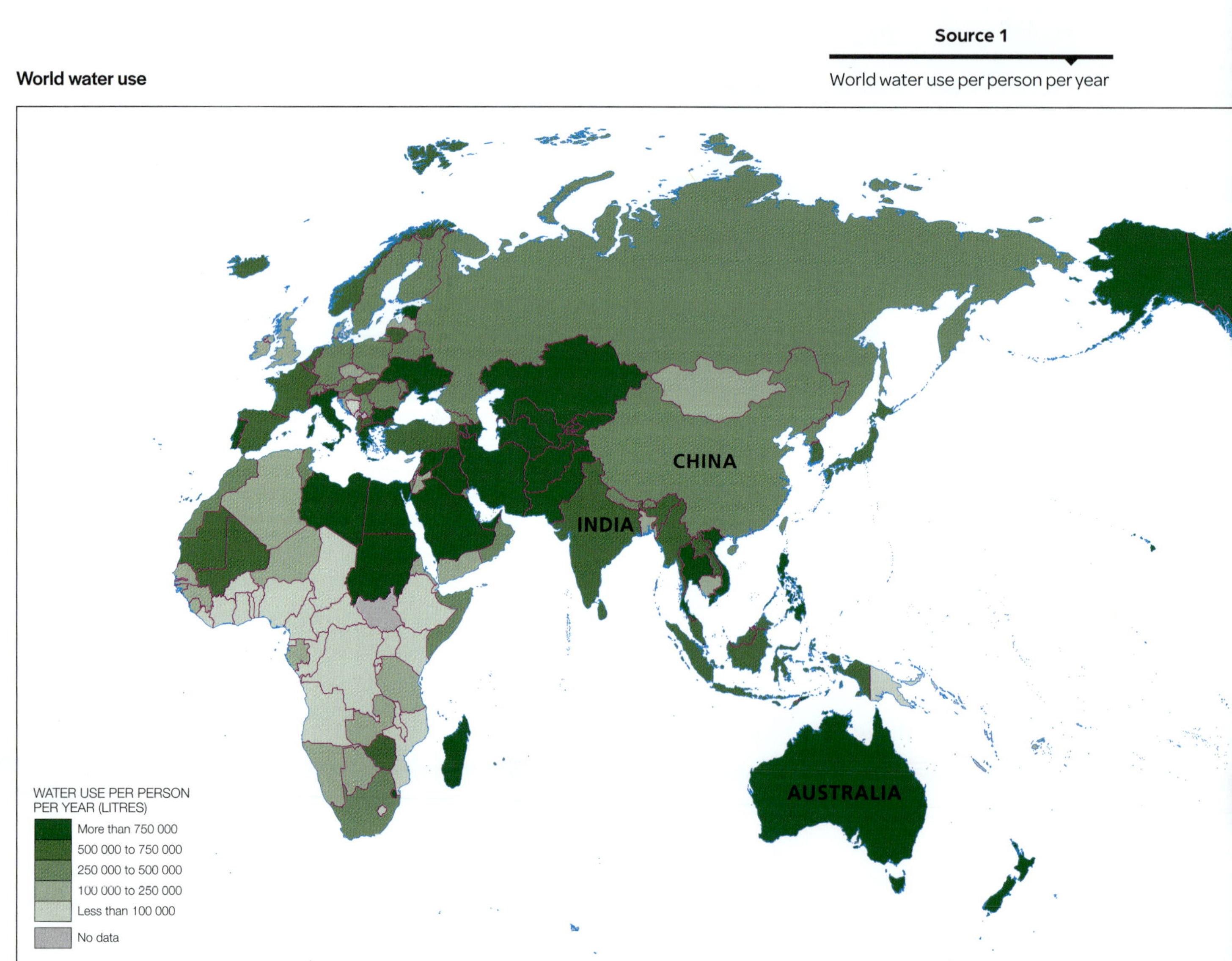

World water use

Nearly 4000 cubic kilometres of fresh water is taken from the world's lakes, rivers and aquifers each year. This amount of water would fill Victoria's Port Phillip Bay 25 times! World water withdrawals have doubled in the last 50 years. Most of the water used by humans is eventually returned to the environment – but the quality of this water may be less than when it was originally taken from the environment.

India, China and the USA consume 34 per cent of the water withdrawn for human use. India and China both use a lot of water to grow their food because they have the largest populations in the world.

The USA has the world's highest use of water per person, at 1550 cubic metres per year. Water used to produce meat – feeding the cattle that eventually become meat – accounts for one-third of American water use (see page 60). In India, where meat consumption is much lower, water use is only 1089 cubic metres per person per year.

Source: Matilda Education Australia

Source 2

The world's largest water consumers

Learning Ladder G2.4

Show what you know

1 How have world water withdrawals changed in the last 50 years?

2 Source 2: Which three countries consume more water than any others, and why?

Analyse data

Step 1: I can use geographic terminology to interpret data

3 Source 1: Compare the difference in water use per person in the northern and southern hemispheres.

Step 2: I can describe patterns and trends

4 Source 1: Use the PQE method to describe the global pattern of water use per person.

Step 3: I can explain the reasons behind a trend or spatial distribution

5 Source 1: Look at the map of water withdrawals per capita.

- a How do Australia and New Zealand compare with the rest of the world?
- b Which continent has the smallest water withdrawals per head of population and why?
- c As a class, discuss why some countries use more water per person than others.

Step 4: I can analyse relationships between different data

6 Compare Source 1 with Source 2. Give reasons to explain why India is high in one category and low in the other.

PQE, page 156

What happens when the water runs out?

Cape Town's water supply running dry

Friday, 23 February 2018

Cape Town in South Africa will become the first major city to run out of water. If drought-saving rains don't arrive, the taps will be turned off on 'Day Zero': 4 June 2018. Despite harsh water restrictions, dam levels are predicted to decline to critically low levels, with the city's largest water reservoir looking more like a desert.

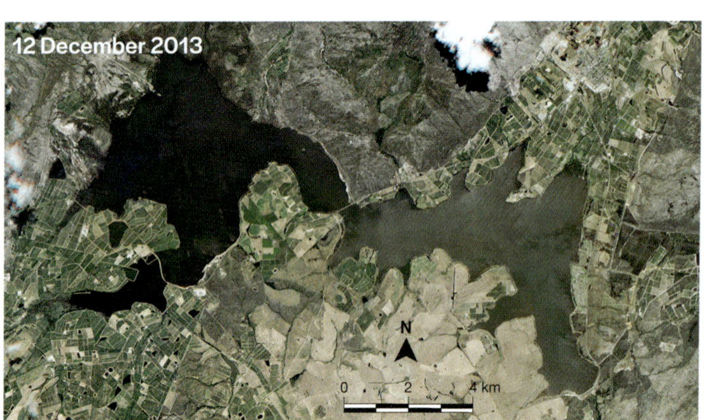

Source: Planet Labs Inc.

Source: Planet Labs Inc.

Source 1

Satellite images of Theewaterskloof Dam in 2013, and during drought in 2018

Source 2

Theewaterskloof Dam supplies 41 per cent of Cape Town's water. It has been turned into a desert after the worst drought in a century.

Water crisis

On 23 February 2018, Cape Town had just 100 days of water left. Day Zero for Cape Town was 4 June, when the city's taps would run dry, cutting off water to its 4 million residents. At the time, the city's largest water supply, the Theewaterskloof Dam, held just 12 per cent of its massive 480 000 million cubic metre capacity.

At the start of 2018, Cape Town residents faced Level 6B water restrictions, which meant a limit of just 50 litres of water per person, per day. Queues formed at natural springs around Cape Town as people filled up water containers to build up their emergency supplies. Shops sold out of bottled water.

If Day Zero arrived, all water to homes would have been shut off and diverted to 200 distribution centres, where people would have to collect their daily rations of 25 litres of water.

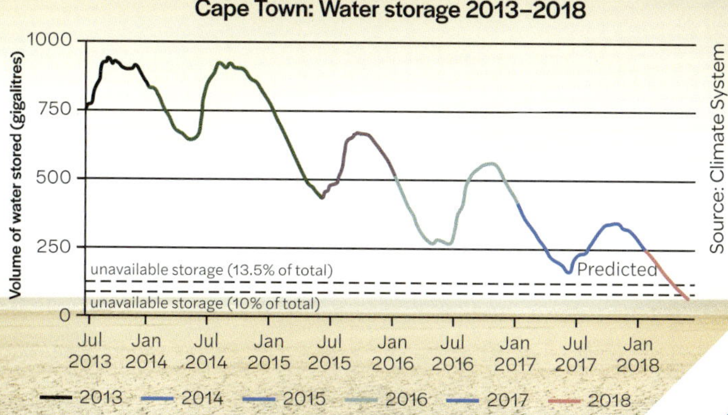

Source 3

This graph shows Cape Town's water storage levels during the period 2013–2018.

Thankfully, Day Zero never eventuated. In response to the crisis, the people of Cape Town became extremely water-wise: households using too much water were fined and limits on water use for agriculture were introduced. Finally, in June 2018, the region experienced average rainfall for the first time in four years. However, Cape Town's water reserves are still critically low.

Causes of the problem

The twin problems behind Cape Town's water crisis were drought and massive population growth. Since 1915, Cape Town's population has grown by nearly 80 per cent, but dam water storage has increased by only 15 per cent.

The year 2017 was the driest on record for Cape Town, and 2015–2017 was the driest three-year period since 1933. In May 2017, the drought was declared the city's worst in a century.

Learning Ladder **G2.5**

Show what you know

1 What was Day Zero, and how did the threat of it come about?

2 What were the two major causes of Cape Town's water crisis?

3 Source 1: Use the satellite images to draw a map of Theewaterskloof Dam with a key that shows the changes that took place between 2013 and 2018. Include the following in your key: very deep water, deep water, shallow water and no water.

Geographical challenge

Step 1: I can identify responses to a geographical challenge

4 What action was taken to respond to the crisis?

Step 2: I can compare responses to a geographical challenge

5 What action could have been taken earlier to avert the water crisis?

Step 3: I can compare strategies for a geographical challenge

6 Search online to compare the current Cape Town water restrictions with those in place in Melbourne. What are the differences?

Step 4: I can evaluate alternatives for a geographical challenge

7 As a class, develop two strategies that could be undertaken to reduce the impacts of water shortages in Cape Town.

Satellite images, page 167

economics+business

How much water is used to make goods?

Every day you consume 3496 litres of water without being aware of it. Most of this water is hidden in the food you eat and the products you use. Geographers call this virtual water. It is the responsibility of every consumer and every business to consider virtual water when making economic decisions.

3781 litres of water are used in the life cycle of one pair of jeans

68% growing cotton

23% washing jeans

7% jeans production

2% other

Source 1

Creating jeans is a water-hungry process

Virtual water

Although it is invisible, millions of litres of water go into making the products we use every day. We don't physically drink this **virtual water**, but it is used in the production of almost every product we use.

The water footprint of a product is the total amount of water used in each step of producing and transporting that product. To make a pair of jeans, cotton needs to be grown, spun and weaved into denim. Water is used during the production process to create steam for power, to cool machinery and to flush and clean equipment. More water is required to make packaging and to transport the jeans overseas.

Source 2

Water footprint of various foods and drinks (in litres)

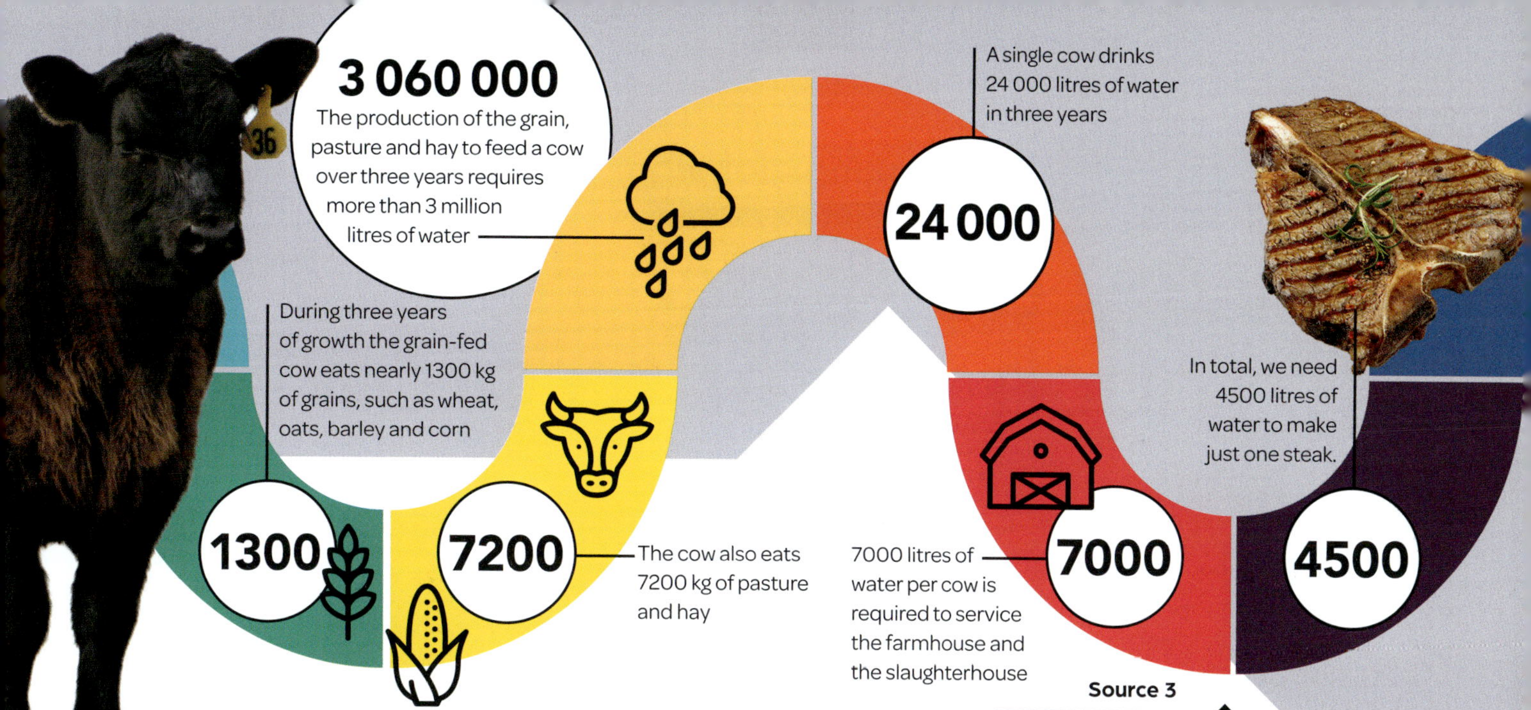

3 060 000
The production of the grain, pasture and hay to feed a cow over three years requires more than 3 million litres of water
A single cow drinks 24 000 litres of water in three years
24 000
During three years of growth the grain-fed cow eats nearly 1300 kg of grains, such as wheat, oats, barley and corn
In total, we need 4500 litres of water to make just one steak.
1300
7200
The cow also eats 7200 kg of pasture and hay
7000 litres of water per cow is required to service the farmhouse and the slaughterhouse
7000
4500
36

Source 3

4500 litres of water are used to make one beef steak

Water for farming

In most parts of the world, more than 70 per cent of fresh water is used for agriculture. By 2050, the world's population will have grown to 9 billion people, and that will require a 50 per cent increase in farm production and a 15 per cent increase in water use.

Of all water used for agricultural production in Australia, 90 per cent is used for irrigation of crops and pastures. Irrigation is the supply of water to farmland that doesn't receive enough rain to raise cattle or grow crops.

Water is required in the production of all foods – one-third of all water used in agriculture is for growing grain to feed livestock. Beef is the most 'water hungry' meat to produce, needing 4500 litres of virtual water to produce just one steak.

70 Apple One
50 Orange One
2500 Coconut One small
720 Wine One bottle
150 Beer One large bottle
125 Potato 500 g
1830 Lamb One steak
1170 Chicken One breast fillet
200 Egg A single egg

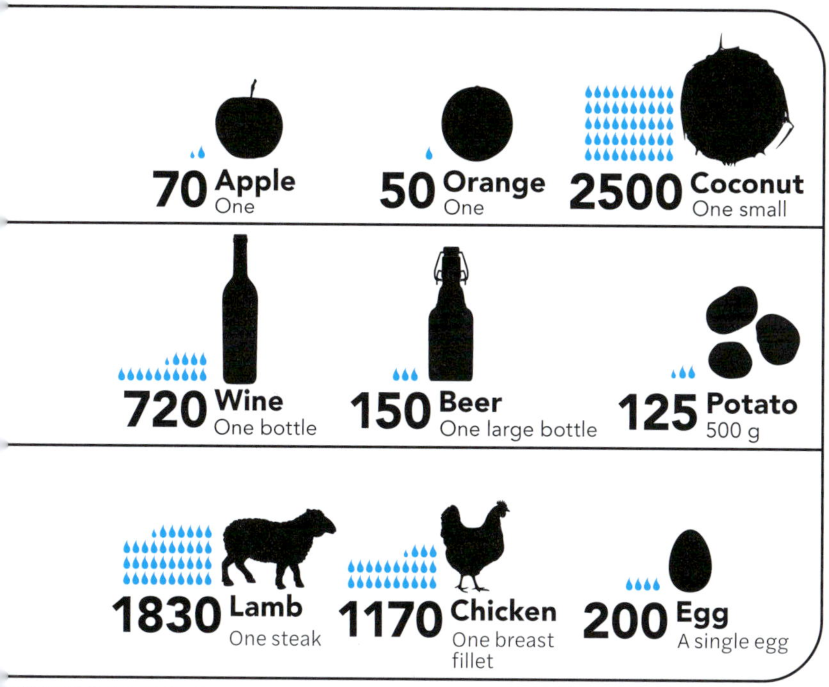

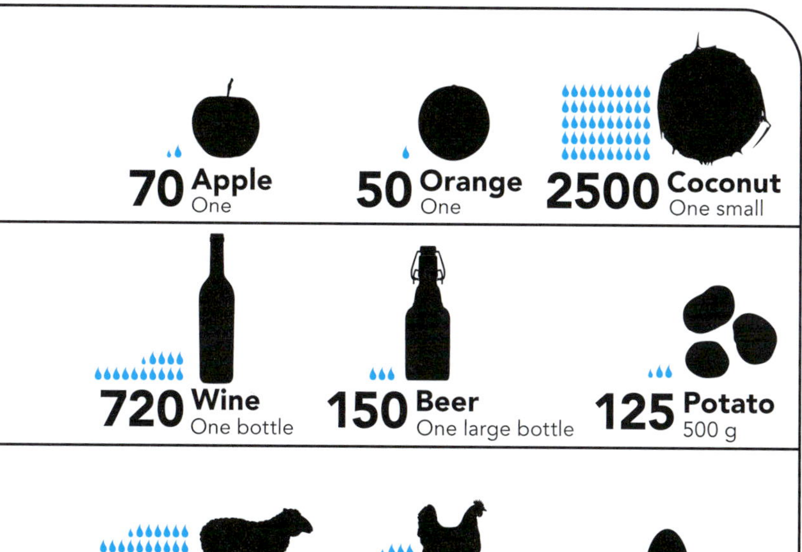

Learning Ladder G2.6

Economics and business

Step 1: I can recognise economic information

1 Source 1: What is the total water footprint of one pair of jeans?

Step 2: I can describe economic issues

2 Why is fresh water vital for industry?

Step 3: I can explain issues in economics

3 Why does it take so much water to produce one beef steak?

Step 4: I can integrate different economic topics

4 Sources 2 and 3: Explain why some beef production requires more virtual water for growth than other farm products.

Step 5: I can evaluate alternatives

5 Source 1: Look at the example of the life cycle of a pair of jeans.
   a Which stage of the jeans-making process uses the most water? Why?
   b Can you suggest an alternative to this stage that would require less water?

6 Calculate your own water footprint using the Water Footprint Network website (http://mea.digital/gh7_g2_1). In what ways could you reduce your footprint?

Why does rice need so much water to grow?

Bali paddies

To grow rice, you need plenty of heat and plenty of water. Rice is well suited to warm, wet, tropical places such as Bali in Indonesia. Rice fields are known as **paddy** fields. They are surrounded by low walls so that the paddies can be flooded with water. In hilly areas, flat rice fields are cut into hillsides as a series of steps, or **terraces**. Paddy rice requires 3400 litres of water to produce just 1 kilogram of the grain.

In Bali, native oxen known as *banteng* are used to plough the muddy fields before planting. After ploughing, the paddies are flooded. Rice seedlings are planted in rows by hand, and rainfall or irrigation is used to keep the paddies flooded for about four months. When the rice plants are a metre tall and turning yellow, they are harvested. To do this, the water in the rice paddies is reduced, and workers use sharp knives to cut the stalks off at ground level. The rice plants are then tied in bundles and left to dry in the sun.

Source 1

Constructing a terraced farm to grow rice

A stone retaining wall is built and backfilled with soil. The soil is flattened to provide a smooth terrace surface.

The terrace is flooded. The depth is controlled by an opening where water can be let out.

Other terraces are built and connected by a water supply.

Source 2

A plan aerial view of terraced rice paddies in Bali, Indonesia. These paddies are ready to be ploughed and planted.

Learning Ladder G2.7

Show what you know

1 Where does the water come from to flood the fields and keep them flooded during the growing season?

2 Why is the water in the paddy fields reduced at harvest time?

Spatial characteristics

Step 1: I can identify and describe spatial characteristics

3 Source 3: Describe the distribution of different kinds of vegetation in this photo.

4 Source 2: Draw the plan aerial view of terraced rice paddies as a map. Include a key to its features.

Step 2: I can explain spatial characteristics

5 Locate Bali on a world map. Describe where Bali is in relation to the equator.

6 Undertake research online to locate Bali's major rice paddy regions. Using the PQE method, describe the distribution of these paddies.

Step 3: I can explain processes influencing places

7 Use the SHEEPT factors to explain why Bali is an ideal place to grow rice.

Step 4: I can predict changes in the characteristics of places

8 Undertake Internet research to discover how rice is grown, sewn and harvested in Australia. Why would harvesting in this manner be different to Bali?

PQE, page 156
SHEEPT, page 158

Source 3

Rice plants grow in neat rows on flooded terraced fields called paddies. When the plants are fully grown, the water in the rice paddy is reduced to allow for easier harvesting.

Is all water safe to drink?

Nearly 800 million people in the world do not have access to a reliable and safe fresh water supply. Millions of people around the world get sick or die from drinking contaminated (or dirty) water. Millions of women and children around the world spend several hours a day collecting and carrying enough water to sustain their families for another day.

The burden of thirst

Even at four in the morning, Binayo can run down the rocks to the river by starlight alone and climb the steep mountain back up to her village with 20 kilograms of water on her back. She has made this journey three times a day for nearly all her 25 years. So has every other woman in her village of Poro, in the Konso district of south-western Ethiopia.

Binayo dropped out of school when she was eight years old, in part because she had to help her mother fetch water from the Toiro River. The water is dirty and unsafe to drink; every year that the ongoing drought continues, the once mighty river grows more exhausted. But it is the only water Poro has ever had.

The task of fetching water defines life for Binayo. She must also help her husband grow cassava and beans in their fields, gather grass for their goats, dry grain and take it to the mill for grinding into flour, cook meals, keep the family compound clean, and take care of her three small sons. None of these jobs is as important or as consuming as the eight hours or so she spends each day fetching water.

Source 1

Gabra women in northern Kenya spend up to five hours each day hauling heavy jerry cans filled with dirty water back to their families.

'In wealthy parts of the world, people turn on a faucet [tap] and out pours abundant, clean water. Yet [in 2010] nearly 900 million people in the world have no access to clean water, and 2.5 billion people have no safe way to dispose of human waste – many defecate in open fields or near the same rivers they drink from. Dirty water and lack of a toilet and proper hygiene kill 3.3 million people around the world annually, most of them children under age five. Here in southern Ethiopia, and in northern Kenya, a lack of rain over the past few years has made even dirty water hard to find.

Where clean water is scarcest, fetching it is almost always women's work. In Konso a man hauls water only during the few weeks following the birth of a baby. Very young boys fetch water, but only up to the age of seven or eight. The rule is enforced fiercely – by men and women. "If the boys are older, people gossip that the woman is lazy," Binayo says. The reputation of a woman in Konso, she says, rests on hard work: "If I sit and stay at home and do nothing, nobody likes me. But if I run up and down to get water, they say I'm a clever woman and work hard."

'The burden of thirst', *National Geographic* April 2010. Tina Rosenberg (text) and Lynn Johnson (photograph).

Waterborne diseases

Fresh water is essential for drinking, cooking, food production and hygiene, as well as for cleanliness. Many countries in the world lack safe water supplies and have poor sanitation (safe disposal of human urine and faeces). People get diarrhoea from drinking and bathing in contaminated water, and this can lead to diseases such as cholera, typhoid fever and dysentery. These illnesses kill 6000 children every day – and 3.4 million people die every year from waterborne diseases.

Water scarcity in Africa

Rapid population growth in Africa has made it even more difficult for African communities to find a reliable source of fresh water. In Africa, **water scarcity** affects one person in every three – and it is getting worse.

Reduced access to water means that people use water of poorer quality. The journeys women take to fetch water for the family become even longer. When they can get water, people store it in their homes – but this provides a breeding ground for mosquitoes, which carry dengue fever, malaria and other diseases.

Better access to safe water

World Vision and other **aid agencies** work with local communities to improve their access to safe water.

Just two protected wells provide the daily water for the 33 000 people of the Mwinilunga community in Zambia. Women and girls walk long distances to fetch water from unprotected wells and contaminated streams. Half of the Mwinilunga children suffer from life-threatening malaria or diseases caused by diarrhoea.

However, family health is improving in the Mwinilunga community as families gain access to safe water for the first time. World Vision has worked with the local community to:

- drill ten boreholes and equip them with pumps
- protect five wells against contamination
- build toilets for homes and schools
- train locals to operate and look after the borehole pumps
- conduct 100 hygiene education sessions and train 20 teachers.

Source 2

Rendille villagers in drought-stricken Northern Kenya in 2010. They are scooping the last remaining water from a tank that was filled by a government truck the night before. The next water delivery won't be for a week.

Source 3

This map shows countries where less than half the population can access clean water. The percentage for each country shows the number of people who have access to safe water supplies.

Learning Ladder G2.8

Show what you know

1 What problems does water scarcity bring?

2 What problems does drinking unsafe water cause?

Geographic challenge

Step 1: I can identify responses to a geographical challenge

3 Read the story 'The burden of thirst' on page 64.

a What is Binayo's most important job? Describe how she performs this task.

b How far away is the Toiro River and how many times does she travel to it each day?

Step 2: I can compare responses to a geographical challenge

4 How has access to safe water for the Mwinilunga community improved?

Step 3: I can compare strategies for a geographical challenge

5 Source 3: Use the map to locate target regions for action to improve safe water.

- In which continent is maintaining a sustainable safe water supply a major issue?
- Which country has the worst access to safe water?
- Which countries neighbouring Australia have difficulties sustaining a safe water supply?

Step 4: I can evaluate alternatives for a geographical challenge

6 Suggest three ways that economically developed countries can assist those with water scarcity.

Who has spiritual links with water?

In many cultures there is a deeper, spiritual connection with rivers, lakes and waterholes. Waterways can be central to religions and cultures and their stories.

Indigenous Australians' connection to water

Australia's Indigenous peoples are connected to the land and water, and are responsible for looking after them. Water places are spiritual for Aboriginal peoples, and their importance has been passed on through song, dance, painting and storytelling.

Many Aboriginal creation stories tell of the Rainbow Serpent as the creator of rivers, lakes and waterholes.

For some Indigenous people, the Rainbow Serpent is the source of all life and the protector of the land and people. According to myth, the Rainbow Serpent moved across the land, shaping the landscape with its powerful body. River beds were formed from the winding tracks made by the serpent.

According to Indigenous law, water places have spiritual significance, along with responsibilities for protecting cultural knowledge. However, Indigenous Australians have been largely left out of the management of their traditional water sources and significant sites.

'Water is the life for us all,' says Karajarri man John 'Dudu' Nangkiriyn. 'It's the main part. If we are gonna lose that, I don't know where we gonna stand. If that water go away, everything will die. That's the power of water.'

The holy Ganges River

The Ganges River in India is perhaps the holiest river in any religion. The Ganges is worshipped by Hindus, who believe they are purified by bathing and praying in the river.

Source 1

Aboriginal people in arid areas can map the location of water in their artwork. Campsites are closely associated with waterholes.

Running Water

Rainbow / Cloud

Waterhole / Campsite

Rain

Source 2

During the Kartik Purnima festival in Patna, India, tens of thousands of Hindus come together on the banks of the Ganges River to bathe and offer prayers.

This sacred river is also used by millions of Indians daily – and it is one of the most polluted rivers in the world.

The Ganges is worshipped as a goddess in Hindu religion, and is referred to as Mother Ganga. Hindus believe that the waters of the Ganges are sacred and can wash the sins of previous lives away.

It is believed that people who die near the Ganges reach heaven, escaping the cycle of rebirth. As a result, cremating a dead body on the banks of the Ganges – or spreading the ashes of the deceased in the water – is believed to lead to heaven. The city of Varanasi on the banks of the Ganges is India's spiritual capital. Every year, 32 000 corpses are cremated in Varanasi, and up to 200 tonnes of half-burnt flesh end up in the Ganges, along with the unburned bodies of those Hindus who cannot afford firewood for their cremation.

However, the waters of the Ganges River that Hindus believe are so purifying are also dangerous to their health. It is estimated that two-thirds of the people who regularly use the Ganges for bathing, or for washing clothes or dishes, will suffer a serious illness.

In 2015, 100 000 Indian children died from the effects of diarrhoea, a waterborne disease (see page 66). This death toll was higher than for any other country in the world.

Learning Ladder G2.9

Show what you know

1 How have Indigenous Australians passed on the spiritual connection to water over time?

2 Why do you think Indigenous Australians want a say in how water resources are managed?

Spatial characteristics

Step 1: I can identify and describe spatial characteristics

3 The concept of place looks at the significance of places. Why is the Ganges River a significant place?

Step 2: I can explain spatial characteristics

4 Source 1: How do some Aboriginal artworks show the spatial location of water resources?

Step 3: I can explain processes influencing places

5 Source 2 : What social factor is behind Patna being considered a special place?

Step 4: I can predict changes in the characteristics of places

6 As a class, discuss:
   a the impact that the spiritual connection with the Ganges River has on the health of the people who bathe in it.
   b how the Indian government can help to clean up the Ganges River, yet still respect people's spiritual beliefs.

Geographic concepts, page 4
SHEEPT factors, page 158

How do we manage water?

Water resources need to be carefully managed to make sure there are reliable supplies of safe water and a healthy environment. There is great competition for water use from farms, factories and cities.

Maintaining water balance

Water balance describes the flow of water in and out of a system. Managing our water supply requires having an understanding of all of the water that comes in and out of that system (see page 20).

Inputs refer to water coming *into* the system. This includes rainfall, **run-off** into rivers and lakes, and water flowing into underground reserves, as well as additional water piped in from outside the system (see desalination on page 72).

Outputs refer to water going *out* from a system. This includes **evaporation** from land and bodies of water, **transpiration** from plants, and our use of water from surface water and groundwater.

Governments and water authorities need to estimate the size of future **water resources** in a region. To do this, they have to calculate all of the inputs and all of the outputs, both natural and for human use.

Reliable water supply

Managing our precious **fresh water** supply is not easy, because:

- it is an essential resource for all
- the amount of water available changes depending on rainfall, usage and rate of evaporation
- water resources are shared among many competing users.

Water storages, called **dams** or reservoirs (see page 27), provide a reliable water supply for **irrigation**, **hydroelectricity** (see page 16) and urban use. Water authorities monitor storage levels to make sure that there is enough water available.

Melbourne's water supply system is made up of water supply reservoirs and a **desalination** plant (see page 72). When the water supply drops below 60 per cent, the government will order water from the desalination plant and bring in water restrictions to reduce how much water people use – the outputs – from the water storages.

Source 1

Melbourne's water use per person per day 2000–17

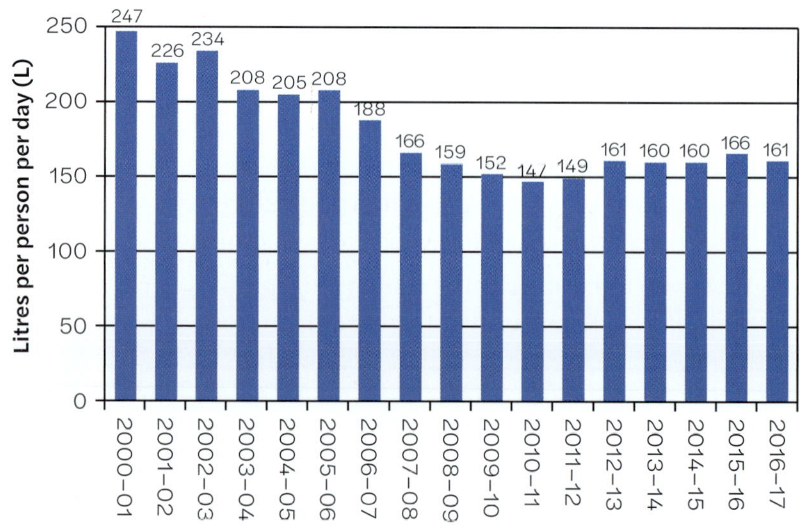

Source: City West Water, Melbourne Water, South East Water & Yarra Valley Water, 2017

Source 2

Water balance

Inputs
1 Precipitation
2 Runoff
3 Groundwater inflow
4 Surface water inflow
5 Water diversions

Outputs
6 Evaporation
7 Transpiration
8 Surface water outflow
9 Groundwater outflow
10 Irrigation
11 Industrial uses
12 Residential uses
13 Water diversions

Competition for Murray–Darling River water

Our largest river system, the Murray-Darling, does not carry much water compared to other rivers in the world. The **drainage basin** that supplies water to the Murray-Darling system receives about 530 000 gigalitres of water as rainfall – but then loses 94 per cent of this to evaporation and transpiration. Because Australia's climate varies so much, the actual amount of water in the river system can vary from less than 7000 gigalitres (in 2006) to almost 118 000 gigalitres (in 1956).

To ensure a regular supply of water from the Murray-Darling system, a series of water storages have been built: four major dams, 16 storage weirs and 15 navigable locks. The amount of water released from these dams depends on the amount of water entering the Murray-Darling system (see page 34).

There are many competing uses for water from the Murray-Darling River system, including irrigation, recreation, navigation, hydroelectricity and water storage, along with water for cities, industry and maintaining the natural environment. Competition also exists between upstream and downstream users of water from the Murray-Darling River system and users in different states.

The Murray River spans three states: Victoria, New South Wales and South Australia. As a downstream user of the Murray River, South Australia is always affected by how the water is used upstream. To ensure fairness in management, the river system is controlled by the independent Murray-Darling Basin Authority.

Learning Ladder G2.10

Show what you know

1 Why are water storages such as dams an important part of water supply?

2 How does the water supply in the Murray-Darling Basin vary?

3 What are competing uses for water from the Murray-Darling River system?

Interconnections

Step 1: I can identify and describe interconnections

4 What water movements in and out of the Murray-Darling Basin cause the water supply to vary so much?

Step 2: I can explain interconnections

5 Source 2: Use the block diagram to look at interconnections in this environment.
   a What are the natural inputs and outputs of a water system?
   b What are the human inputs and outputs?

Step 3: I can identify and explain the implications of interconnections

6 Summarise how our environment maintains a water balance. Use the concepts of interconnection and change in your writing.

Step 4: I can evaluate the implications of significant interconnections

7 Source 1: Look at the graph. Use the PQE method to describe and quantify the pattern. How have changes in water use helped maintain Melbourne's water supply?

PQE, page 156

G2.11

thinking locally

How can we make fresh water?

Australia's largest desalination plant is the Dalyston facility in Victoria. A desalination plant removes salt and impurities from seawater and turns it into fresh water.

Planning for the **desalination** plant started in 2007, when Melbourne's dam water storages fell below 30 per cent – their lowest level ever. The plant was built to supply Melbourne with water in times of drought. It was completed in 2012, but not used until 2017.

The desalination plant costs over $600 million per year to run and can produce up to 150 gigalitres (150 billion litres) of water each year.

The plant takes in seawater through a pipeline 1.2 kilometres long. Seawater passes through filters and then through two stages of a process called **reverse osmosis**, where it is pushed under high pressure through ultra fine filters (called membranes). Fresh water passes through the membranes and leaves the salt and other impurities behind.

Drinking water is then pumped via an 84-kilometre pipeline from the Dalyston plant to Cardinia Reservoir, near Melbourne.

Source 1

Oblique aerial view of the Dalyston desalination plant. The long intake and outlet tunnels are buried beneath the ground and are not visible in this image.

1 Seawater intake
2 Feedwater pipe delivers seawater to filter
3 Multistage filters remove solids
4 Microfilter removes remaining small solids
5 Seawater passes through reverse osmosis process to produce fresh water
6 Water is stored
7 Water is pumped into water supplies
8 Outlet tunnel transports concentrated seawater back to the ocean

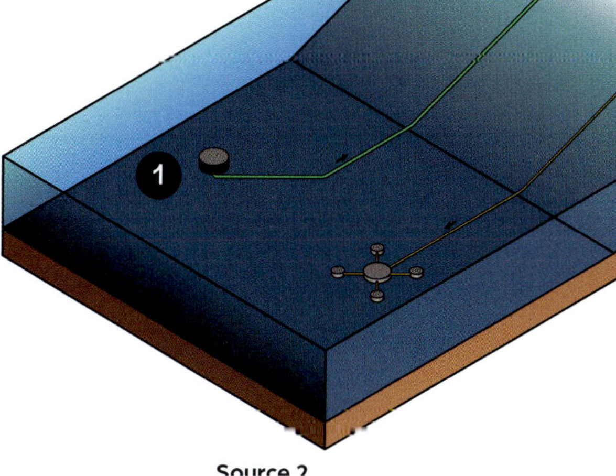

Source 2

How a reverse-osmosis water desalination plant works

Source 3

Total volume of water in Melbourne water storages 1996–2016

Melbourne: Water storage 1996–2016

Full supply level

1996
2012
2014
2016
2005
2010
2006
2007
2008

Total volume (gigalitres)

1750 1650 1550 1450 1350 1250 1150 1050 950 850 750 650 550 450

Jan Feb Mar Apr May Jun Jul Aug Sep Oct Nov Dec

Source: State Government of Victoria, Department of Environment, Land, Water and Planning, 2017

Learning Ladder G2.11

Show what you know

1 Why was the desalination plant built near the ocean?

2 Source 2: Using the symbols on this source, outline the process of desalination.

Analyse data

Step 1: I can use geographic terminology to interpret data

3 Source 1: Why don't the intake and outlet pipes show on the oblique aerial photograph?

Step 2: I can describe patterns and trends

4 Source 3: Look at the graph. Use the data to answer these questions.
   a At what time of year do water storages peak?
   b When were the lowest storage levels recorded?

Step 3: I can explain the reasons behind a trend or spatial distribution

5 Source 3: Look at the graph.
   a Why do you think plans to build a desalination plant began in 2007?
   b Why do you think the desalination plant was not used until 2017?

Step 4: I can analyse relationships between different data

6 Many people living near Dalyston protested against the plant being built. As a class, discuss the issues they might have raised.

HOW TO

Flow diagrams, page 163
Graphing, page 164

How can we save water at home?

Most Australians are lucky to have instant access to clean, safe water at home. Per person, Australians are one of the biggest users of water in the world – but we can all help save water in the home with a few simple changes.

Domestic water use

Most Australian homes have good access to a reliable supply of clean water. As the country's supply of fresh water is low and varies a lot, Australian governments have encouraged water conservation and introduced water restrictions during times of drought.

Domestic use of water only accounts for 12 per cent of Australia's total water use, but Australia has one of the highest rates of water consumption per person in the world. The average daily water use is 340 litres per person, or 900 litres per household. Average water consumption ranges from 100 litres per person in some coastal areas to more than 800 litres per person in the **arid** inland areas.

Saving water in the home

More than 1.7 million Australian households have installed rainwater tanks – saving both water and money. Water collected in rainwater tanks is usually used for watering the garden and flushing toilets. It also provides an alternative supply of water during times of drought.

To improve water management in the home, we can capture water, **recycle** water or use less water. We can select water-saving appliances such as low-flush toilets and water-efficient showerheads to reduce the amount of water we use.

Source 1

Water-saving measures at home help to save our valuable water. Between water restrictions and changes in water use, the amount of water used by Australian households fell 12 per cent between 2003 and 2009 – despite a population increase of 7.7 per cent.

2 Roof
- Install rain water catchment system
- Connect gutters to water tanks

3 Bathroom
- Use water-saving shower head
- Install low-flush toilet

1 Laundry
- Install water saving appliances
- Install grey water system

4 Kitchen
- Install water saving appliances
- Use low-flow taps

6 External
- Ensure pipes are in good order
- Fix leaks quickly
- Install and check water meter to measure usage

5 Garden
- Use drip irrigation
- Use drought-resistant plants
- Obtain water from rain tanks

TOP TIPS FOR SAVING WATER

Install a dual-flush toilet and do not dispose of rubbish in toilets.

Check and repair all leaks.

Do not leave water running when brushing your teeth, shaving or washing dishes.

Install watersaving showerheads and limit showers to four minutes.

Use full loads in the dishwasher and washing machine. Use an eco setting where possible.

Other ways you can save water include watering the garden in the early morning and late evening, installing a rain water tank, and using buckets of water for cleaning.

Source 2

Tips for saving water

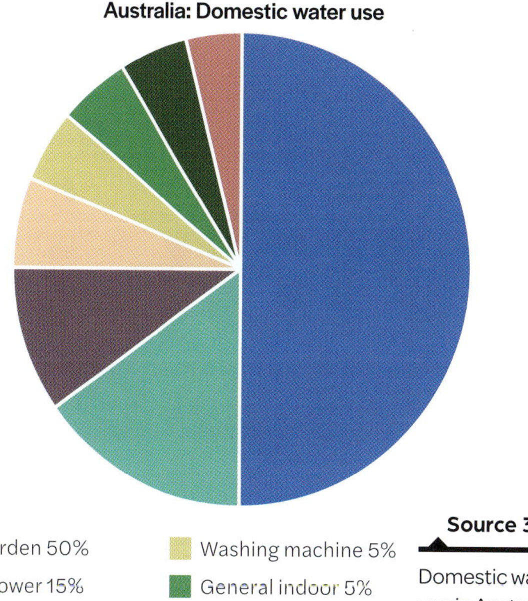

Source 3

Domestic water use in Australia

Learning Ladder G2.12

Show what you know

1 How much of Australia's total water use occurs in our homes?

2 How did Australians reduce water usage by 12 per cent between 2003 and 2009?

Analyse data

Step 1: I can use geographic terminology to interpret data

3 What does the blue section of Source 3 show?

Step 2: I can describe patterns and trends

4 Source 3: Examine the pie chart.
   a Where is half of domestic water used? List three ways we can reduce domestic water use.
   b Which is the most water-hungry appliance inside the house? How can we reduce water use in this area?

Step 3: I can explain the reasons behind a trend or spatial distribution

5 'Water is a valuable resource that should not be used carelessly.' Support this statement using data from this spread.

Step 4: I can analyse relationships between different data

6 Create a checklist of methods to investigate water use at school. Use one or more of these methods to conduct a survey of water use at your school.

Masterclass

Learning Ladder

Work at the level that is right for you or level-up for a learning challenge!

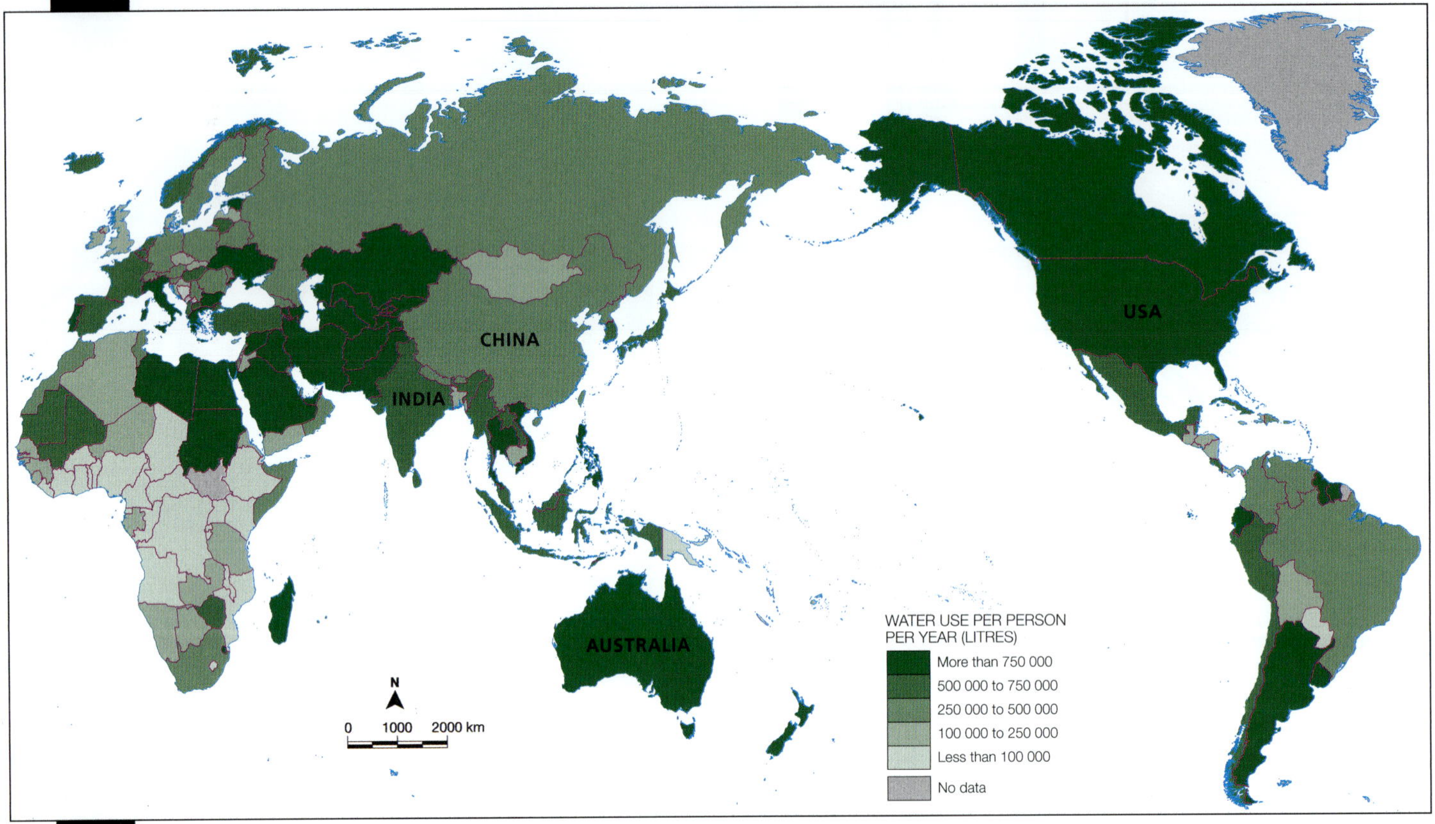

CHINA
INDIA
USA
AUSTRALIA
N
0 1000 2000 km
WATER USE PER PERSON PER YEAR (LITRES)
More than 750 000
500 000 to 750 000
250 000 to 500 000
100 000 to 250 000
Less than 100 000
No data

Source 1

World water use

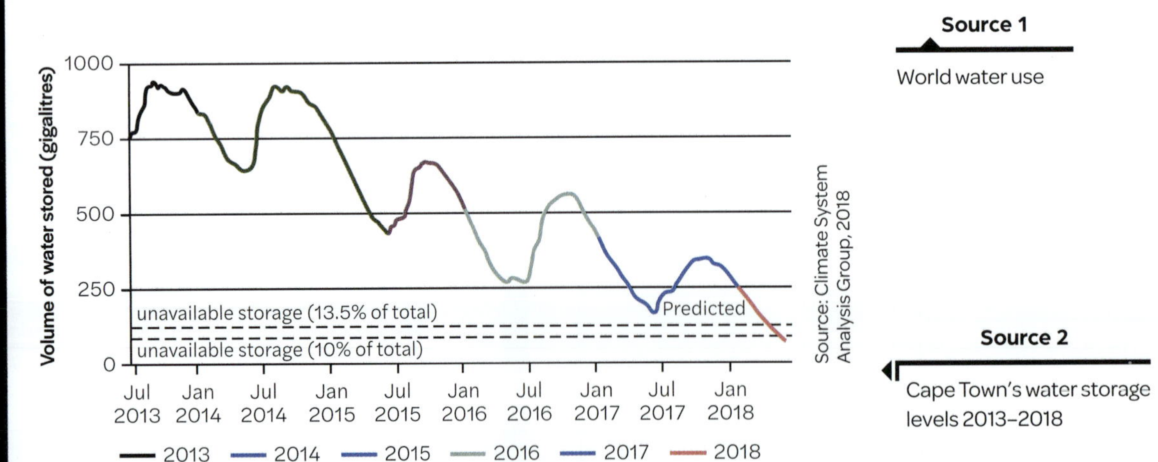

Source 2

Cape Town's water storage levels 2013–2018

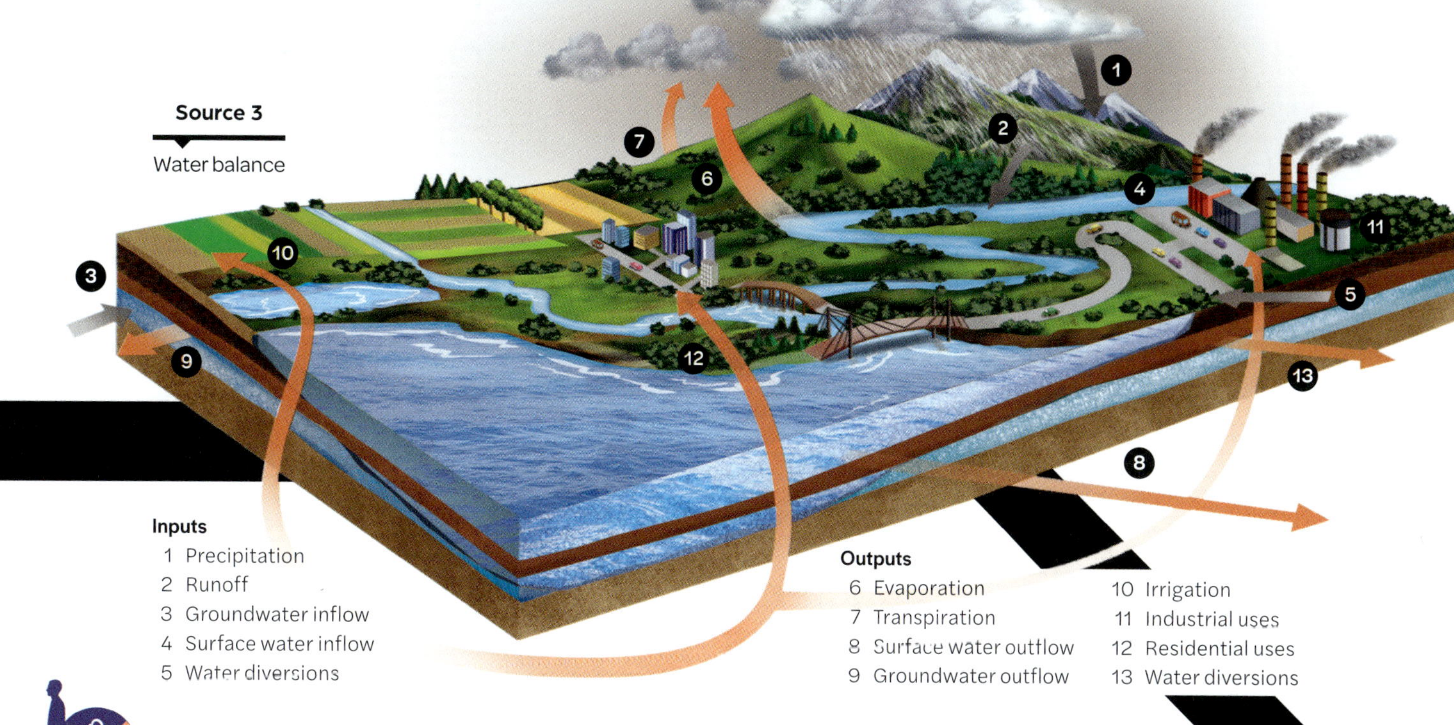

Step 1

a **I can identify and describe spatial characteristics**

Look at Source 2. Identify the place shown in this source.

b **I can identify and describe interconnections**

Source 3: What are the natural inputs and outputs of a water system?

c **I can identify responses to a geographical challenge**

How did the government respond to the geographical challenge shown in Source 2?

d **I can collect, record and display data in simple forms**

Look at Source 2. What does the line represent?

e **I can use geographic terminology to interpret data**

Look at Source 1. Do countries in the northern or southern hemisphere generally use more water?

Step 2

a **I can explain spatial characteristics**

Water is said to connect places. Identify what the words 'connect' and 'place' mean in this context.

b **I can explain interconnections**

How is Cardinia Reservoir interconnected with the creation of fresh water at the Dalyston desalination plant 84 kilometres away (page 72)?

c **I can compare responses to a geographical challenge**

Compare Source 1 on page 65 with Source 2 on page 66. How do these responses to the challenge of providing water for the family differ?

d **I can recognise and use different types of data**

List a series of data collection methods that could be used to identify whether water in a particular place is safe for humans to drink.

e **I can describe patterns and trends**

Describe the trend for the volume of stored water in Cape Town using Source 2.

Step 3

a **I can explain processes influencing places**

Explain the interconnection between spirituality and water sources in Australia (page 68).

b **I can identify and explain the implications of interconnections**

Explain implications of the interconnection between spirituality and water sources in India (page 69).

c **I can compare strategies for a geographical challenge**

Compare water saving and storing strategies in Cape Town with those in Melbourne. What are the differences (pages 58 and 70)?

d **I can choose, collect and display appropriate data**

Create a step-by-step guide explaining how to create a photo essay to display information gathered during fieldwork.

e **I can explain the reasons behind a trend or spatial distribution**

Source 1: Using the SHEEPT factors explain the global pattern of water withdrawals.

Masterclass

Step 4

a I can predict changes in the characteristics of places

Predict how the Ganges River in India may change in the future based on current water uses (page 68).

b I can evaluate the implications of significant interconnections

In the last decade Kenya has faced many years of below average rainfall, leading to lower food harvests and high prices for food. What are the implications for the people of Kenya of the interconnection of these three factors (see pages 65–67)?

c I can evaluate alternatives for a geographical challenge

Suggest three ways that economically developed countries can assist those with water scarcity.

d I can use data to support claims

A student would like to explore how water connects people within their local community. Suggest two methods that would assist them in their investigation. Provide some reasons for your response.

e I can analyse relationships between different data

Compare the relationship between Sources 2 and 3. Which factors in Source 3 changed or stayed the same to produce the situation in Source 2?

Step 5

a I can analyse the impact of change on places

Suggest five ways we can sustainably use our fresh water sources.

b I can explore spatial association and interconnections

Find an online map of Australian land use from a reliable source. Compare the map to the map of Australia's average annual rainfall on page 50. Describe the spatial association between land use and rainfall regions with less than 250 mm per year.

c I can plan action to tackle a geographical challenge

Australia has the lowest volume of water in rivers of any inhabited continent in the world. Why is this the case and what action could be taken to improve it?

d I can evaluate data

How relevant is climate data when investigating how much water we use and have access to in Australia?

e I can draw conclusions from analysing collected data

Explain why some beef production requires more virtual water for growth than other farm products.

Capstone

How can I understand water for life?

In this chapter, you have learnt a lot about water for life. Now you can put your new knowledge and understanding together for the capstone project to show what you know and what you think.

In the world of building, a capstone is an element that finishes off an arch or tops off a building or wall. That is what the capstone project will offer you, too: a chance to top off and bring together your learning in interesting, critical and creative ways. You can complete this project yourself, or your teacher can make it a class task or a homework task.

Scan this QR code to find the capstone project online.

mea.digital/GHV7_G2

G3

Living in Australia

How can I understand living in Australia?

Australia is classified as a **more economically developed country (MEDC)**. This means that we generally have good access to education, healthcare, transport, open spaces and other key elements that make our country liveable. In 2019, three of our cities – Melbourne, Perth and Adelaide – were all ranked in the top 10 most liveable cities in the world.

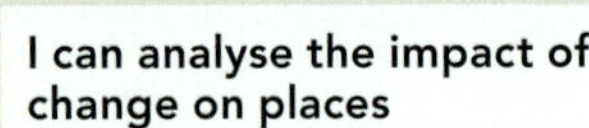

learning ladder

	Spatial characteristics	Interconnections	Geographical challenge
step 5	**I can analyse the impact of change on places** I can analyse and evaluate the implications of growth or decline of places and calculate the impact on people and environments.	**I can explore spatial association and interconnections** I can compare distribution patterns and the interconnections between them; e.g. the importance of community of place for people in rural areas.	**I can plan action to tackle a geographical challenge** I can frame questions, evaluate findings, plan actions and predict outcomes to tackle a population-based geographical challenge.
step 4	**I can predict changes in the characteristics of places** I can predict changes in the characteristics of places over time due to population growth.	**I can evaluate the implications of significant interconnections** I can identify, analyse and explain key population-based interconnections within and between places, and evaluate their implications over time and at different scales.	**I can evaluate alternatives for a geographical challenge** I can weigh up alternative views and strategies on a population-based geographical challenge using environmental, social and economic criteria.
step 3	**I can explain processes influencing places** I can explain the series of actions leading to change in a place, such as the decline of rural towns.	**I can identify and explain the implications of interconnections** I can identify, analyse and explain population-based interconnections and explain their implications.	**I can compare strategies for a geographical challenge** I can compare strategies for a geographical challenge, taking into account a range of factors and predicting the likely outcomes.
step 2	**I can explain spatial characteristics** I can identify concepts of Space, Place, Interconnection, Change, Environment, Sustainability and Scale (SPICESS) when I read about living in Australia.	**I can explain interconnections** I can describe and explain interconnections and their effects, such as different farming activities in different climates and environments.	**I can compare responses to a geographical challenge** I can identify and compare responses to a geographical challenge and describe its impact on different groups.
step 1	**I can identify and describe spatial characteristics** I can talk about spatial characteristics at a range of scales; e.g. Australian cities.	**I can identify and describe interconnections** I can identify and explain simple interconnections involved in phenomena such as people with similar interests coming together in community groups.	**I can identify responses to a geographical challenge** I can find responses to a geographical challenge such as population growth and understand the expected effects.

Source 1

Melbourne's central business district (CBD) is surrounded by parklands and recreational spaces.

I can evaluate data
I can determine whether data presented about human settlements is reliable and assess whether the methods I used in the field or classroom were helpful in answering a population-based research question.

I can use data to support claims
I can select or collect the most appropriate data and create specialist maps and information using ICT to support investigations into human settlements.

I can choose, collect and display appropriate data
I can select useful sources of human data and represent them to conform with geographic conventions.

I can recognise and use different types of data
I can define the terms primary, secondary, qualitative and quantitative data and represent data in more complex forms.

I can collect, record and display data in simple forms
I can identify that maps and graphs use symbols, colours and other graphics to represent data.

Collect, record and display data

I can draw conclusions from analysing collected data
I can summarise findings and use collected data to support key patterns and trends I have identified for population-based research.

I can analyse relationships between different data
I can use multiple data sources, overlays and GIS to find links and relationships that exist in patterns of human settlement on Earth.

I can explain the reasons behind a trend or spatial distribution
I can identify Social, Historical, Economic, Environmental, Political and Technological (SHEEPT) factors to help me explain patterns in data.

I can describe patterns and trends
I can identify Patterns, Quantify them and point out Exceptions (PQE) to describe the patterns I see.

I can use geographic terminology to interpret data
I can identify increases, decreases or other key trends on a map, graph or chart about human settlements.

Analyse data

Spatial characteristics

1 What place is shown in Source 1?

Interconnections

2 *Interconnection* is the idea that two things or phenomena are related, interact or are linked in some way. Suggest why most major cities of the world are interconnected with rivers.

Geographical challenge

3 As a class, brainstorm key factors that make your local community liveable. Are there any elements that restrict the liveability of your area?

Collect, record and display data

4 Annual world liveability rankings are based on survey results.
 a Make a list of the types of questions you think should be asked in a world liveability survey.
 b Is this primary or secondary data collection?

Analyse data

5 Sketch the photo of Melbourne's CBD and surrounds in your notebook. Label it with elements that make this capital city one of the most liveable in the world.

Where did the first Australians live?

Aboriginal and Torres Strait Islanders are the first peoples of Australia. Most early Aboriginal people lived along the coast and in the river valleys, where the resources they needed to survive were plentiful. Today, Indigenous Australians live in all parts of Australia: in cities, in suburbs and in regional and remote areas.

Finding resources

Early Indigenous Australians looked for places that were **liveable** and had the water and food they needed to survive and thrive.

The highest **density** of Indigenous Australians was along the coastline – which means that was where the most people lived. Those people who chose to live inland were located mainly along the banks of the Murray River and its tributaries.

Fresh water was the most important resource for survival. Early Indigenous peoples hunted animals that were attracted to sources of water, and built stone traps to catch fish and eels. Ancient fish traps used by Wailwan people on the Barwon River at Brewarrina, New South Wales are 40 000 years old – which makes them the oldest human-made structures on Earth.

Living in the desert

Further inland, resources were scarcer and Aboriginal populations were less dense. In desert regions, early Aboriginal peoples changed where they lived according to the seasons. They followed the summer monsoonal rains and the movement of animals, or the natural cycles of fruiting plants.

Aboriginal desert peoples knew the locations of permanent waterholes and followed birds to other available water sources. They dug up the roots of trees known to hold water – and even caught a species of water-holding frog and drained the water from it to get the water they needed.

Connection with the land

For Indigenous Australians, liveabilty means much more than whether an area can supply food and water.

Indigenous Australians have a strong sense of identity that comes from belonging to a region. The deep relationship between Indigenous Australians and the land is often described as **connection to Country**.

Source 1

The Wailwan people built fish traps on the Barwon River at Brewarrina, New South Wales, about 40 000 years ago.

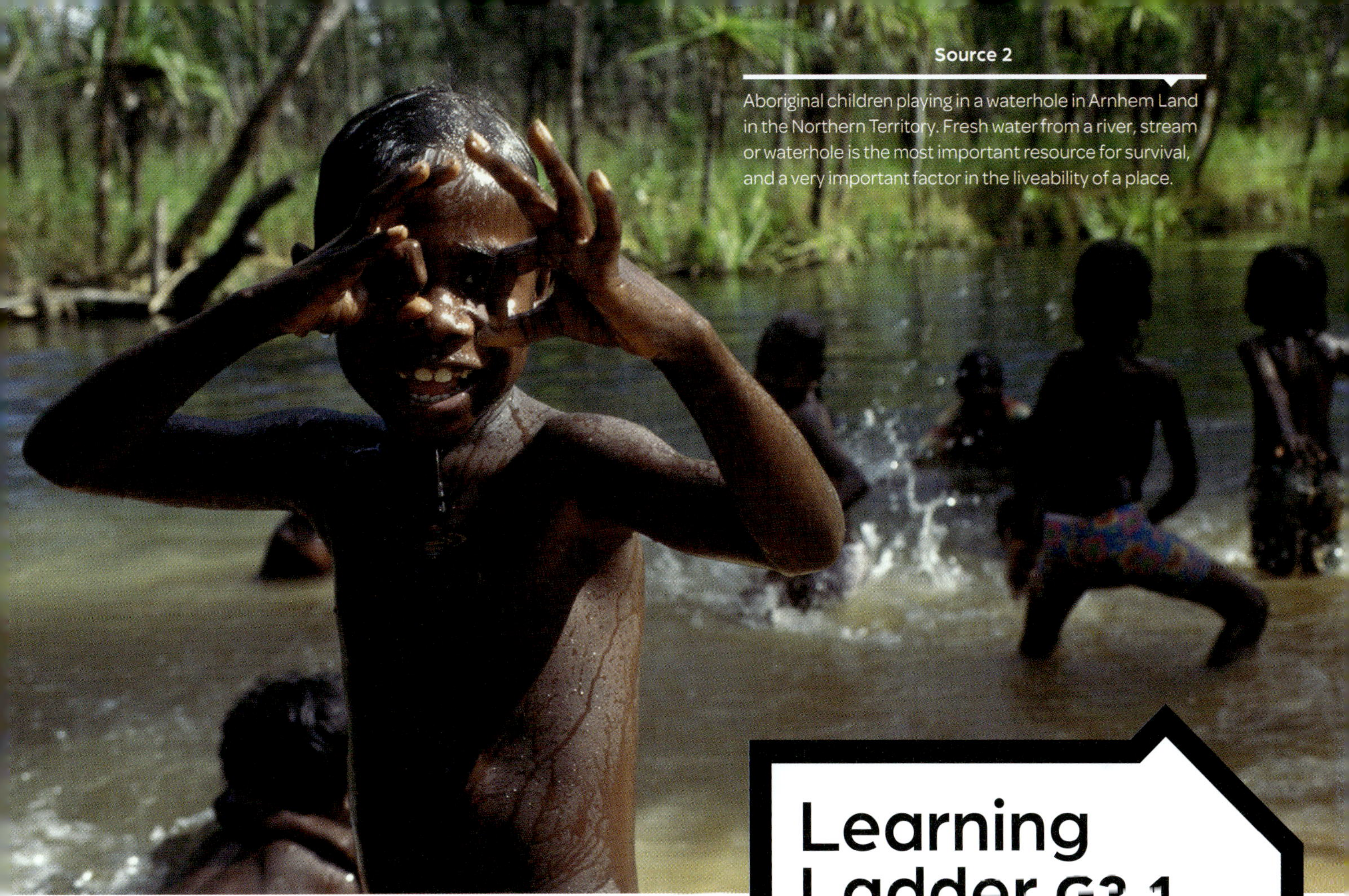

Source 2

Aboriginal children playing in a waterhole in Arnhem Land in the Northern Territory. Fresh water from a river, stream or waterhole is the most important resource for survival, and a very important factor in the liveability of a place.

Population density of Aboriginal Australians before 1788

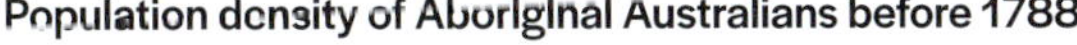

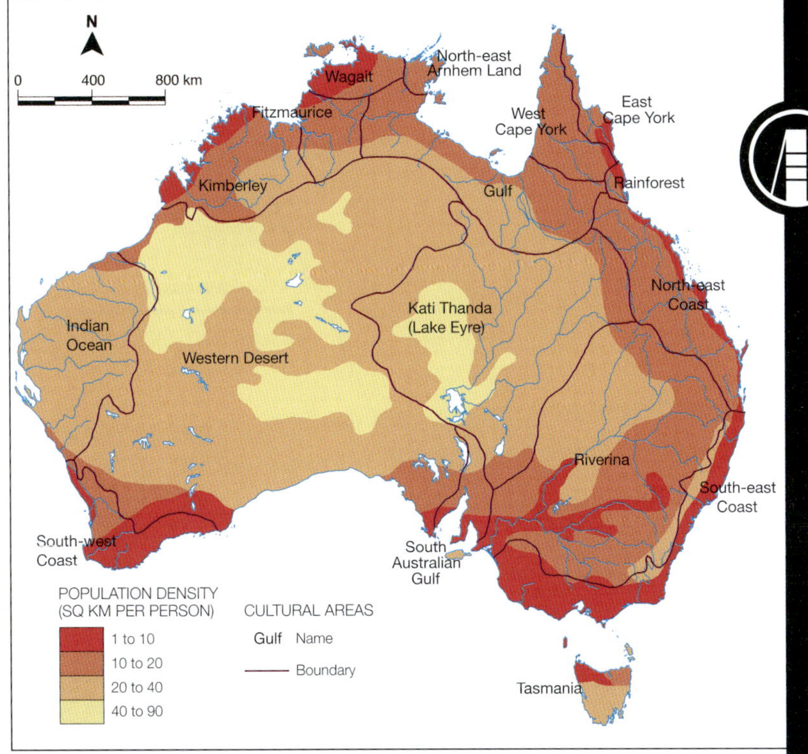

Source: Matilda Education Australia

Source 3

Population density of Aboriginal Australians before 1788

Learning Ladder G3.1

Show what you know

1 List the important factors that contributed to a place's liveability for early Indigenous Australians.

Spatial characteristics

Step 1: I can identify and describe spatial characteristics

2 Source 3: Identify the regions of Australia that had the densest populations of Indigenous Australians before the arrival of Europeans in 1788.

Step 2: I can explain spatial characteristics

3 Describe why a sense of connection to the land is important for Indigenous Australians.

Step 3: I can explain processes influencing places

4 Source 3: Which inland environments had the greatest populations before 1788 and why were they important resources for Aboriginal Australians?

Step 4: I can predict changes in the characteristics of places

5 Look at Source 1 on page 30 of the History section. Describe the cultural borders of Indigenous Australians, and compare this to our state and territory boundaries today.

Where do we live today?

Australia is mainly desert, and is one of the world's most sparsely populated countries. Nine out of ten Australians live in urban areas along the coastline.

Australia's population distribution

In area, Australia is the sixth largest country in the world, but it is only the 53rd largest in terms of population. In summary: large area, not many people. So even though Australia's total population reached 25 million people in 2018, it was still the third most sparsely populated country in the world. The average population density is just 3.3 people per square kilometre.

Australia's low **population density** – the number of people per square kilometre – is due to the desert land that makes up much of Australia's interior. Seventy per cent of Australia receives less than 500 mm of rainfall per year and is classified as desert or semi-desert. Very few Australians live in the arid centre of the continent.

Eighty-nine per cent of Australians live in **urban** areas along the coastline, with 40 per cent of all Australians living in either Sydney or Melbourne. Sydney is Australia's largest city, with a population of 5.1 million. Melbourne is growing faster than Sydney and looks set to take over as Australia's largest city by 2054.

	Growth[1]	2014	2034[2]	2054[2]
		Population (,000)	Population (,000)	Population (,000)
Adelaide	1.06%	1304.6	1604.7	1843.7
Brisbane	1.92%	2274.6	3340.6	4406.6
Canberra	1.70%	386.0	N/A	N/A
Darwin	2.30%	140.4	175.5	212.6
Hobart	0.67%	219.2	250.4	266.6
Melbourne	1.95%	4440.3	6238.1	7982.4
Perth	3.05%	2021.2	3451.8	4923.0
Sydney	1.50%	4840.6	6432.6	7966.4

1 Annual population growth rate between 2009 and 2014

2 Estimates based on medium levels of fertility, overseas migration, life expectancy and interstate migration flows

Source 1

Expected growth of Australia's capital cities over time

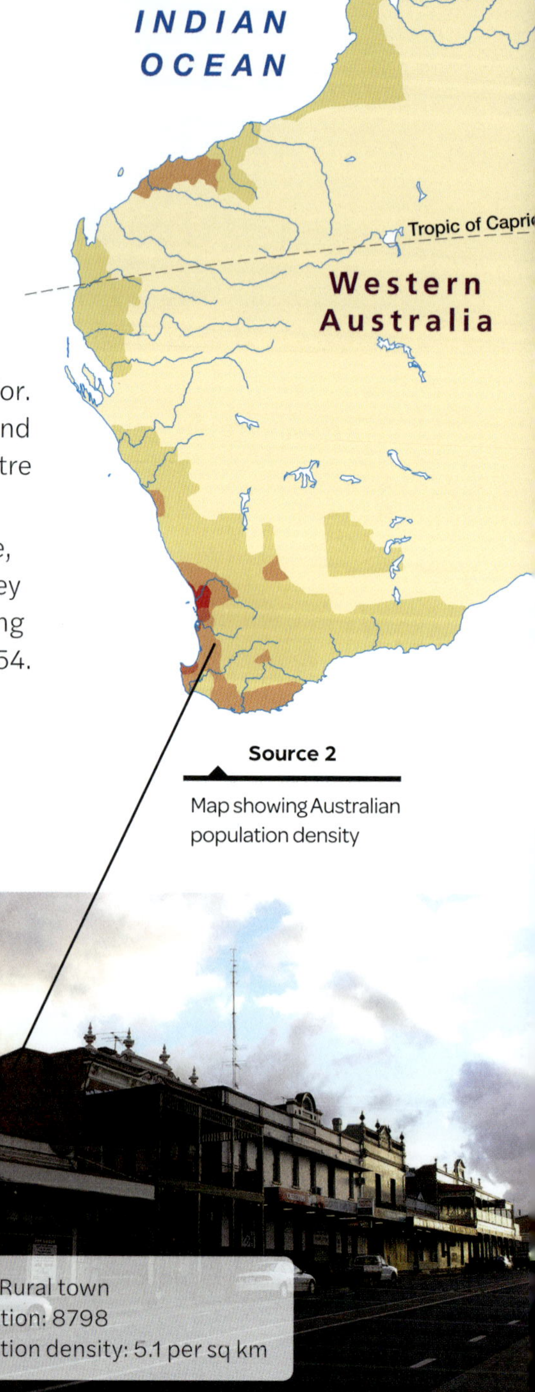

Source 2

Map showing Australian population density

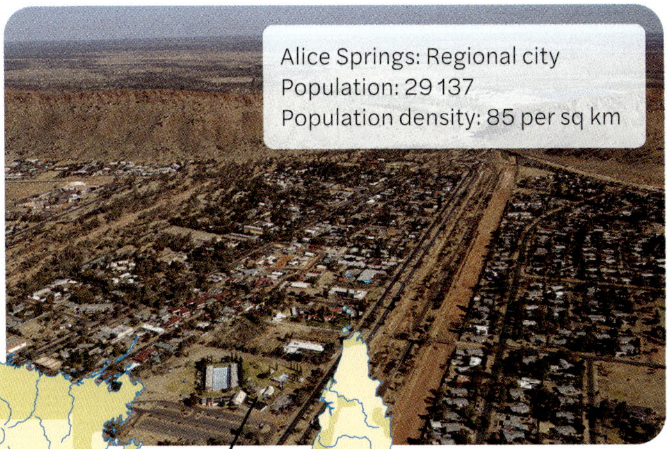

Alice Springs: Regional city
Population: 29 137
Population density: 85 per sq km

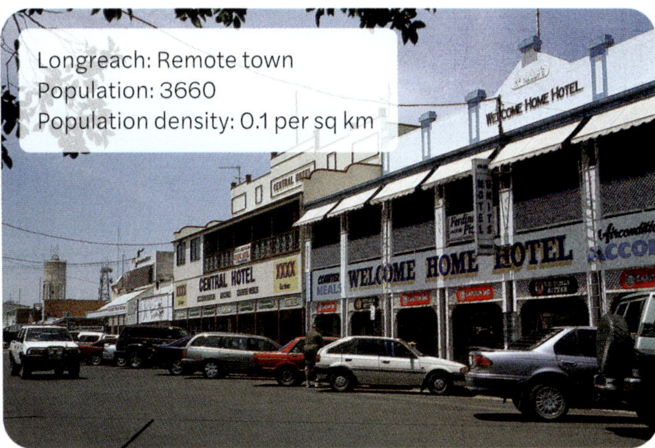

Longreach: Remote town
Population: 3660
Population density: 0.1 per sq km

PACIFIC OCEAN
Northern Territory
Queensland
Kati Thanda–Lake Eyre
South Australia
Darling River
New South Wales
Murray River
Victoria
ACT
Tasmania
N
250 500 km

POPULATION DENSITY (PEOPLE PER SQ KM)
Over 100
10 to 100
1 to 10
0.1 to 1
Under 0.1

Source: Matilda Education Australia

Sydney: Major city
Population: 5 131 326
Population density: 407 per sq km

Learning Ladder G3.2

Show what you know

1 Why does Australia have such a small population if we have one of the biggest countries?

2 How does climate influence liveability in different regions of Australia?

Spatial characteristics

Step 1: I can identify and describe spatial characteristics

3 Source 1: Use the table to rank Australia's capital cities based on their predicted population in 2034.

Step 2: I can explain spatial characteristics

4 Source 2: Use PQE to describe the distribution of people in Australia per square kilometre.

Step 3: I can explain processes influencing places

5 Sources 1 and 2: Use SHEEPT to explain Australia's **population distribution**.

Step 4: I can predict changes in the characteristics of places

6 Source 1: Use the data in the table to create a graph showing how the population is expected to grow between 2014 and 2054 in Australia's eight capital cities.

7 Using the graph you created in question 6, describe those cities that are expected to grow the fastest and slowest over the 40-year period. Suggest reasons to support the trends.

HOW TO
Using PQE, page 156
Using SHEEPT, page 158

Why do we live where we do?

Early Indigenous Australians made decisions on where to live based on natural factors they needed to survive. European settlers had to make these decisions too. Towns and cities in Australia have developed in these areas. Today, the liveability of a place also includes access to services and work, and the quality of the environment.

Natural factors

Liveability in Australia depends on water, climate and flat land.

Fresh water

Access to fresh water is the most important factor in the liveability of a place. People originally settled near rivers and lakes, and in areas of high rainfall. Today, modern technology means that food and water are available right across Australia.

Settlements in Australia are mainly located around the coastline, where the highest rainfall occurs. The biggest towns and cities are mainly located along the east coast, between the Great Dividing Range and the coastline.

Climate

Climate is a very important factor for liveability. Most Australians choose to live in those regions that have a mild climate, without extremes of heat or cold. Other people prefer hotter or cooler climates.

Many retired Australians prefer the warmer coastal climates of Queensland's Sunshine Coast and Gold Coast.

Flat land

Flat or gently sloping lands are the most densely populated areas in Australia. Most Australian towns and cities have developed on the flatter coastal plains, and along mountain valleys. That's because farms and transport networks are easier to develop on flat land, and it is easier and cheaper to build on than hilly land.

The most liveable areas of Melbourne are on the flat coastal plain surrounding Port Phillip Bay, as shown in Source 1.

Source 1

Melbourne's metropolitan area spreads out across the flat coastal plain between Port Phillip Bay and the Dandenong Ranges.

Australian average annual rainfall

N
0
500
1000 km
Darwin
Alice Springs
Tropic of Capricorn
Brisbane
Perth
Sydney
Adelaide
Canberra
Melbourne
Hobart
AVERAGE ANNUAL RAINFALL (mm)
Over 2000
1500 to 2000
1000 to 1500
500 to 1000
250 to 500
Under 250
Semi-desert
Desert

Source: Matilda Education Australia

Source 2

Australia's average annual rainfall

Australian average annual daily maximum temperature

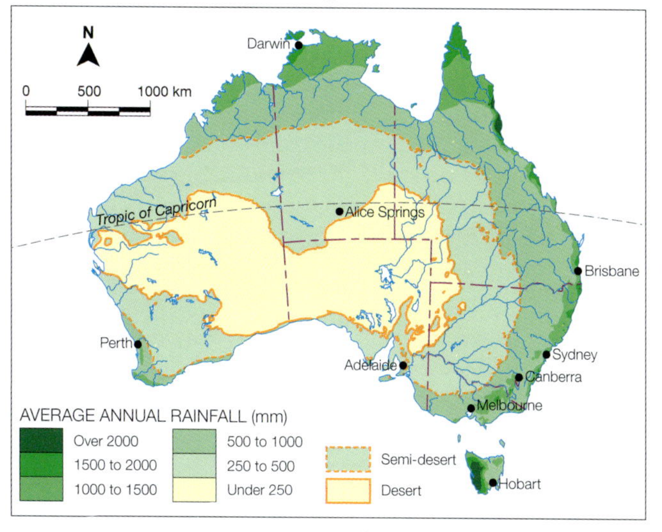

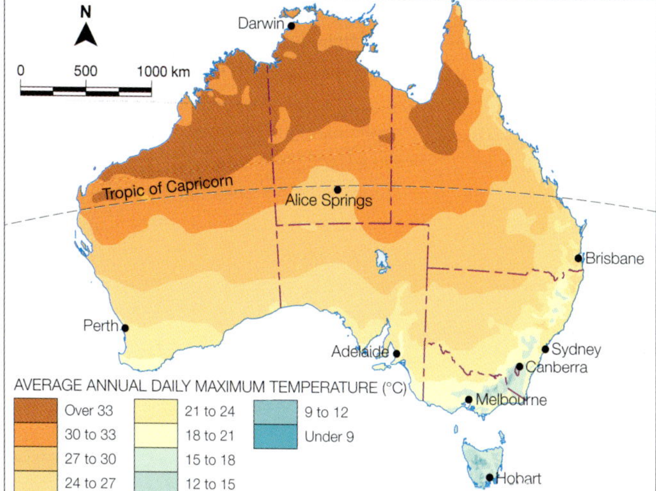

N
0
500
1000 km
Darwin
Tropic of Capricorn
Alice Springs
Brisbane
Perth
Adelaide
Sydney
Canberra
Melbourne
Hobart
AVERAGE ANNUAL DAILY MAXIMUM TEMPERATURE (°C)
Over 33
30 to 33
27 to 30
24 to 27
21 to 24
18 to 21
15 to 18
12 to 15
9 to 12
Under 9

Source: Matilda Education Australia

Source 3

Average annual daily maximum temperatures

Physical and human environments map of Australia

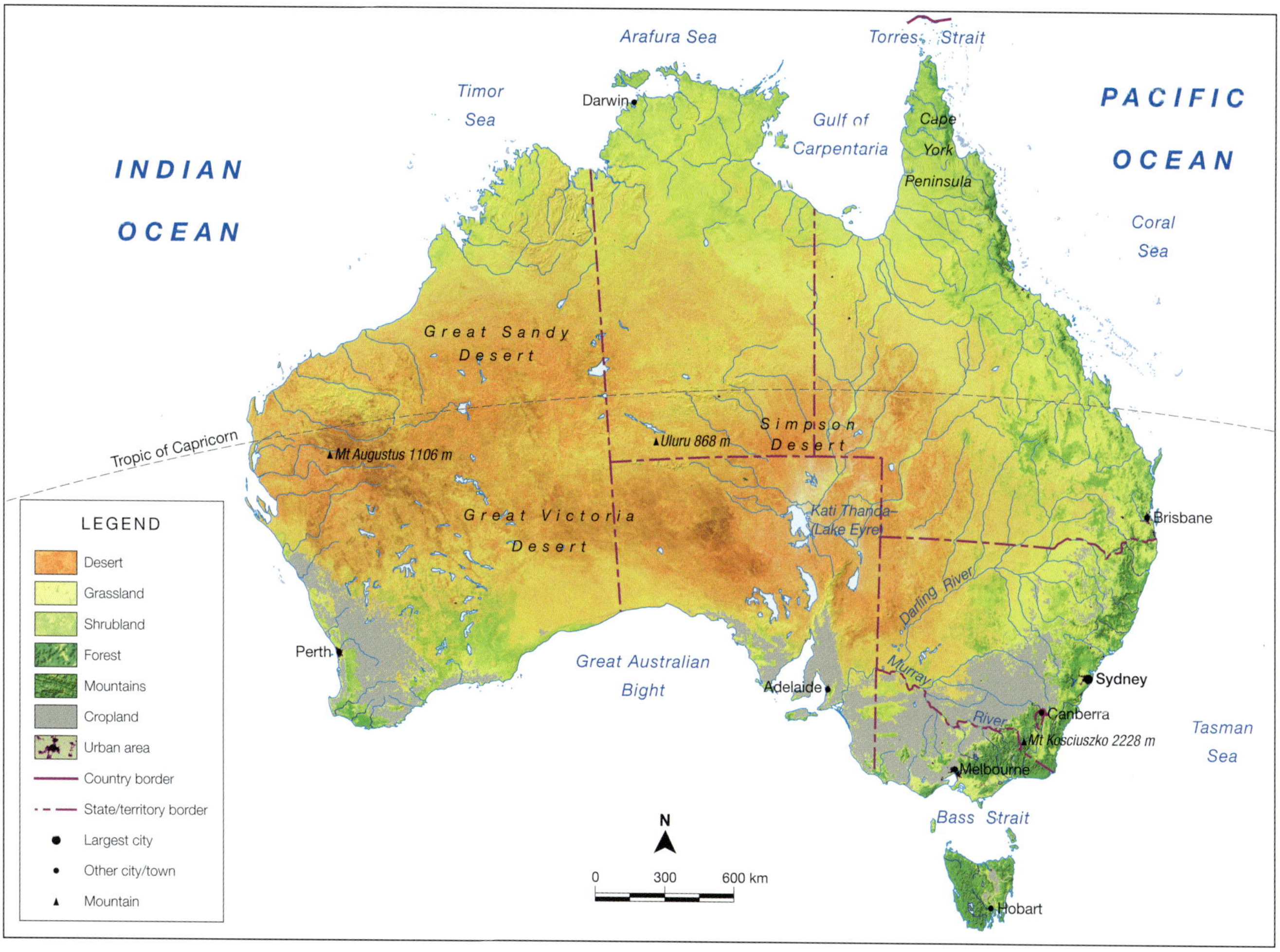

Arafura Sea
Torres Strait
Timor Sea
Darwin
Gulf of Carpentaria
Cape York Peninsula
PACIFIC OCEAN
INDIAN OCEAN
Coral Sea
Great Sandy Desert
Tropic of Capricorn
Mt Augustus 1106 m
Uluru 868 m
Simpson Desert
Kati Thanda–Lake Eyre
Great Victoria Desert
Brisbane
Darling River
Perth
Great Australian Bight
Adelaide
Murray River
Sydney
Canberra
Mt Kosciuszko 2228 m
Tasman Sea
Melbourne
Bass Strait
Hobart
N
0
300
600 km
LEGEND
Desert
Grassland
Shrubland
Forest
Mountains
Cropland
Urban area
Country border
State/territory border
Largest city
Other city/town
Mountain

Source: Matilda Education Australia

Source 4

This map shows different environments, including where people choose to live.

Human factors

Where people choose to live also depends on human factors – such as access to services and facilities, the quality of the environment, safety, and closeness to friends and family. A person's ideas about liveability change depending on their age and the stage of life they are at. School-age children might rate the liveability of an area highly if they are close to school, friends, public transport and facilities such as sporting grounds and shops. Adults with families might need schools, healthcare, affordable housing, outdoor recreation areas and access to jobs. How liveable a place is depends on what is important to you.

Employment

People of working age need to live in a place where they can access employment. The most jobs are generally in capital cities and larger towns, so many young people move from country towns to cities to study and to find employment. Mining towns with very few facilities and services can also be a magnet for workers because they offer high-paying jobs.

Services

Liveabilty increases with good access to services. Local shopping centres give people access to supermarkets, shops and restaurants, as well as services such as doctors and hairdressers. Good access to public transport and an efficient road system help people to move around. Other community facilities that can improve liveability include playgrounds, sports arenas, parks, skate ramps and bike tracks.

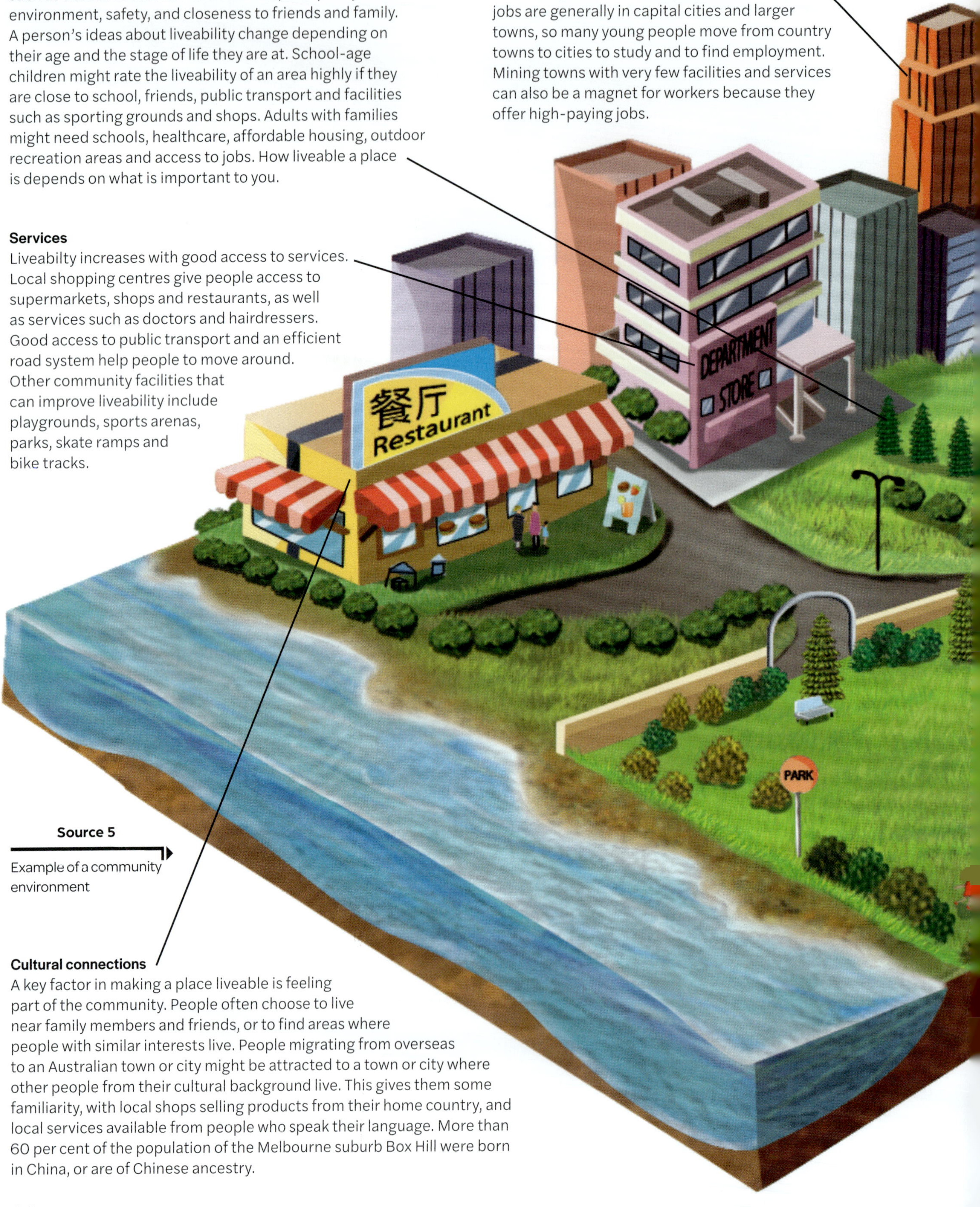

Source 5

Example of a community environment

Cultural connections

A key factor in making a place liveable is feeling part of the community. People often choose to live near family members and friends, or to find areas where people with similar interests live. People migrating from overseas to an Australian town or city might be attracted to a town or city where other people from their cultural background live. This gives them some familiarity, with local shops selling products from their home country, and local services available from people who speak their language. More than 60 per cent of the population of the Melbourne suburb Box Hill were born in China, or are of Chinese ancestry.

Housing
Selecting a home is based on finding the most liveable area where you can afford to buy or rent. Some key decisions include how close the housing is to facilities such as shops, transport and parks. The size and type of the home selected will differ – some people will want a large home with a garden, where other people will want a low-maintenance apartment.

Safety
A very important factor in the liveability of a place is that people feel safe in their homes and within their communities. Areas with lower crime rates are the most liveable. Safety is even more important in countries that suffer from war and conflict.

Environment
Environmental quality impacts on the liveability of places, and the quality of life for people living there. Factors that contribute to the quality of the environment are the climate, the quality of the air and water, the availability of parks and open spaces, and how it looks – the **aesthetics**. Increasing numbers of Australians are choosing to move to coastal areas or rural areas to enjoy the natural environment and leave the pressures of urban life behind.

Learning Ladder G3.3

Show what you know

1 Why is living next to a fresh water source no longer necessary in order for a place to be liveable?

Analyse data

Step 1: I can use geographic terminology to interpret data

2 Source 1: Suggest two reasons why most Australian cities developed on the flatter coastal plains.

Step 2: I can describe patterns and trends

3 Sources 2 and 3. Why do fewer Australians choose to live in central Australia?

Step 3: I can explain the reasons behind a trend or spatial distribution

4 Source 5: List the factors that make living in a community more liveable. Rank your list according to the factors that make your community liveable.

Step 4: I can analyse relationships between different data

5 Compare Source 4 on page 87 with Source 2 on page 84. How is population density interconnected with the physical features of the land?

What's life like in Tully?

Tully is a town in far north Queensland. It is built on flat land in a valley between some of the highest peaks in the Great Dividing Range.

Climate and liveability

Tully has a tropical climate, with maximum temperatures above 24°C throughout the year. It also has the second-highest rainfall of any town in Australia – with more than four metres of rainfall each year! The warm climate is part of the liveability of Tully for its residents – along with the World Heritage-listed rainforest of Tully Gorge, the tropical islands and nearby Mission Beach.

Tully was established in the 1920s as a sugarcane-growing town, because it has the perfect climate for growing tropical crops such as sugar cane and bananas. In harvesting season, the Tully sugar mill employs more than 12 per cent of the town's population of 2400 people. About 325 growers in the region supply the mill with sugarcane, which is transported to the mill via 200 kilometres of special 'cane train' railway.

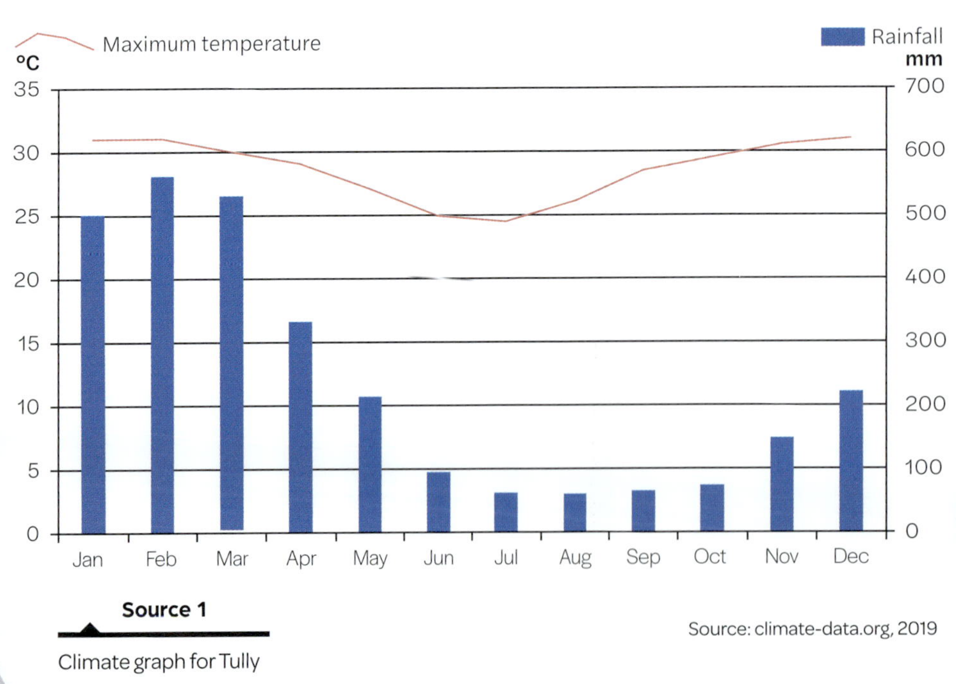

Source 1

Climate graph for Tully

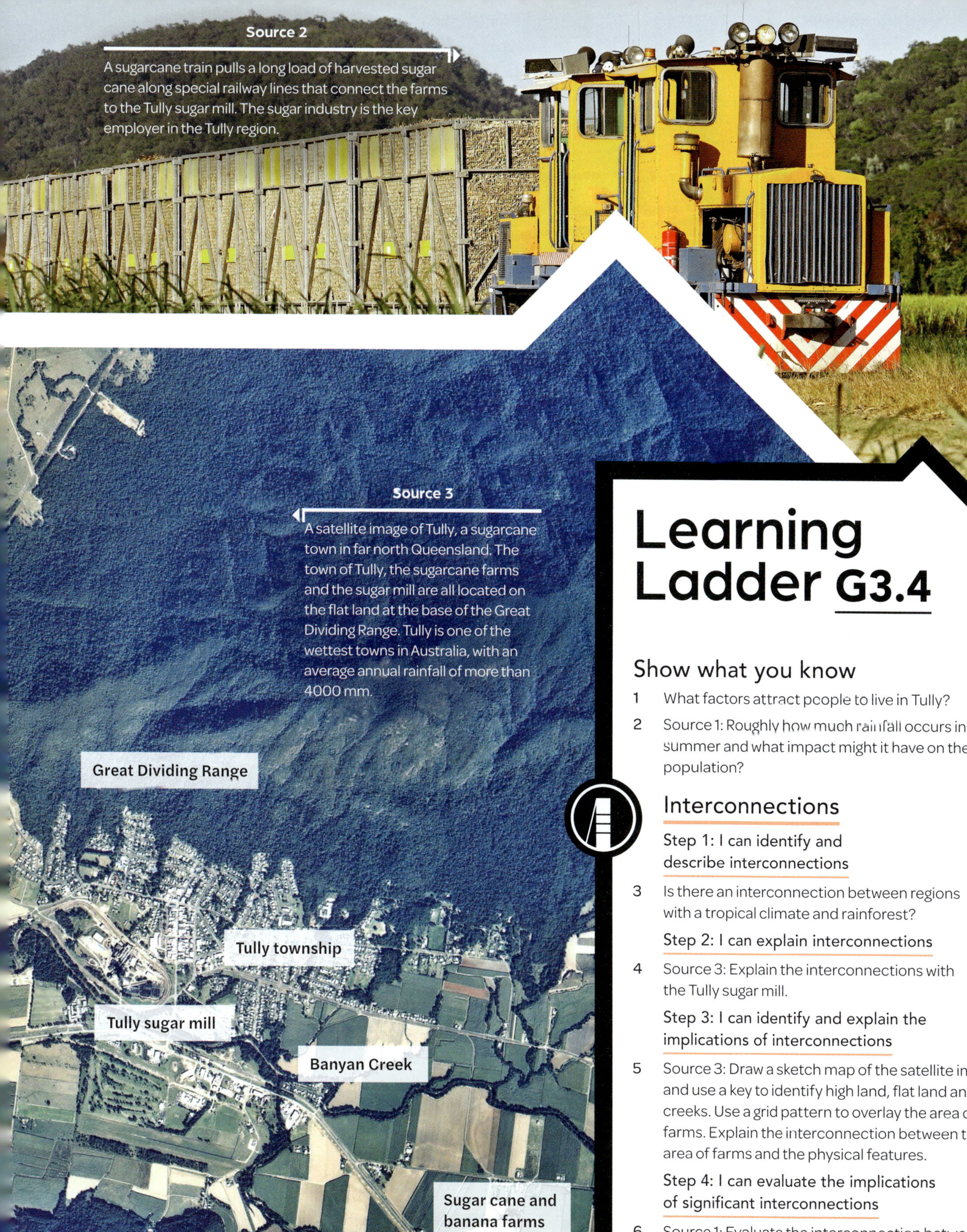

Source 2

A sugarcane train pulls a long load of harvested sugar cane along special railway lines that connect the farms to the Tully sugar mill. The sugar industry is the key employer in the Tully region.

Source 3

A satellite image of Tully, a sugarcane town in far north Queensland. The town of Tully, the sugarcane farms and the sugar mill are all located on the flat land at the base of the Great Dividing Range. Tully is one of the wettest towns in Australia, with an average annual rainfall of more than 4000 mm.

Learning Ladder G3.4

Show what you know

1 What factors attract people to live in Tully?

2 Source 1: Roughly how much rainfall occurs in summer and what impact might it have on the population?

Interconnections

Step 1: I can identify and describe interconnections

3 Is there an interconnection between regions with a tropical climate and rainforest?

Step 2: I can explain interconnections

4 Source 3: Explain the interconnections with the Tully sugar mill.

Step 3: I can identify and explain the implications of interconnections

5 Source 3: Draw a sketch map of the satellite image and use a key to identify high land, flat land and creeks. Use a grid pattern to overlay the area of farms. Explain the interconnection between the area of farms and the physical features.

Step 4: I can evaluate the implications of significant interconnections

6 Source 1: Evaluate the interconnection between climate and growing sugarcane and bananas. Why do farmers choose to grow sugarcane and bananas in the Tully region?

HOW TO

Satellite images, page 167

How do places change over time?

Growth areas on the **rural–urban fringe** of major cities are attracting thousands of new residents – at the same time as many small rural towns are losing population and services and becoming less liveable.

A simple way to measure the liveability of a place is to calculate whether it is growing or declining. Fast growing areas show there is demand – people want to live there. The fastest growing places in Australia are new housing estates on the edges of capital cities.

Source 1

Percentage of population growth by region, 2006–2016

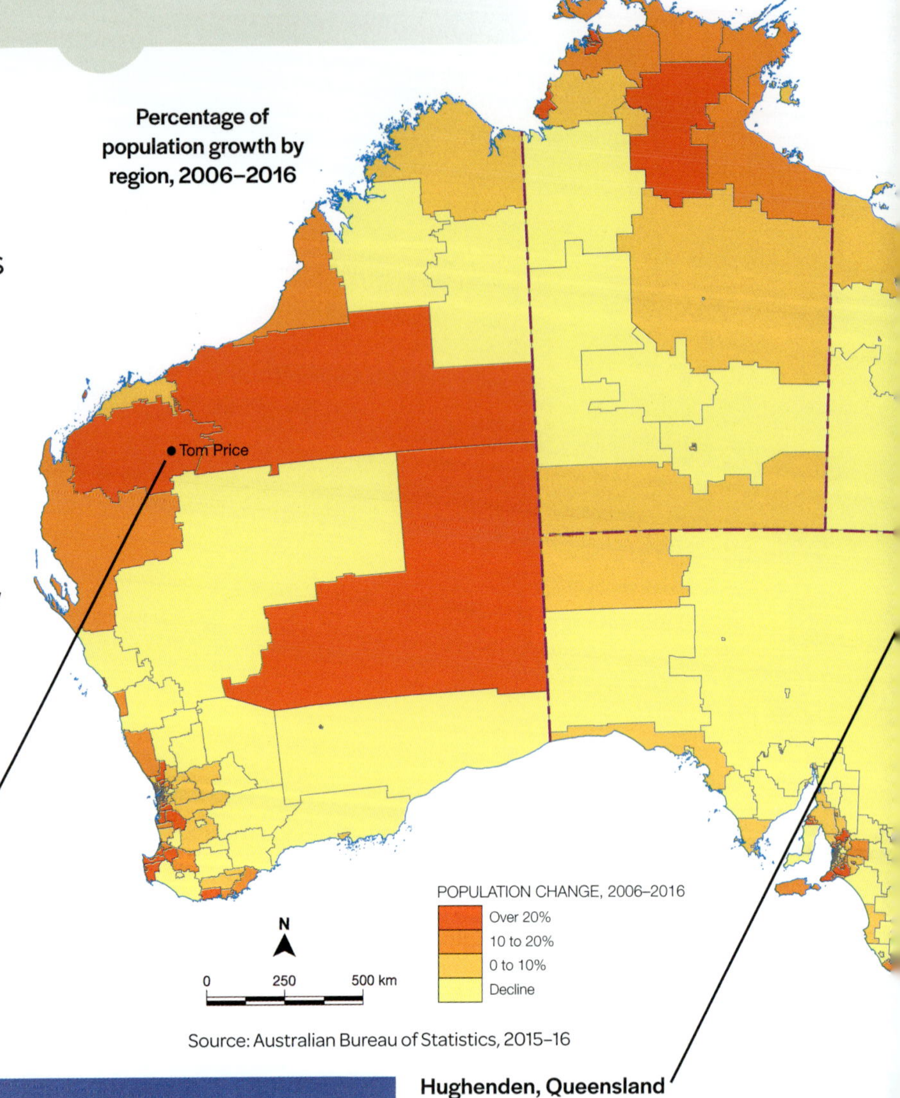

Source: Australian Bureau of Statistics, 2015–16

Tom Price

The mining industry has driven growth in places such as Tom Price, a town in Western Australia. The population of Tom Price grew by 85 per cent during the 2006–2011 mining boom. Then, as the mining boom ended, the population declined by 50 per cent between 2011 and 2016.

Hughenden, Queensland

Hughenden is a small town in outback Queensland. It has lost half its population in the last 50 years. Hughenden is a service centre for the surrounding agricultural areas, but businesses have closed as people have left the town – and one of the businesses to close was the Grand Hotel. The town is hopeful that a recent discovery of dinosaur bones will bring tourists to the area.

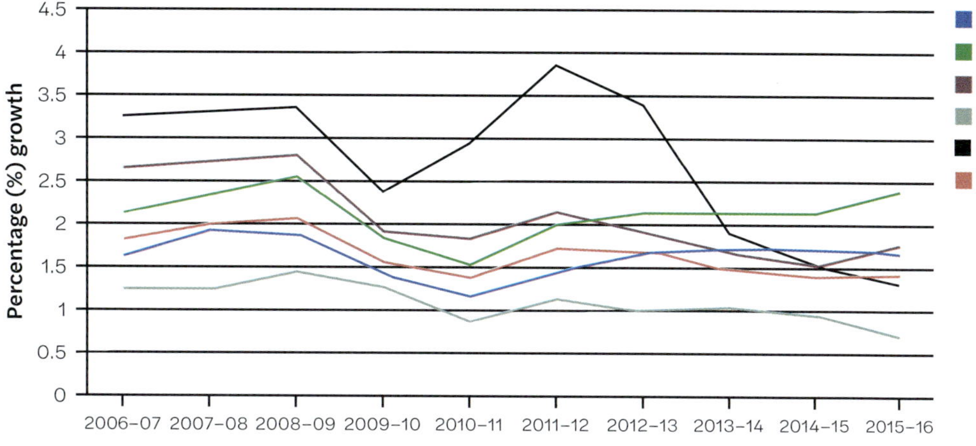

Source 2

Population growth in Australian capital cities, 2006–2016

Source: Australian Bureau of Statistics, 2019

Growth and decline can also be linked to work opportunities, such as the rapid increase and decrease of populations in mining towns like Tom Price, in Western Australia. Small rural towns more than 100 kilometres from capital cities represent many of Australia's declining populations. Tough farming conditions have forced many farmers off the land; in turn, this forces shops, schools and hospitals in nearby country towns to close.

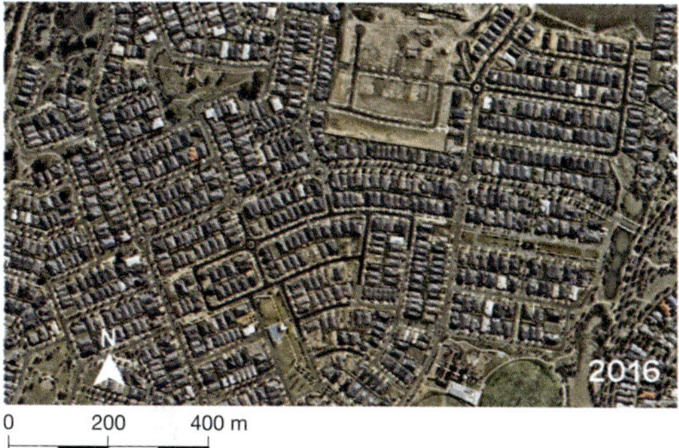

Doreen, Victoria

The population of Doreen in Melbourne's outer north doubled between 2011 and 2016, as new housing estates were build on what had been farmland. New services such as supermarkets, shopping centres, schools and medical centres have been built to support the growing population.

Learning Ladder G3.5

Show what you know

1 Why are places with growing populations more liveable and those with declining populations less liveable?

2 Why did the population of Mount Tom Price change so rapidly in the decade between 2006 and 2016?

3 Where are the fastest growing places in Australia and why are people attracted to these places?

Collect, record and display data

Step 1: I can collect, record and display data in simple forms

4 What do the lines on the graph in Source 2 show?

Step 2: I can recognise and use different types of data

5 Source 2: Is this line graph a source of primary or secondary data? Justify your response.

Step 3: I can choose, collect and display appropriate data

6 List the primary and secondary methods you could use to collect appropriate data investigating whether the population of your local suburb is growing or shrinking.

Step 4: I can use data to support claims

7 Look up Google Earth Engine (http://mea.digital/gh7_g3_1). How has Google used data to show changes to Las Vegas? How effective is it?

thinking locally

What is the impact of Nyah West's declining population?

Nyah West is a small town in north-western Victoria, near the New South Wales border. It was established in 1915 when the railway was extended from Swan Hill, and it has since supported the surrounding irrigated farms that produce wine, dried fruit, vegetables and wool.

Many farmers sold up and moved away after a long drought and falling prices for dried fruit. Like many small rural towns in Australia, Nyah West is losing its people to larger cities such as Melbourne and to the nearby 'sponge city' of Swan Hill, which soaks up the populations of smaller surrounding towns as they lose their services. Nyah West lost several government services and all of its banks in the 1980s and 1990s. These are all now available in the growing town of Swan Hill, just 20 minutes away.

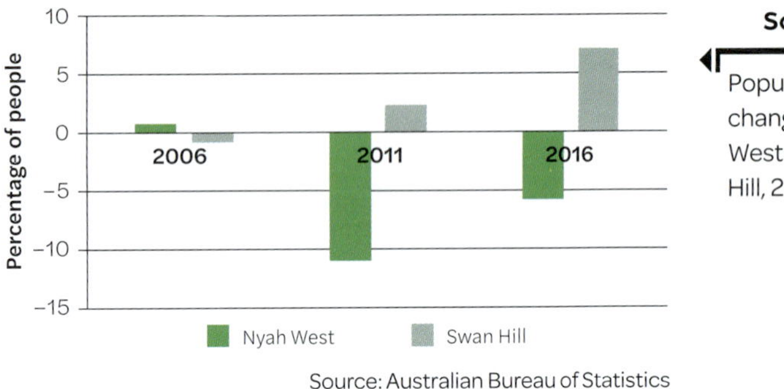

Source 1

Population change in Nyah West and Swan Hill, 2006–2016

Age groups living in Nyah West and Victoria, 2016

Percentage of people: 40, 35, 30, 25, 20, 15, 10, 5, 0

Age group: 0–19, 20–39, 40–59, 60+

Nyah West | Victoria

Source: Australian Bureau of Statistics

Source 2

Age groups living in Nyah West and Victoria, 2016

Source 3

Swan Hill offers a wide range of services for its residents and the local region.

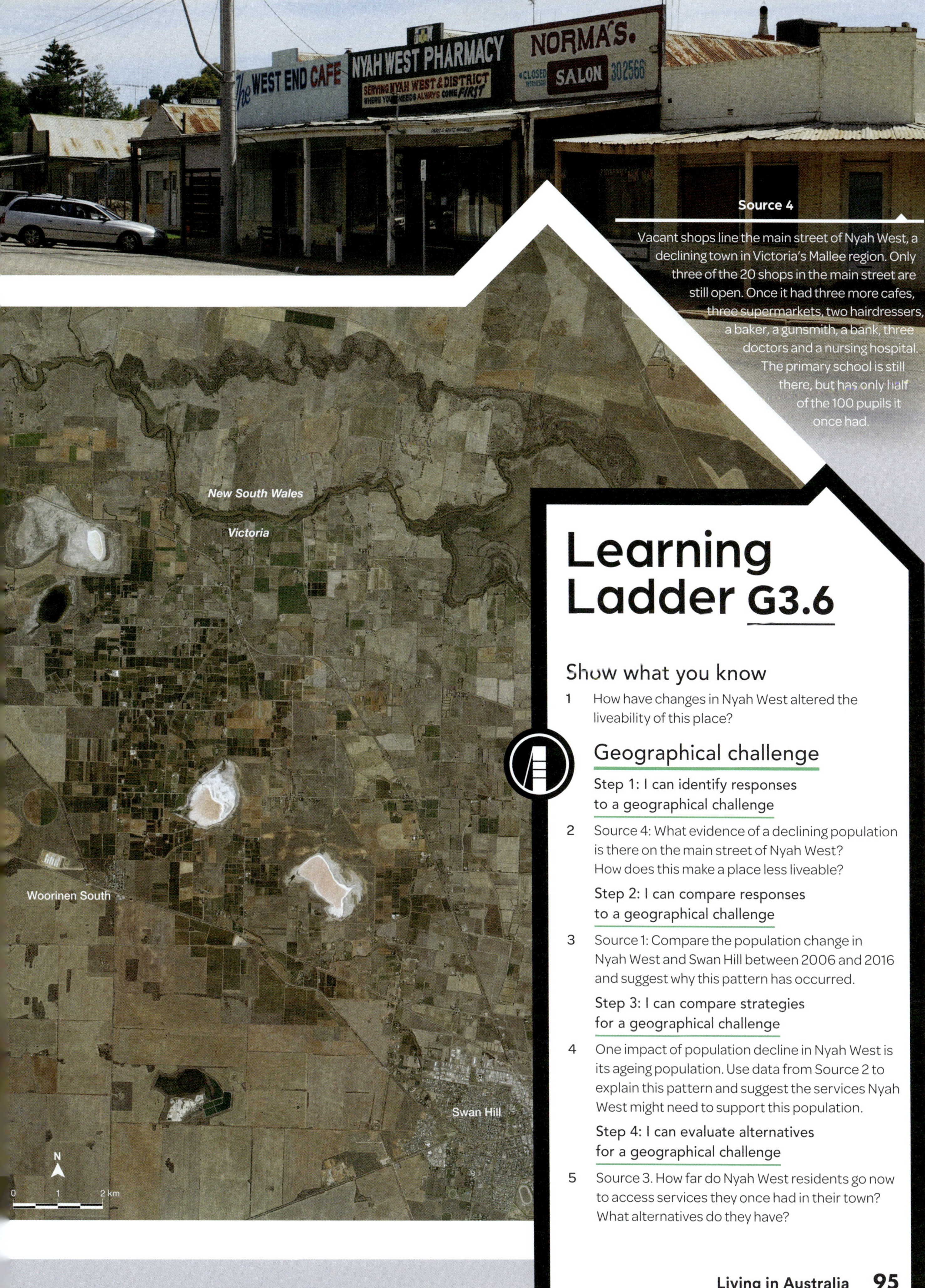

Source 4

Vacant shops line the main street of Nyah West, a declining town in Victoria's Mallee region. Only three of the 20 shops in the main street are still open. Once it had three more cafes, three supermarkets, two hairdressers, a baker, a gunsmith, a bank, three doctors and a nursing hospital. The primary school is still there, but has only half of the 100 pupils it once had.

Learning Ladder G3.6

Show what you know

1 How have changes in Nyah West altered the liveability of this place?

Geographical challenge

Step 1: I can identify responses to a geographical challenge

2 Source 4: What evidence of a declining population is there on the main street of Nyah West? How does this make a place less liveable?

Step 2: I can compare responses to a geographical challenge

3 Source 1: Compare the population change in Nyah West and Swan Hill between 2006 and 2016 and suggest why this pattern has occurred.

Step 3: I can compare strategies for a geographical challenge

4 One impact of population decline in Nyah West is its ageing population. Use data from Source 2 to explain this pattern and suggest the services Nyah West might need to support this population.

Step 4: I can evaluate alternatives for a geographical challenge

5 Source 3. How far do Nyah West residents go now to access services they once had in their town? What alternatives do they have?

Why do people live in cities?

Major cities such as Sydney and Melbourne are some of the most liveable in the world. They attract people with job opportunities and a wide range of services available in education, health, entertainment and shopping.

Living in big cities

Australia is one of the most **urbanised** countries in the world – with 89 per cent of our population living in cities. Over 66 per cent of Australians live in the metropolitan area of Australia's eight capital cities.

Large cities attract people to live there because they have:

- jobs – 75 per cent of all jobs are based in Australia's major cities
- education opportunities at a range of public and private schools, universities and training colleges
- a large range of shops and businesses
- entertainment at large sporting arenas, theatres, cinemas, museums and art galleries
- transport infrastructure such as airports, major highways and railway lines
- a large range of health services, including specialists and large hospitals.

Liveable capital cities

Australian cities are continually voted the most liveable cities in the world. In 2018 and 2019, Melbourne was listed as the second most liveable city in the world after being the most liveable city for seven years.

Source 1

Cyclists on a bike path heading towards the Melbourne CBD. Open space, walking trails and bike paths make busy cities more liveable.

Source 2

Major cities such as Melbourne offer large entertainment venues, such as the Melbourne Cricket Ground and Rod Laver Arena, along with a range of art galleries, theatres, cinemas and restaurants.

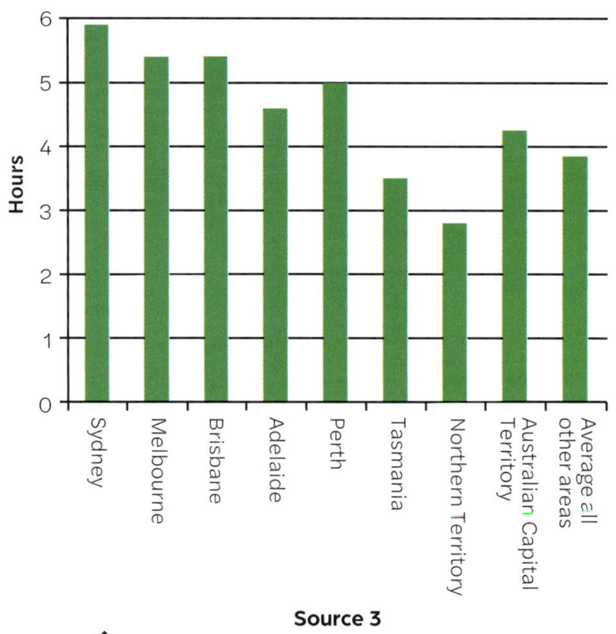

Source 3

As a city grows, more people use its infrastructure, such as transport. This leads to increased traffic and travel times, which can make a city less liveable.

The Global Liveability Index ranks 140 cities in the world according to these factors:

- Stability – the threat of terror, war and crime
- Health care availability and quality
- Culture and environment
- Education availability and quality
- Infrastructure – quality of housing, transportation, water, energy and telecommunications.

Australian cities score well in liveability studies because they are smaller and less crowded than many other large cities around the world. Australian cities also have good education and healthcare, and relatively low crime rates.

Maintaining liveability

As cities grow, open spaces are replaced with housing – this happens particularly on the edges of cities, where new suburbs appear on what was previously farmland. Then schools, shopping centres, medical centres, transportation, sporting facilities and other services need to be provided to meet demand from residents in new suburbs.

As city populations rise, more pressure is put on **infrastructure** and services. On average, Sydney and Melbourne drivers now spend one hour and 10 minutes commuting to and from work each day. By 2031 commute times are predicted to be two hours a day.

Learning Ladder G3.7

Geographical challenge

Step 1: I can identify responses to a geographical challenge

1 Work with a partner to create a list of the pros and cons of living in a city environment. Share your list with the class. Did you have different ideas to other groups?

Step 2: I can compare responses to a geographical challenge

2 Give two reasons why Melbourne was ranked second-most liveable city in the world in 2018.

Step 3: I can compare strategies for a geographical challenge

3 Source 3: Which capital city has the longest weekly work commute? What strategies could be used here to improve the liveability of this place?

Step 4: I can evaluate alternatives for a geographical challenge

4 Source 1: What examples are given here to improve the liveability of large cities. Why is it sometimes difficult to supply these in large cities?

G3.8

economics+business

How do we plan for population growth?

Melbourne is Australia's most liveable city – but it is under pressure from record population growth. This is an economic issue. City planners have come up with a range of solutions, each with its own costs and benefits.

Record population growth

Melbourne is one of the fastest growing cities in the **developed** world. In 2017 it grew at a rate of 2.7 per cent, which is double the growth rate of most cities in advanced countries. It is a growth rate usually associated with cities in **less economically developed countries (LEDC)**, such as those in South America and Africa.

According to figures from the Australian Bureau of Statistics, Melbourne's population grew more than 2.5 per cent each year between 2011 and 2017. In the same period, Sydney grew by 1.8 per cent.

In 2017, Melbourne added 123 000 people – and since 2011 it has increased its population by 650 000 people.

Melbourne's economic problem

Population growth is an **economic problem**. As more and more people come to live in Melbourne, the need for essential services such as housing and water is increased.

Needs are items that people require for survival, such as food, water, clothing and housing. Needs are different to **wants**, which are products or services that are not critical for survival, such as more public transport and better roads to reduce travel times.

Population change in Melbourne 2005–2015

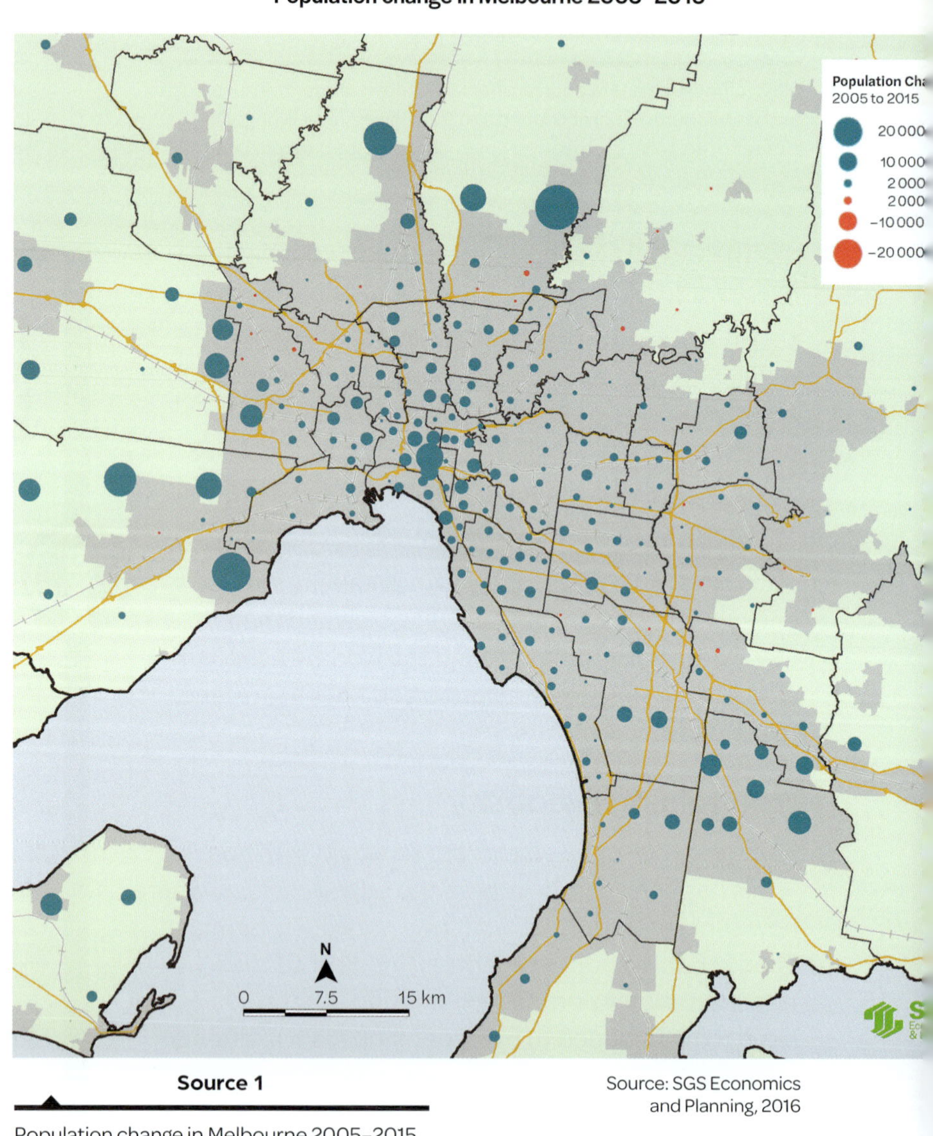

Source 1

Population change in Melbourne 2005–2015

Source: SGS Economics and Planning, 2016

Source 2

Melbourne is one of the fastest growing cities in the developed world – which makes it difficult for transport systems to cope.

In 2018, Vienna was declared the world's most liveable city, replacing Melbourne. Vienna's smaller population of 1.8 million is a big factor in its liveability – cities with higher populations face greater challenges.

Melbourne's fast urban growth has a negative effect on its liveability. As urban areas grow into what used to be farmland – called urban sprawl – the liveability of cities is threatened. Melbourne's fast population growth has led to high prices for homes and lengthy traffic jams.

Planning for the future

Economists work with governments and businesses to help solve economic problems, such as rapid population growth. They calculate the needs for Melbourne's growing population, such as the number of homes required. Economists prioritise the different wants of consumers – such as transport, schools, medical services and green space – and suggest ways to satisfy as many wants as possible.

There is often more than one way to solve an economic problem, and each solution will have its own **costs** and **benefits**. Some suggested solutions for Melbourne are:

- reducing the number of migrants coming to Australia to remove pressure on resources
- increasing the density of new housing developments, so more homes can be built in a smaller area
- encouraging people to move from urban to country areas to relieve urban congestion – which is called **decentralisation**.

Learning Ladder G3.8

Economics and business

Step 1: I can recognise economic information

1 Why is population growth an economic problem?

Step 2: I can describe economic issues

2 Source 1: Look carefully at the map showing the change in Melbourne's population.
   a Use compass directions to explain the fastest growing population areas.
   b What do most of the areas with the largest population change have in common?

Step 3: I can explain issues in economics

3 Source 1: What additional problems might economists face when planning for growth of new suburbs on the outer rim of the city, compared to growth in the inner suburbs?

Step 4: I can integrate different economic topics

4 In what ways is Melbourne's population growth affecting its liveability?

Step 5: I can evaluate alternatives

5 List three solutions to solve the economic problem of population growth in Melbourne. Search online to find one example of a plan to ease the problems of overpopulation in Melbourne.

Why is a *community of place* important?

Living in a community brings people together. Being part of a community helps people feel more connected and supported – and makes a place more liveable.

Connected communities

The most **liveable** places are those where everyone feels connected to the local **community**. A community is a group of people who have something in common such as a neighbourhood, a workplace, a school, a sporting club, a returned soldiers league or a language group.

We are all members of many different communities that make our lives more enjoyable. For example, over 6 million Australians are actively involved in sporting communities. They might play sport at an elite level or just for fun, or they might be a spectator or follow a particular sporting team.

Sports practised by boys and girls aged 6–13 years

Base: *Australian children 6–13*

Boys
Girls

Percentage (%) of participants
0
10
20
30
40
50
60
70

Swimming
Soccer
Bicycling
Cricket
Tennis
AFL
Netball

Sport

Source: Roy Morgan Young Australians Survey, 2015

Source 1

Sports practised by Australian boys and girls aged 6–13 years

Community of place

A **community of place** is a community of people who come together because of where they live, work, visit or spend time.

People living in remote or rural areas need the support and sense of belonging to a local community to overcome the isolation of where they live. Places such as schools, churches, sporting clubs, cafes and hotels are important meeting points for country people. Community events – such as rodeos, agricultural shows, festivals, sporting events and social celebrations – are important to help overcome the loneliness felt by many people on the land.

Source 2

Shri Shiva Vishnu Hindu temple at Carrum Downs is a popular community venue for Melbourne's Indian community.

Source 3

People connect with each other when they join up to share a common interest, such as sport. If people feel connected through communities, the place they live in becomes more liveable. Boys and girls have a variety of sports they can be involved in, and their participation in sports has changed over time.

Learning Ladder G3.9

Show what you know

1 Define the term 'community of place'.

2 How does 'community of place' increase the liveability of an area?

Interconnections

Step 1: I can identify and describe interconnections

3 Discuss how living in a community helps people interconnect.

Step 2: I can explain interconnections

4 Source 3: Suggest two reasons why sport enables interconnection.

Step 3: I can identify and explain the implications of interconnections

5 List the local community groups in your area. As a class, discuss how these groups contribute to the local community of place.

Step 4: I can evaluate the implications of significant interconnections

6 Source 4: Why are opportunities for interconnection through a community of place even more important for people in rural communities?

Cultural communities

People are often attracted to a place where they know there are other people from the same cultural background. Half of Australia's population was either born overseas or has a parent who was born overseas. So some people find specific rural towns – and specific suburbs in some cities – are more liveable for them because they know there are other people living there who share their language and culture.

Cultural communities often share the same religious beliefs. Melbourne has Australia's largest Indian community, so Hinduism, which is India's biggest religion, has become Melbourne's fastest growing religion. Belonging to a religious community helps make places more liveable for new Australians who have migrated from another country.

Source 4

Agricultural shows are a way of bringing rural communities together. Local, regional and state shows give an opportunity to showcase livestock and produce. For many people in remote areas, the local show is the highlight of the year – and the focus for the whole community.

What are social values?

Liveability is improved when people feel connected to a place, community or society. Australia is a **multicultural** and **multi-faith** country. This means that everyone has the freedom and the right to practise and share their cultural beliefs, free from discrimination or censorship.

Values

While each individual has their own beliefs, Australians share the values of respect, compassion, equality, inclusion and freedom. These values guide our behaviour, and help us to live in harmony. They are reflected in the Australian **Constitution** and in our laws.

As Australia has developed over time, its people, governments and the legal system have promoted these values:

- **Freedom** to make our own decisions within the boundaries of the law. Australians enjoy freedoms that some other people don't have, such as the freedom to openly express opinions in public and the freedom to practise different religions.
- *Equality* for everyone, without **discrimination**. All Australians deserve a 'fair go' to access work, education and healthcare, regardless of their sex, race or wealth.
- *Compassion* towards people that need our help. For example, compassion towards refugees who may have fled from difficult circumstances, such as war, and who arrive with different values, beliefs and language.
- *Inclusion* of every person that makes up our society, no matter what they look like or how they choose to live. The most liveable places are those where everyone feels connected to the community, rather than living in isolation.

Source 1

A key Australian value is freedom of speech. Australian citizens can speak openly and protest peacefully when they want to let their views be known.

Source 2

A mosque built by the Cyprus Turkish Islamic community in Sunshine, Victoria

Religion Australian population 2016 (%)

Catholic	22.6
Anglican	13.3
Other Christian	16.3
Islam	2.6
Buddhism	2.4
Hinduism	1.9
Sikhism	0.5
Judaism	0.4
Other religion	0.4
No religion	30.1
Declined to answer	9.5

Source: Australian Bureau of Statistics, 2016

Source 3

Major religions in Australia in 2016

- *Responsibility* for our actions, and making sure that others do not suffer as a result. All Australians should understand what is right and wrong – laws are in place to make sure that people act responsibly (see page 65 of the History section).

Religion

One of the freedoms enjoyed in Australia is the opportunity to practise whatever religion you choose, or no religion at all. Australia has developed as a multi-faith society as people from many different cultures have brought their customs, beliefs and religions.

Australia is also a **secular** country, where the church is separate from the government and does not directly influence the decisions of government. In a secular nation the religious beliefs of one group of people are not forced onto those of other faiths. Australia is able to support people to choose their own religious beliefs or to be **atheists** (people who do not have a religion).

Nearly a quarter of the world's countries are non-secular, where the legal and government systems are based on religious teachings. Most of these countries are in northern Africa and south-east Asia. Some examples of non-secular nations are Iran (Shia Islam), Iraq (Islam), Israel (Judaism), Malaysia (Sunni Islam), Cambodia (Buddhism) and England (Protestantism).

Civics and citizenship

Step 1: I can identify topics about society

1 Source 2: Look at the photo of the mosque.
   a How has the Turkish community influenced the character of Sunshine?
   b How does the mosque help the Islamic people in Sunshine feel connected to the community?

Step 2: I can describe societal issues

2 What is inclusion? Why is it an important Australian value for communities?

Step 3: I can explain issues in society

3 What does it mean to an Australian to be given a 'fair go'?

Step 4: I can explain different points of view

4 Source 1: Look at the photo of the protesters.
   a Which key value in Australian society are these people using to stage a peaceful protest?
   b Which values do they think are being ignored by other Australians?

Step 5: I can analyse issues in society

5 Discuss this question: Are Australians free to do whatever they want to?

How do environments affect liveability?

The quality of the environment is an important factor in the liveability of a place. In large cities, liveability improves with access to clean air and open space. To find clean air and open space, people are increasingly moving away from cities to coastal or rural areas so they can enjoy a more liveable environment.

Sea change

More people are choosing to live in coastal towns, where they can enjoy the natural environment as well as swimming, surfing, fishing and walking along the beach. The decision to move to a coastal location is known as a **sea change**. Increasingly, large numbers of people are moving to coastal areas to get away from the pressures of big city life. Apart from capital cities, coastal communities are the fastest growing regions in Australia.

Tree change

When people move from the city to a rural area, it is known as a **tree change**. A tree change is usually made in search of a more **liveable** environment. Country towns offer people a better balance of work and leisure (called 'work–life balance'), a lower cost of housing and living, clean air and a slower pace of life.

For people considering a tree change, the lower cost of homes in rural areas is tempting. A buyer might get a large home on a block of land for the price of an apartment in a capital city. More than 30 per cent of the people moving to country areas say the cost of living was their main reason to move.

Source 2

Adelaide is consistently voted one of the most liveable cities in the world. The central business district is surrounded by parklands for Adelaide's residents to enjoy.

Population change in Victorian coastal regions 2006–2016

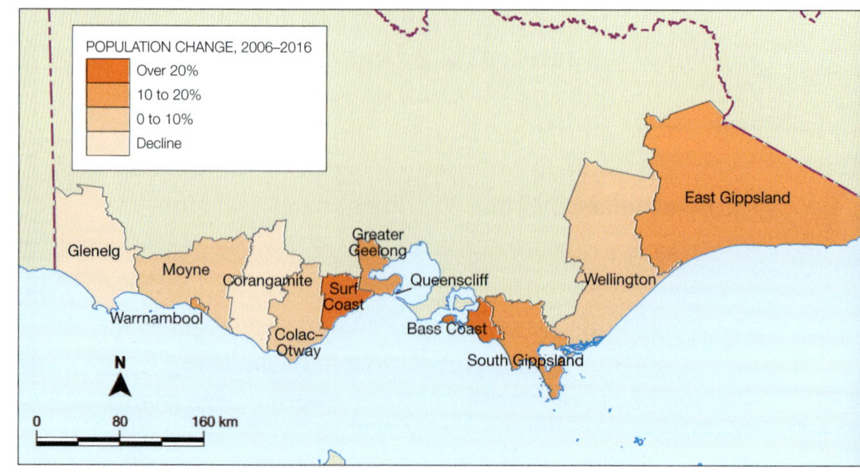

Source: Matilda Education Australia

Source 1

Population change in Victorian coastal regions 2006-2016

Source 3

Residents of Victoria's Surf Coast enjoy a pre-dawn surf. Some of them will then head off to work locally. Others will be driving to jobs in Geelong (20 minutes) or Melbourne (90 minutes). More and more people are making a 'sea change' to enjoy the coastal environment and the activities available there.

With improvements to internet connections across Australia, more residents can make a tree change and work from home to minimise their travel time and increase their quality of living.

A key motivation for a person to make a tree change to a rural area is to find a better living environment. The aesthetic appeal of rolling green pastures with grazing animals and fresh air appeals to many people.

Green belts and wedges

Green belts and **green wedges** are undeveloped areas within or around cities that have been set aside for open space. Green belts surround the outer limits of cities, and green wedges run through urban areas.

These green zones help make major cities more liveable, as they:

- protect natural and rural environments
- improve air and water quality
- give city residents access to rural and natural environments
- provide bushwalking, camping and riding areas close to cities
- provide a connected habitat for native animals and plants.

Learning Ladder G3.11

Show what you know

1 What is a sea change? What is a tree change? Why do people choose to make them?

Collect, record and display data

Step 1: I can collect, record and display data in simple forms

2 Source 1 is a choropleth map. What do these types of maps show?

Step 2: I can recognise and use different types of data

3 Source 4 is an oblique aerial photograph. How is this form of data useful in understanding places?

Step 3: I can choose, collect and display appropriate data

4 As a class, create a list of questions to determine whether people in your class would prefer a tree change or a sea change. Survey the whole class and graph the results.

Step 4: I can use data to support claims

5 Refer to the data you collected in question 4. Provide some possible explanations as to why your class prefers a sea change or tree change.

Source 4

Birregurra is a small rural town with a population of 828 people. It is located in the foothills of the Otway Ranges, about 130 km south-west of Melbourne. Birregurra is surrounded by grazing land and is close to coastal attractions such as Port Campbell.

HOW TO

Surveys, page 150
Graphing, page 164

Why is Torquay booming?

Torquay is a coastal town just 90 minutes from Melbourne. Between 2001 and 2016, its population grew by 130 per cent – putting it at the centre of one of the biggest population booms in Australia.

Families are choosing to make a sea change to enjoy the coastal environment of Torquay, with many people commuting to Melbourne each day for work. As the population grows, new services such as schools and shopping centres are being built.

Source 1a

Oblique aerial view of Torquay in 2006 when Torquay's population was just 6695

2006 Torquay

Torquay's age-sex profile between 2006 and 2016

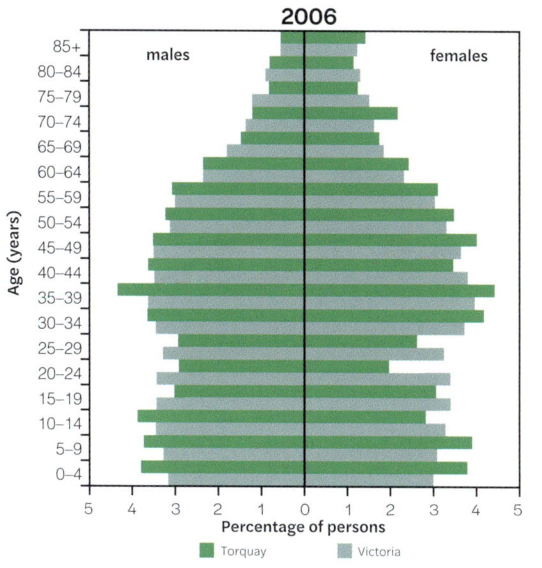

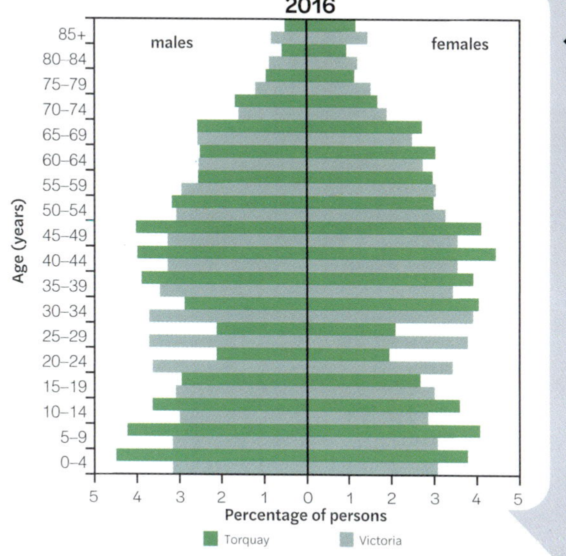

Source: Australian Bureau of Statistics, 2016

Source 2

Population pyramids showing Torquay's changing age-sex profile between 2006 and 2016

Source 1b

Oblique aerial view of Torquay in 2016 when the population had grown to more than 13 000

Learning Ladder G3.12

Show what you know

1. Why are more people making the sea change to Torquay?
2. What new services does a community require with a fast-growing population?

Analyse data

Step 1: I can use geographic terminology to interpret data

3. On a blank map of Victoria, locate Torquay and describe its relative location.

Step 2: I can describe patterns and trends

4. Source 2: Use the age-sex pyramids to describe how the population structure in Torquay has changed over time.

Step 3: I can explain the reasons behind a trend or spatial distribution

5. Sources 1a and 1b: Compare the two images from 2001 and 2016. As a class, using SHEEPT, discuss key changes that have occurred in Torquay and suggest reasons why the growth is occurring.

Step 4: I can analyse relationships between different data

6. Source 2: How does the population profile for Torquay compare to the average for Victoria in 2016? Based on this information, suggest the types of services that need to be built to support the Torquay community.

HOW TO

SHEEPT, page 158
Population pyramids, page 166
Satellite images, page 167

Masterclass

Learning Ladder

Overall liveability ranking of Melbourne suburbs

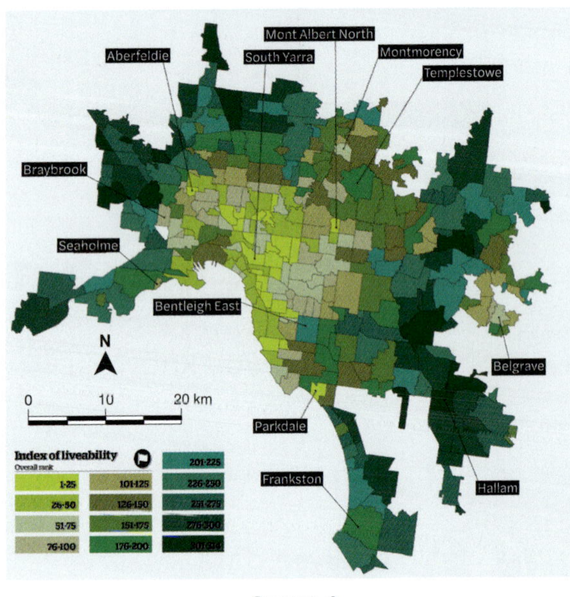

Source 1

Choropleth map showing liveability of Melbourne suburbs

Number of trees in Victorian suburbs

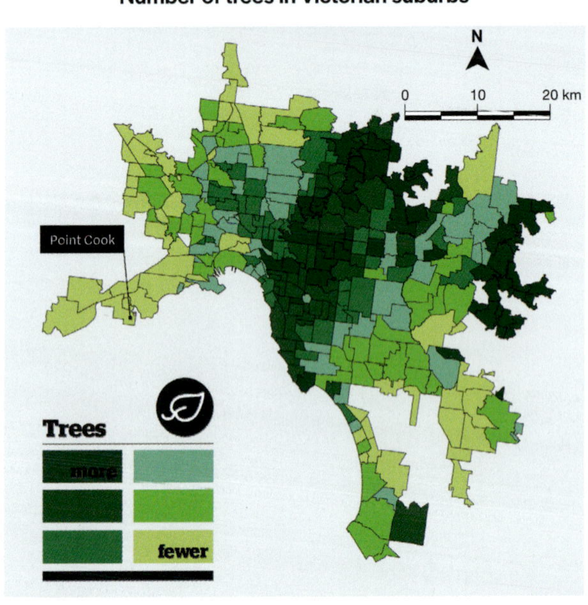

Source 2

Choropleth map showing amounts of trees in Melbourne suburbs

Liveability score for Australia's major cities

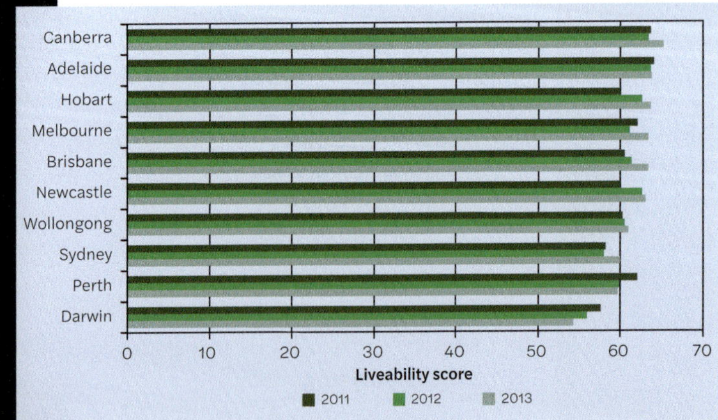

Source 3

Liveability scores of Australian cities

Suburb	2015–16 population	2015–16 population growth	2015–16 growth rate	Area (km²)
South Morang	64 354	4971	8.40	59.70
Cranbourne East	23 901	4956	26.20	42.90
Craigieburn – Mickleham	52 848	4491	9.30	106.10
Point Cook	50 774	3512	7.40	36.90
Epping	42 236	3226	8.30	84.90
Tarneit	36 961	2872	8.40	62.10
Truganina	23 222	2863	14.10	27.00
Melbourne – Hoddle Grid	35 447	1975	5.90	2.40
Melton South	24 341	1831	8.10	111.00
Beaconsfield – Officer	13 605	1817	15.40	41.70

Source: Australian Bureau of Statistics

Source 4

Population change over time for a range of Victorian suburbs

Work at the level that is right for you or level-up for a learning challenge!

Step 1

a I can identify and describe spatial characteristics

Explain why the environment is an important factor when determining local liveability.

b I can identify and describe interconnections

How do communities help enable interconnections between people (see pages 100–101)?

c I can identify responses to a geographical challenge

List three responses to population decline in rural communities (see pages 94–95)?

d I can collect, record and display data in simple forms

List the data collected to create the table in Source 4.

e I can use geographic terminology to interpret data

Look at Source 1. Describe the location of the least liveable suburbs in Melbourne.

Step 2

a I can explain spatial characteristics

Refer to Source 3. Explain why Darwin's liveability score has changed over time.

b I can explain interconnections

Is there an interconnection between Melbourne's most liveable suburbs (Source 1) and suburbs with more trees (Source 2)?

c I can compare responses to a geographical challenge

Compare responses from urban planners working to improve the liveability of large cities. Which do you think is the most important (see pages 98–99)?

d I can recognise and use different types of data

Source 3 is an example of a clustered bar graph. Explain how this type of graph works.

e I can describe patterns and trends

What do the suburbs in Source 4 have in common?

Step 3

a I can explain processes influencing places

Refer to Source 1. Using data, outline the liveability ranking of South Yarra and Hallam and suggest possible reasons for the difference.

b I can identify and explain the implications of interconnections

Explain the interconnection between the area of farms and the physical features in Tully (pages 90–91).

c I can compare strategies for a geographical challenge

List three solutions to solve the geographical challenge of population growth in Melbourne (see pages 98–99).

d I can choose, collect and display appropriate data

Using the data from Source 4, graph the population growth rate of 10 Victorian suburbs for 2015–2016.

e I can explain the reasons behind a trend or spatial distribution

Use SHEEPT to explain the pattern shown in Source 1 on page 92.

Step 4

a I can predict changes in the characteristics of places

Describe the capital cities in Australia that are expected to grow the fastest and slowest by 2054 and suggest reasons to support the trends (Source 1, page 84).

Masterclass

b **I can evaluate the implications of significant interconnections**

Why are opportunities for interconnection through a community of place even more important for people with different cultures and languages (see pages 100–01)?

c **I can evaluate alternatives for a geographical challenge**

How can those suburbs shown as least liveable in Source 1 improve their liveability rating?

d **I can use data to support claims**

Local residents in Torquay are pushing for new aged care facilities. Does the data in Source 2 on page 107 support these claims?

e **I can analyse relationships between different data**

Why is Parkdale a good example to support the hypothesis that liveable suburbs have more trees?

step 5 Step 5

a **I can analyse the impact of change on places**

Compare the satellite images of Torquay in 2006 and 2016 on pages 106 and 107. What impact has population change had in this area?

b **I can explore spatial association and interconnections**

Source 1 page 92: What is the spatial association between Australian regions with declining populations and inland rural towns?

c **I can plan action to tackle a geographical challenge**

Using Nyah West as an example (pages 94–95), list the actions the town could take to improve its liveability.

d **I can evaluate data**

What additional information would be useful to test the reliability of the data used to build Source 3?

e **I can draw conclusions from analysing collected data**

Using a map of Melbourne to locate the suburbs in Source 4, identify the region of Melbourne that will most urgently require require infrastructure such as roads , schools and hospitals. Justify your decision.

Capstone

How can I understand living in Australia?

In this chapter, you have learnt a lot about living in Australia. Now you can put your new knowledge and understanding together for the capstone project to show what you know and what you think.

In the world of building, a capstone is an element that finishes off an arch or tops off a building or wall. That is what the capstone project will offer you, too: a chance to top off and bring together your learning in interesting, critical and creative ways. You can complete this project yourself, or your teacher can make it a class task or a homework task.

Scan this QR code to find the capstone project online.

mea.digital/GHV7_G3

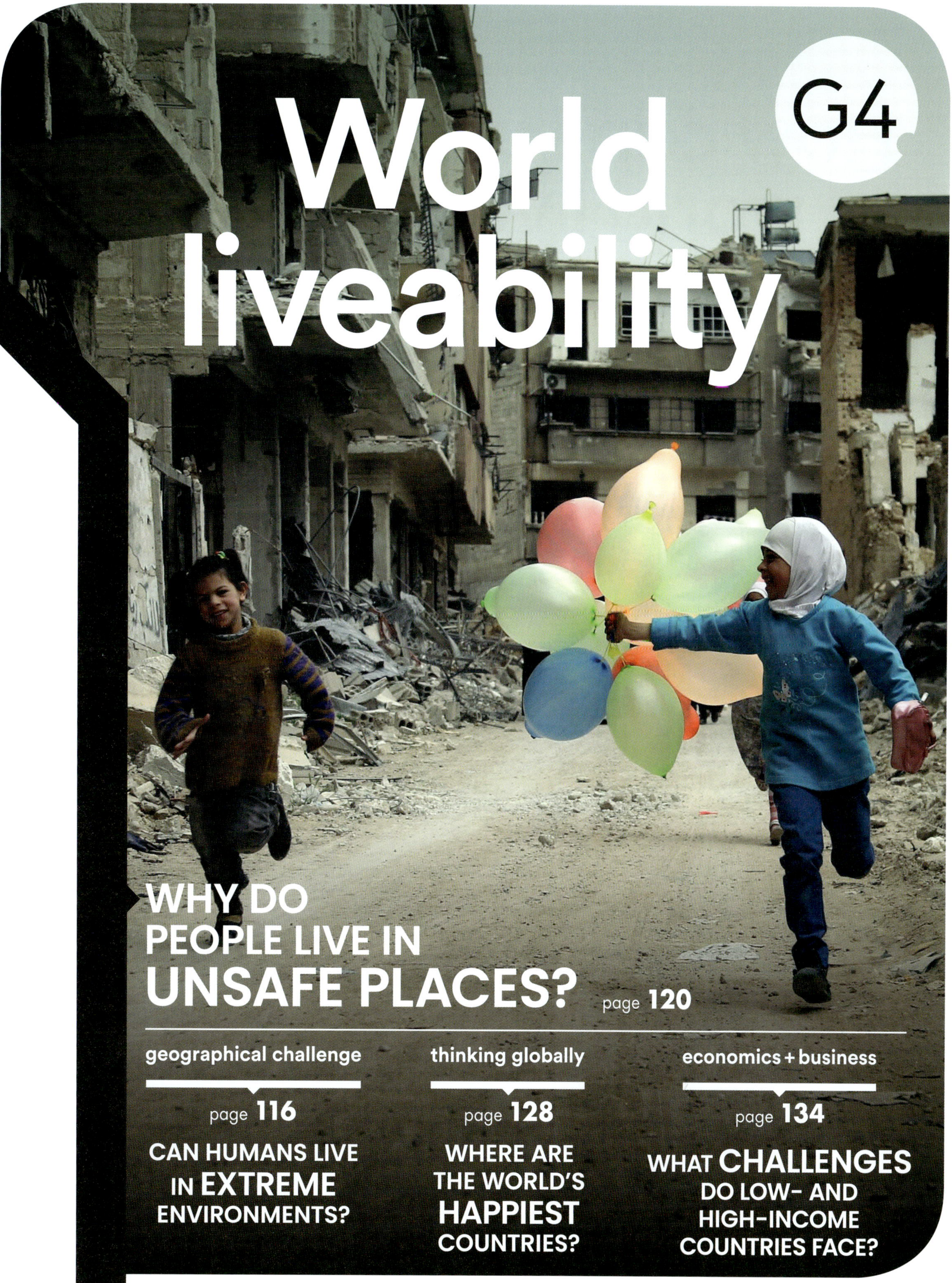

G4

World liveability

WHY DO PEOPLE LIVE IN UNSAFE PLACES?

page 120

geographical challenge

page 116

CAN HUMANS LIVE IN EXTREME ENVIRONMENTS?

thinking globally

page 128

WHERE ARE THE WORLD'S HAPPIEST COUNTRIES?

economics + business

page 134

WHAT CHALLENGES DO LOW- AND HIGH-INCOME COUNTRIES FACE?

How can I understand world liveability?

Liveability varies between regions. High-income countries tend to have access to resources, high-quality education and advanced health care, while low-income countries may not have the technology and infrastructure to support their growing populations.

learning ladder

Step	Spatial characteristics	Interconnections	Geographical challenge
step 5	**I can analyse the impact of change on places** I can analyse and evaluate the implications of factors such as safety and calculate the impact on people and environments.	**I can explore spatial association and interconnections** I can compare distribution patterns and the interconnections between them; e.g. the safety of places and patterns of liveability.	**I can plan action to tackle a geographical challenge** I can frame questions, evaluate findings, plan actions and predict outcomes to tackle a liveability-based geographical challenge.
step 4	**I can predict changes in the characteristics of places** I can predict changes in the characteristics of places over time due to factors such as natural hazards.	**I can evaluate the implications of significant interconnections** I can identify, analyse and explain key liveability-based interconnections within and between places, and evaluate their implications over time and at different scales.	**I can evaluate alternatives for a geographical challenge** I can weigh up alternative views and strategies on a liveability-based geographical challenge using environmental, social and economic criteria.
step 3	**I can explain processes influencing places** I can explain the series of actions leading to change in a place, such as people being forced to become refugees.	**I can identify and explain the implications of interconnections** I can identify, analyse and explain liveability-based interconnections and explain their implications.	**I can compare strategies for a geographical challenge** I can compare strategies for a liveability-based geographical challenge using environmental, social and economic criteria.
step 2	**I can explain spatial characteristics** I can identify concepts of Space, Place, Interconnection, Change, Environment, Sustainability and Scale (SPICESS) when I read about world liveability.	**I can explain interconnections** I can describe and explain interconnections and their effects, such as malnutrition and infant deaths.	**I can compare responses to a geographical challenge** I can identify and compare responses to a geographical challenge and describe its impact on different groups.
step 1	**I can identify and describe spatial characteristics** I can talk about spatial characteristics at a range of scales; e.g. extreme environments.	**I can identify and describe interconnections** I can identify and explain simple interconnections involved in phenomena such as environmental quality and liveability.	**I can identify responses to a geographical challenge** I can find responses to a geographical challenge such as adapting to extreme environments.

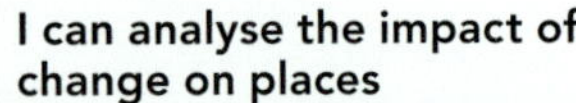
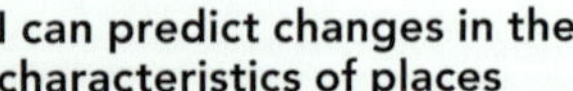
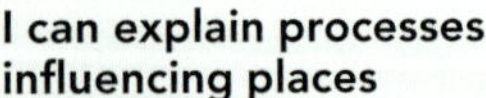

Source 1

An example of inequality in liveability in Mumbai, India. In the background we see large skyscrapers, parklands and road networks, and in the foreground are slums characterised by their tarp roofs and cramped living conditions.

Warm up

Collect, record and display data	Analyse data
I can evaluate data I can determine whether data presented about human settlements is reliable and assess whether the methods I used in the field or classroom were helpful in answering a liveability-based research question.	**I can draw conclusions from analysing collected data** I can summarise findings and use collected data to support key patterns and trends I have identified for liveability-based research.
I can use data to support claims I can select or collect the most appropriate data and create specialist maps and information using ICT to support investigations into liveability.	**I can analyse relationships between different data** I can use multiple data sources, overlays and GIS to find links and relationships that exist in patterns of liveability on Earth.
I can choose, collect and display appropriate data I can select useful sources of liveability data and represent them to conform with geographic conventions.	**I can explain the reasons behind a trend or spatial distribution** I can identify Social, Historical, Economic, Environmental, Political and Technological (SHEEPT) factors to help me explain patterns in data.
I can recognise and use different types of data I can define the terms primary, secondary, qualitative and quantitative data and represent data in more complex forms.	**I can describe patterns and trends** I can identify Patterns, Quantify them and point out Exceptions (PQE) to describe the patterns I see.
I can collect, record and display data in simple forms I can identify that maps and graphs use symbols, colours and other graphics to represent data.	**I can use geographic terminology to interpret data** I can identify increases, decreases or other key trends on a map, graph or chart about factors of liveability.

Spatial characteristics

1 Source 1: Comment on how liveability can differ in one place.

Interconnections

2 Source 1: Suggest why the world's largest slums are found in countries where people from poor rural areas are migrating to large cities.

Geographical challenge

3 Source 1: What challenges do you think the growth of urban populations and the development of slums present for governments?

Collect, record and display data

4 Source 1: This is a ground level photograph. Ground level and oblique aerial photographs have a foreground and a background. Why is some information hidden in these photographs?

Analyse data

5 Source 1: Sketch one of the shanty homes shown in this slum in Mumbai and label the materials used to build it. Why do you think this strip of land was left open for new migrants to the city to claim for their makeshift homes?

Where do people live in our world?

Ninety per cent of the world's population lives on just 10 per cent of all the available land on Earth. Some places on Earth – such as deserts, high mountains and polar regions – are inhospitable, and it is difficult for humans to live there.

Living on Earth

The most **liveable** places on Earth are between the Arctic Circle and the Antarctic Circle, where the temperatures are warm enough to grow food.

Antarctica is the only continent that does not have a permanent population. It is the coldest place on Earth. Scientists spend time on Antarctica – but mainly during the summer months, when there are more opportunities to work outside.

The world's population is 7.8 billion people, and we inhabit six of the seven continents. Asia has the highest population, at 4.5 billion people. Sixty per cent of the world's population lives in Asia, and 36 per cent of the world's people live in just two countries: China and India.

Population density

The **population density** of a region is a measure of how crowded that place is. It is calculated by dividing the population by the area of the region. For example, the continent of Australia has a population of 25 000 000, and an area of 7 692 024 square kilometres, so its population density is 3.25 persons per square kilometre.

Population density varies from continent to continent. Australia has the lowest population density of all the populated continents. Asia has the world's highest population density – with 146 people for every square kilometre of land. On average, the world's population density is about 15 people for every square kilometre

Source 1

World **population distribution** dot map

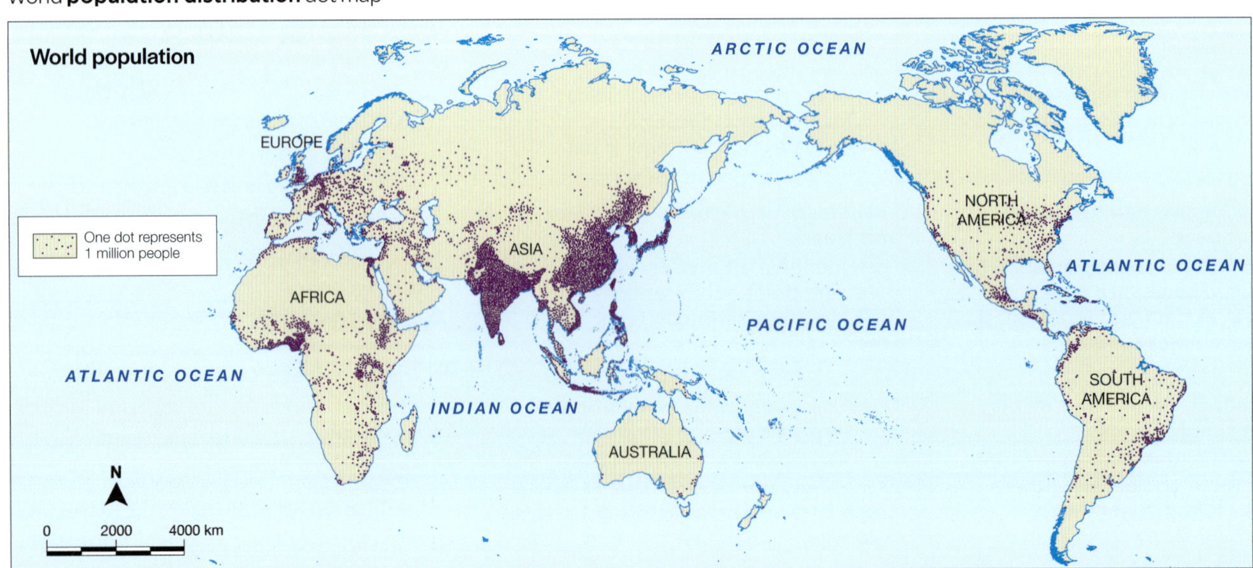

Source: Matilda Education Australia

Source 2

World environments shown on a satellite image

of land on Earth. However, about 90 per cent of the Earth's population live on just 10 per cent of the land – mainly along coastlines and near sources of freshwater.

Least liveable regions

Over half of Earth's land mass consists of regions with harsh conditions that make it difficult for humans to survive. This is why **polar** regions, **deserts** and high mountains have very low population densities (see pages 116–117).

Every continent has a region (or regions) with low population density:

- Africa: northern half contains the Sahara desert
- Europe: far north is in the frozen Arctic Circle
- Asia: central Asia has the Himalayan Mountains and desert
- Australia: all of central Australia is desert or semi-desert
- North America: northern tip is in the Arctic, plus deserts in south-west
- South America: northern half contains the Amazon rainforest.

Learning Ladder G4.1

Show what you know

1 Define the term 'liveability'.

2 Why isn't there a permanent human population on Antarctica?

3 Brainstorm a list of basic things that humans require to live in a region.

Spatial characteristics

Step 1: I can identify and describe spatial characteristics

4 Which continent has the highest population density?

Step 2: I can explain spatial characteristics

5 How does Antarctica's environment contribute to its low population density?

Step 3: I can explain processes influencing places

6 Source 2: Identify the environments that provide the least liveable conditions on Earth. Provide one example for each.

Step 4: I can predict changes in the characteristics of places

7 Source 1: How will the pattern shown on this map change as the world's population grows?

Can humans live in extreme environments?

Some of the least liveable places on Earth are mountains, deserts and polar regions. Over time, some people have adapted well to living in these harsh environments. Their bodies have become used to extremes in temperature, or less water or oxygen.

Difficult environments

Some environments have very harsh conditions that make it difficult for humans to survive. They may be extremely hot or cold, or there may be no water or little oxygen to breathe. Polar regions, deserts and high mountains are examples of extreme environments where the few humans who live there have had to adapt to survive.

Polar regions

The Arctic and Antarctica are the Earth's two **polar** regions. The Arctic is the cold area around the North Pole. One indigenous people who live there are called the Inuit. The Inuit have adapted their diet, clothing, transport and housing to help them live in the extreme cold.

Antarctica is the coldest, highest and windiest place on Earth. It is the only continent that has no permanent human population. Scientists from Australia, New Zealand and many other countries work and live in scientific stations on Antarctica where even in summer the temperature can drop to −50°C.

Buildings on Antarctic stations are heated to 18°C. When working outside people have to wear lots of layers topped with a thick, quilted freezer suit. **Hypothermia** (lowering of body temperature below normal) and **frostbite** (freezing of the skin and other tissues) are both dangerous risks.

Antarctic air is so dry that a person's body can lose water just through breathing. A person needs to drink six to eight litres of water per day to stay hydrated while on Antarctica.

Desert regions

Deserts are very dry areas that receive less than 250 millimetres of rain a year. Most people think that deserts are hot places, and in the day many deserts are. But overnight, deserts are cold. Deserts known as 'cold deserts' also have relatively low temperatures during the day.

Places without a water supply are difficult to live in. People such as the Bedouins of the Middle East and North Africa are **nomadic** – they herd animals around in search of the most productive land. In Chile's Atacama Desert – the world's driest place – the people of the village of Chungungo use fine nets to capture moisture from the sea fogs that pass over the desert. The water droplets drip into large containers and pipes that supply the village with water.

Mountain regions

People living in the high mountains of the world need to adapt to cold temperatures and rugged terrain. Living in mountains also offers unique challenges for breathing. Many visitors to high altitudes suffer mountain sickness caused by the oxygen-thin air. Symptoms range from shortness of breath, dizziness and headaches to coma or death.

Ethnic groups that have lived in high altitudes for many generations, such as Tibetans and Sherpas in the Himalayas and Quechuan in the Andes mountains, have developed a genetic change in their blood that allows them to comfortably live and work at high altitudes with less oxygen.

Source 1

Traditional Inuit people live in the Arctic areas of Alaska, Canada, Siberia and Greenland. The Inuit learnt to make warm winter homes, known as igloos, from snow and ice. They used animal skins and furs to make warm clothing. The Inuit people are unable to grow their own food because the Arctic climate is so harsh, and mostly eat meat from seals, whales, polar bears and caribou. Originally, Inuit used dog sleds for transport – but today they use snowmobiles.

Source 2

Bedouin tribes live in desert areas in North Africa and the Middle East. They have herds of camels and other animals that are well adapted to desert conditions. Their traditional nomadic lifestyle means they frequently move to find scarce resources in the dry desert environment. Bedouin tents are designed to allow air to circulate. Animal hair is used to insulate the tents so they stay cool during the hot days but warm during the freezing nights.

Learning Ladder G4.2

Show what you know

1 Define the term 'extreme environment'.

Geographical challenge

Step 1: I can identify responses to a geographical challenge

2 Source 1: How have the Inuit adapted to life in their polar environment?

Step 2: I can compare responses to a geographical challenge

3 Sources 1 and 2: Compare how the Inuit and Bedouins have responded to living in extreme environments.

Step 3: I can compare strategies for a geographical challenge

4 Compare the strategies for working and living in polar regions used by the Inuit and scientists working in Antarctica.

Step 4: I can evaluate alternatives for a geographical challenge

5 List three adaptations humans have made to live in extreme environments.

How is safety interconnected with liveability?

Being safe is a key ingredient of a liveable place. Many people in the world live in unsafe places with the threat of volcanoes, earthquakes and tsunamis. Others live with conflict, or are forced to escape war zones as refugees.

Looking for safety

Along with having food, water and shelter, feeling safe is the most important factor in the **liveability** of a place. Many of the world's least liveable cities are found in war-torn countries such as Syria (see page 122), Afghanistan and South Sudan.

In regions of conflict, people are forced to leave their homes. They seek safety, food and shelter in camps that have been set up by welfare organisations. Some people are forced to flee their homes and move to camps somewhere else in their country – these are called **internally displaced persons**. Many threatened people flee to other countries as **refugees** in order to escape war, violence or a natural disaster. More than 1 million refugees from South Sudan are now living across the southern border in Uganda.

Safe and unsafe countries

The Global Peace Index ranks countries according to their level of peacefulness. In 2018, the countries rated as the most peaceful were Australia, New Zealand, Canada and Japan, as well as some countries in Europe. Russia, the Middle East and North Africa were rated among the world's least peaceful regions.

The following measures (or criteria) are used to measure peacefulness.

- *Safety and security:* amount of crime, numbers of people jailed, the threat of terrorism, the number of refugees.
- *Ongoing conflict:* number of conflicts and casualties.
- *Militarisation:* military spending, and numbers of military personnel.

Source 1

Global Peace Index 2018

Source: Institute for Economics and Peace, 2019

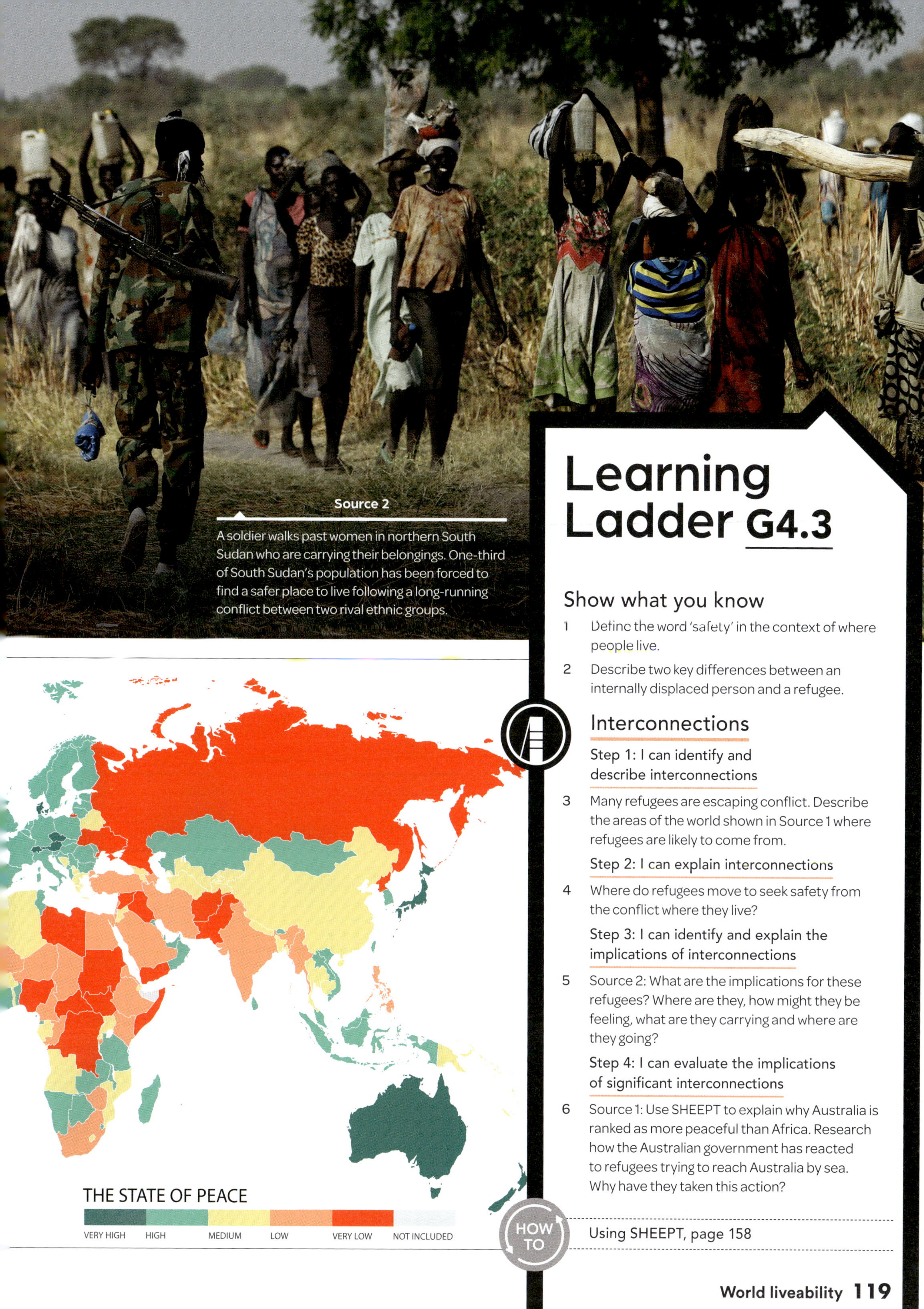

Source 2

A soldier walks past women in northern South Sudan who are carrying their belongings. One-third of South Sudan's population has been forced to find a safer place to live following a long-running conflict between two rival ethnic groups.

Learning Ladder G4.3

Show what you know

1 Define the word 'safety' in the context of where people live.

2 Describe two key differences between an internally displaced person and a refugee.

Interconnections

Step 1: I can identify and describe interconnections

3 Many refugees are escaping conflict. Describe the areas of the world shown in Source 1 where refugees are likely to come from.

Step 2: I can explain interconnections

4 Where do refugees move to seek safety from the conflict where they live?

Step 3: I can identify and explain the implications of interconnections

5 Source 2: What are the implications for these refugees? Where are they, how might they be feeling, what are they carrying and where are they going?

Step 4: I can evaluate the implications of significant interconnections

6 Source 1: Use SHEEPT to explain why Australia is ranked as more peaceful than Africa. Research how the Australian government has reacted to refugees trying to reach Australia by sea. Why have they taken this action?

HOW TO

Using SHEEPT, page 158

Why do people live in unsafe places?

The threat of natural hazards can make some areas of Earth less safe than others to live in. For example, floods occur in low-lying areas near water, and earthquakes and volcanoes tend to occur near the boundaries of **tectonic plates**. Fear of crime forces people to avoid places where they feel threatened.

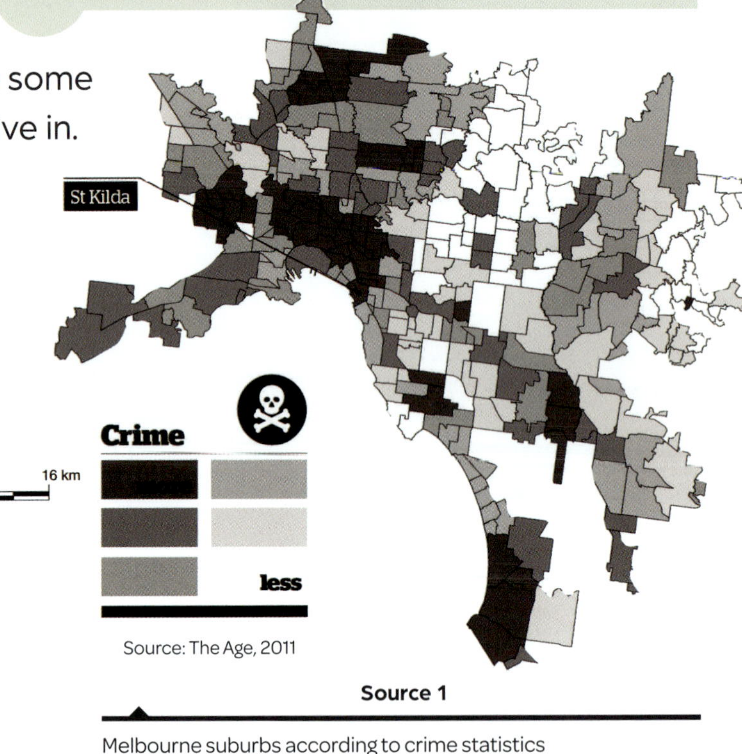

Source 1

Melbourne suburbs according to crime statistics

Living in a hazardous place

Hundreds of millions of people around the world choose to live in areas that are prone to **natural hazards**. For example, areas with volcanoes are hazardous, but they attract many people because the rich soil is good for farming.

Because so many people have chosen to live in hazardous zones, natural events such as volcanic eruptions, earthquakes, tsunamis and cyclones can quickly turn into **natural disasters**. In Australia the most common natural disasters are bushfires, floods and cyclones.

Making places safe from crime

The most liveable places are where people feel safe from injury, abuse, theft and damage. Australian cities are regarded as safe by world standards, so they perform well in liveability rankings (see page 97). Governments are continually looking at ways to make places safer. They measure safety by surveying communities and studying crime statistics. Maps can show the areas where most crimes occur and police can be moved to areas of high crime.

Source 2

The dots and triangles on this map indicate major volcanic eruptions, earthquakes and tsunamis.

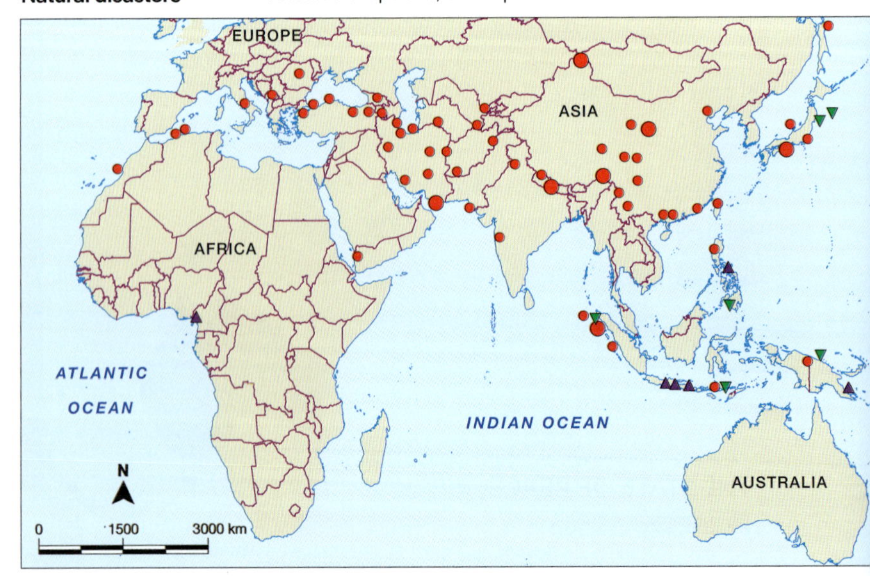

Source 3

An aerial view showing the complete destruction of Palu in Sulawesi, Indonesia. The area was devastated in 2018 by an earthquake and tsunami that claimed more than 2000 lives.

Different measures are put in place to make public spaces safer. For example, improved lighting and building design can help make public areas more visible. Police patrols and surveillance cameras help prevent crime, and make residents feel safer. These strategies help to reduce the number of crimes because if places are more visible and a police presence is more obvious, then any criminal activity is also easier to see, too.

Source: Matilda Education Australia

Learning Ladder G4.4

Show what you know

1 What factors cause a location to be 'unsafe'? Hint: Use SHEEPT to help you answer the question.

Analyse data

Step 1: I can use geographic terminology to interpret data

2 Source 3: Describe what has happened here.

Step 2: I can describe patterns and trends

3 Source 1: Use PQE to describe the distribution of crime across Melbourne.

Step 3: I can explain the reasons behind a trend or spatial distribution

4 Source 2: What natural hazards are shown here and which areas of the world are affected by them? Why do people choose to live in these regions even though they are hazardous?

Step 4: I can analyse relationships between different data

5 Source 1: Suggest what action could be taken in St Kilda to make it a safer place.

PQE, page 156
SHEEPT, page 158

Why is Syria an unsafe place to live?

From 2013 to 2019, Syria's capital Damascus has been rated as the world's least liveable city in the annual survey compiled by the Economist Intelligence Unit. Conflict in Syria has made this Middle Eastern country one of the most unsafe regions in the world.

In 2011, Syrian people who were peacefully protesting against their government were shot at by the Syrian army. This led to civil war, as opposition groups armed themselves to fight back. By 2019, the fighting had killed nearly half a million people, injured more than 1 million and forced about 12 million people from their homes – roughly two-thirds of the Syrian population.

More than 6 million people have fled their homes and are displaced within Syria. More than 5 million **refugees** have crossed the borders of neighbouring countries looking for a safe environment, with 3.5 million Syrian refugees crossing the border into Turkey alone.

Thousands of Syrians flee their country every day. The risks on the journey are high, with families walking long distances at night to avoid being shot or kidnapped.

Source 1

Territories held by opposing forces in Syria in April 2019

Control of Syria April 2019 Source: Matilda Education Australia

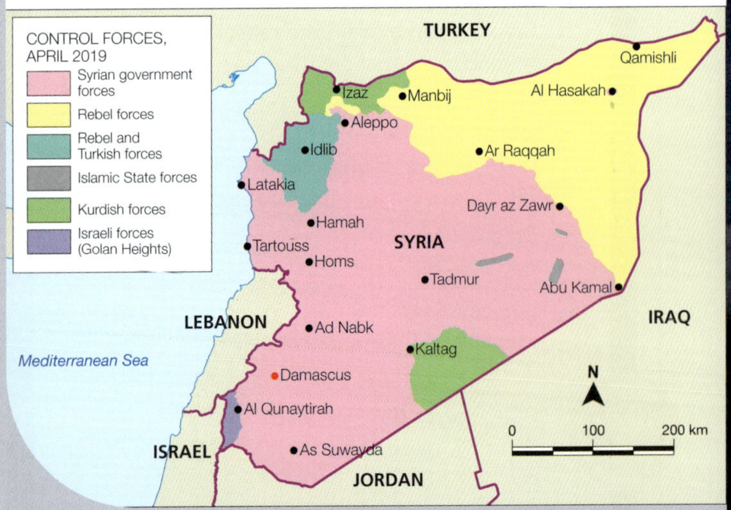

Source 2

The explosion of a suicide car bomb rocks the Syrian city of Kobani, near the border with Turkey. Since 2011, fighting in Syria has killed nearly half a million people and forced 12 million people from their homes. From 2013 to 2019, the Syrian capital Damascus was named the world's least liveable city.

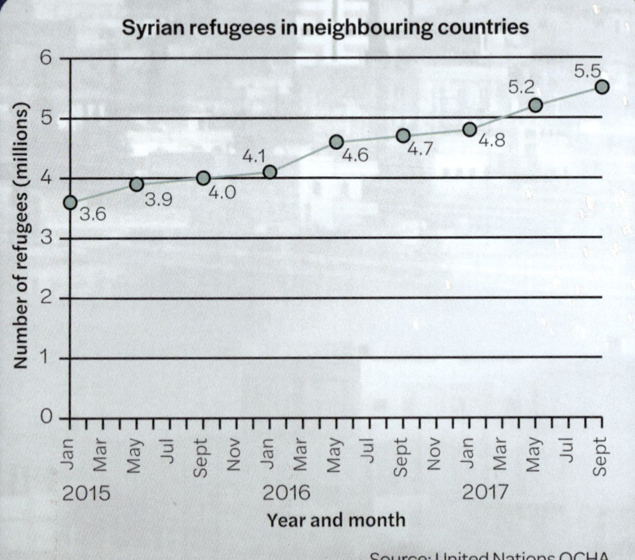

Source 3

More than 12 million Syrians have been displaced from their homes. This includes about 5.5 million refugees who have been forced to seek safety in neighbouring countries, out of a total 6.3 million Syrian refugees worldwide.

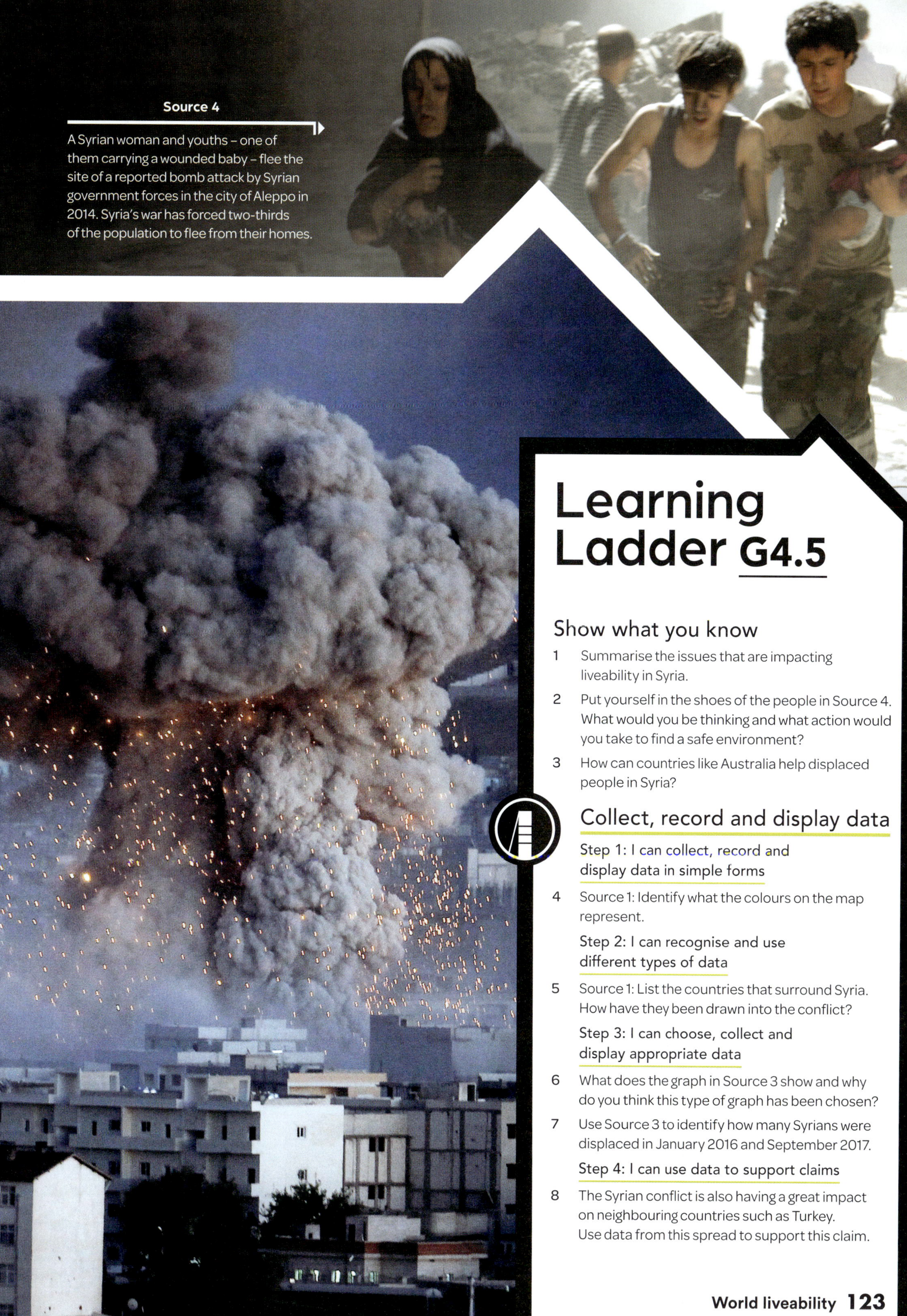

Source 4

A Syrian woman and youths – one of them carrying a wounded baby – flee the site of a reported bomb attack by Syrian government forces in the city of Aleppo in 2014. Syria's war has forced two-thirds of the population to flee from their homes.

Learning Ladder G4.5

Show what you know

1. Summarise the issues that are impacting liveability in Syria.
2. Put yourself in the shoes of the people in Source 4. What would you be thinking and what action would you take to find a safe environment?
3. How can countries like Australia help displaced people in Syria?

Collect, record and display data

Step 1: I can collect, record and display data in simple forms

4. Source 1: Identify what the colours on the map represent.

Step 2: I can recognise and use different types of data

5. Source 1: List the countries that surround Syria. How have they been drawn into the conflict?

Step 3: I can choose, collect and display appropriate data

6. What does the graph in Source 3 show and why do you think this type of graph has been chosen?
7. Use Source 3 to identify how many Syrians were displaced in January 2016 and September 2017.

Step 4: I can use data to support claims

8. The Syrian conflict is also having a great impact on neighbouring countries such as Turkey. Use data from this spread to support this claim.

Are environments and health interconnected with liveability?

The most liveable places are those where the quality of the environment is high, and people have food to eat and the medical services they need to help them live a healthy life.

Environmental quality

The quality of the environment is a key factor that determines how liveable a place is. Characteristics such as clean air and clean water, low noise levels, and access to open spaces make a place healthier and more liveable.

Source 1

New Delhi, November 2017: Indian schoolchildren cover their faces with handkerchiefs as they walk to school through heavy smog. Other commuters wore masks or covered their mouths with scarves. The government shut all the primary schools, as pollution levels were nearly 30 times the World Health Organization safe level.

Air pollution

Many large cities around the world suffer badly from harmful particles in the air, known as **air pollution**. The major sources of outdoor air pollution are:

- exhaust gases from motor vehicles
- power generation – such as oil and coal power plants
- industrial factories, mines and oil refineries
- burning of agricultural waste
- home cooking, heating and lighting.

Source 3

Te Anau in New Zealand is one of the least polluted areas in the world. It is the gateway to the wilderness area of Fiordland National Park.

Source 2

World map showing global pollution levels

Source: Matilda Education Australia

If there are too many harmful particles in the air, people breathe them in when they inhale. This can cause serious health conditions, such as lung cancer, heart disease and respiratory diseases. Each year, around 4.2 million people die because of exposure to air pollution – and 91 per cent of the world's population lives in places where air quality is worse than (or 'exceeds') World Health Organization guidelines.

Air pollution affects both high-income and low-income countries, with the highest levels experienced in South-East Asia. In 2018, India had the 14 most polluted cities in the world. In November 2017, air pollution levels in India's capital, New Delhi, were 30 times beyond the level that is safe to breathe. The Indian government closed schools for three days, stopped all construction, and warned people to stay indoors for their own safety.

To improve the liveability of cities, India and other countries and cities have been trying to reduce their air pollution levels. Some have banned old, polluting cars, or reduced the number of cars on the road. Other countries are investing in clean energy, such as solar power and wind power.

Improving global health

Over the last 30 years there has been a great improvement in the health of the world's population. There has been a reduction by more than 4 million in the number of children dying before they reach one year old.

However, despite the improvements in healthcare, there are still millions of people who don't have a healthy place to live. Wealthy countries have better health opportunities than poorer countries:

- In Africa, children are 15 times more likely to die before the age of five than children in high-income countries, and one child in every 13 dies before they are five.
- Children living in the world's poorest 20 per cent of households are twice as likely to die before their fifth birthday as children in the richest 20 per cent of households.
- Only one per cent of mothers who die during pregnancy or from complications during childbirth live in wealthy countries.

Improving children's health

In 2017, more than 5 million children under the age of five died – and half of these deaths could have been prevented with access to simple and affordable medical care.

Almost half (45 per cent) of all deaths in children under the age of five were because they didn't have enough to eat, or because they did not eat enough of the right things, which is called **malnutrition**. When children suffer from malnutrition, it puts them at greater risk of dying from common infections.

In 2017, there were 51 million underweight children below five years of age – and 16 million of these were severely underweight. More than 50 per cent of all malnourished children live in South Asia, and 25 per cent live in Africa.

Source 4

World hunger map, 2018

Prevalence of undernourishment in the total population (per cent) in 2015–2017

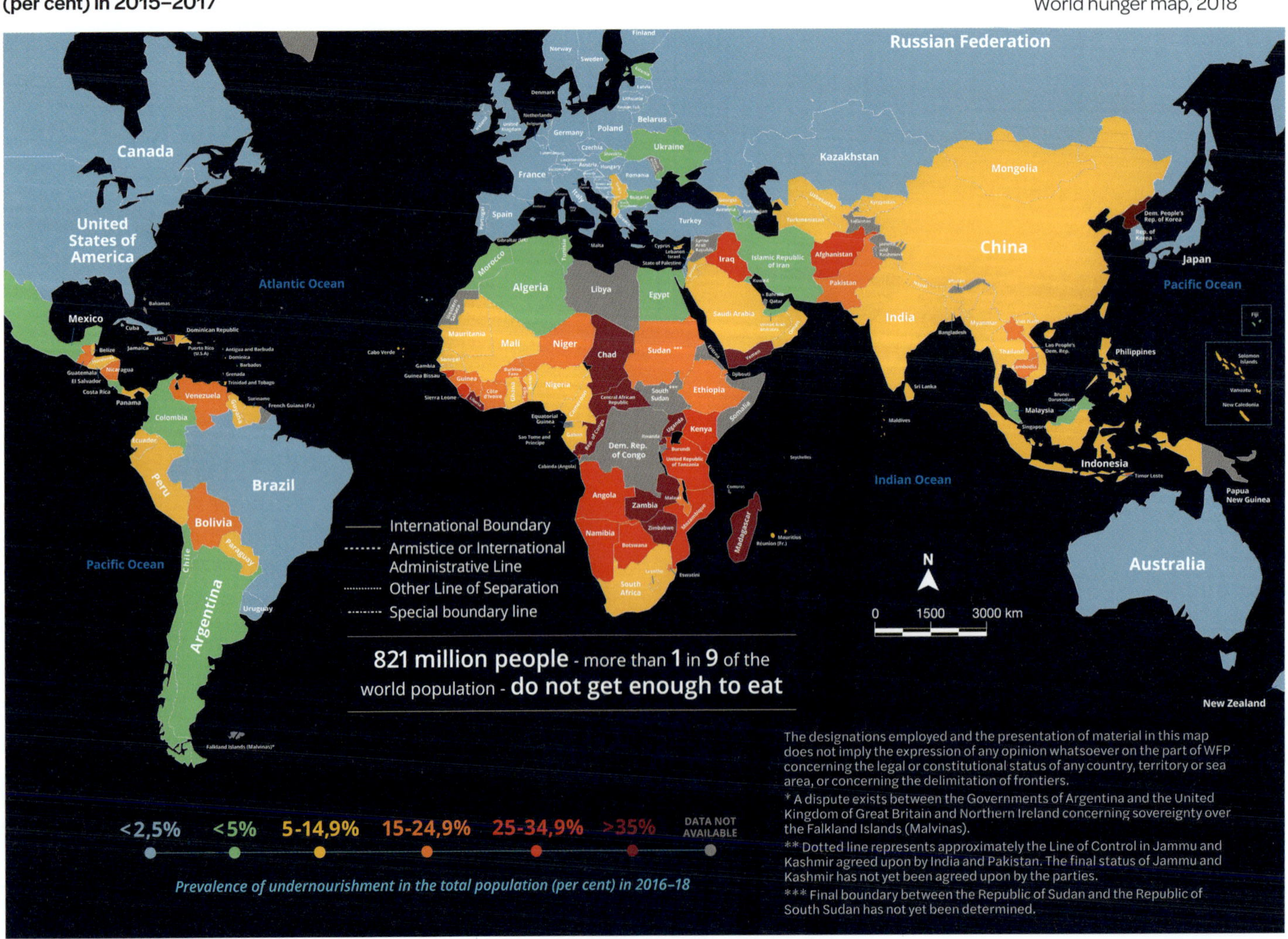

Source: World Food Programme, 2019

Aboriginal health in Australia

In Australia, a man can expect to live for 80 years, and a woman for 85 years. But the **life expectancy** for Indigenous people is much lower. An Aboriginal man can expect to live for only 69 years and an Aboriginal woman for only 74 years. In Australia, which is such a wealthy and liveable country, it is shameful that the health of the Indigenous population lags so far behind.

Aboriginal communities have higher rates of **infant mortality** and suffer more from infectious diseases, such as trachoma, which makes people blind. Australia is the only wealthy country in the world where people suffer from trachoma. Trachoma exists in remote Aboriginal communities that lack clean water and have overcrowded housing. Australia's Aboriginal population is also much more likely to contract diabetes and suffer heart disease than the rest of the population.

Infant mortality, 1985

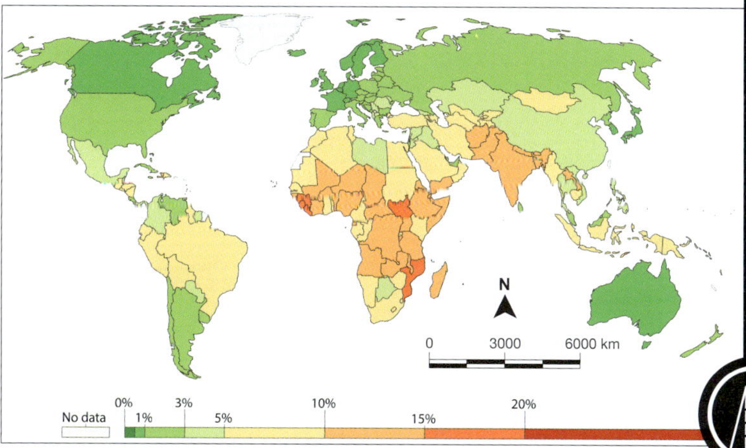

Infant mortality, 2015

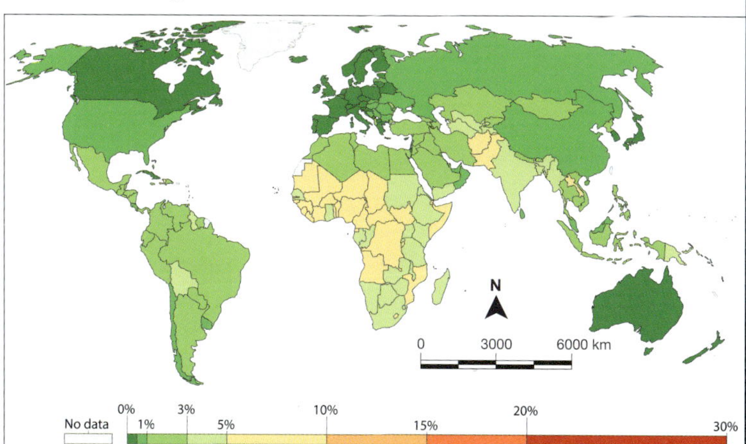

Source: Our World in Data

Source 6

World infant mortality rates, 1985 and 2015. Infant mortality refers to the number of deaths of children who are less than one year old per 1000 live births.

Source 5

The combination of drought and conflict in South Sudan in 2015 led to food shortages and high levels of malnutrition. Here, a two-month-old girl with severe malnutrition lies on a hospital bed next to her mother.

Learning Ladder G4.6

Show what you know

1 Identify where most of the world's most polluted cities are located.

2 Research and outline three things governments can do to help decrease pollution.

3 List some key factors that influence the health of the population.

4 How are wealthy countries different to other countries when it comes to health?

Interconnections

Step 1: I can identify and describe interconnections

5 How is the quality of the environment interconnected with the most liveable places?

Step 2: I can explain interconnections

6 Source 2: Australia has some of the most liveable cities in the world. Is there an interconnection with air pollution to help explain this pattern?

Step 3: I can identify and explain the implications of interconnections

7 Compare sources 4 and 6. Is there an interconnection between hunger and infant mortality?

Step 4: I can evaluate the implications of significant interconnections

8 What is the interconnection between Aboriginal Australians and health? What are the implications for Aboriginal Australians and how can the situation be improved?

Where are the world's happiest countries?

The United Nations annual poll of the happiest countries of the world selected Finland as the winner for 2018 and 2019 – and its Scandinavian neighbours were all in the top 10, too.

World Happiness Report

The happiest countries in the world have been ranked in the World Happiness Report 2019. The 2019 report combines the happiness rankings for 2016–2018, and, Finland ranked number 1, while its Scandinavian neighbours Denmark, Norway, Iceland and Sweden were all in the top 7.

Each year since 2012, the United Nations has measured the quality of life for citizens by surveying over 1000 people in more than 150 countries around the world. The surveys measure how happy citizens say they are – and this is added to data about wealth, life expectancy, social support, generosity, freedom and corruption to arrive at an overall score.

Source 1

The world's happiest countries, as ranked in the World Happiness Report 2019. Zero represents the worst possible life and 10 the best possible life.

The world's happiest countries, 2018

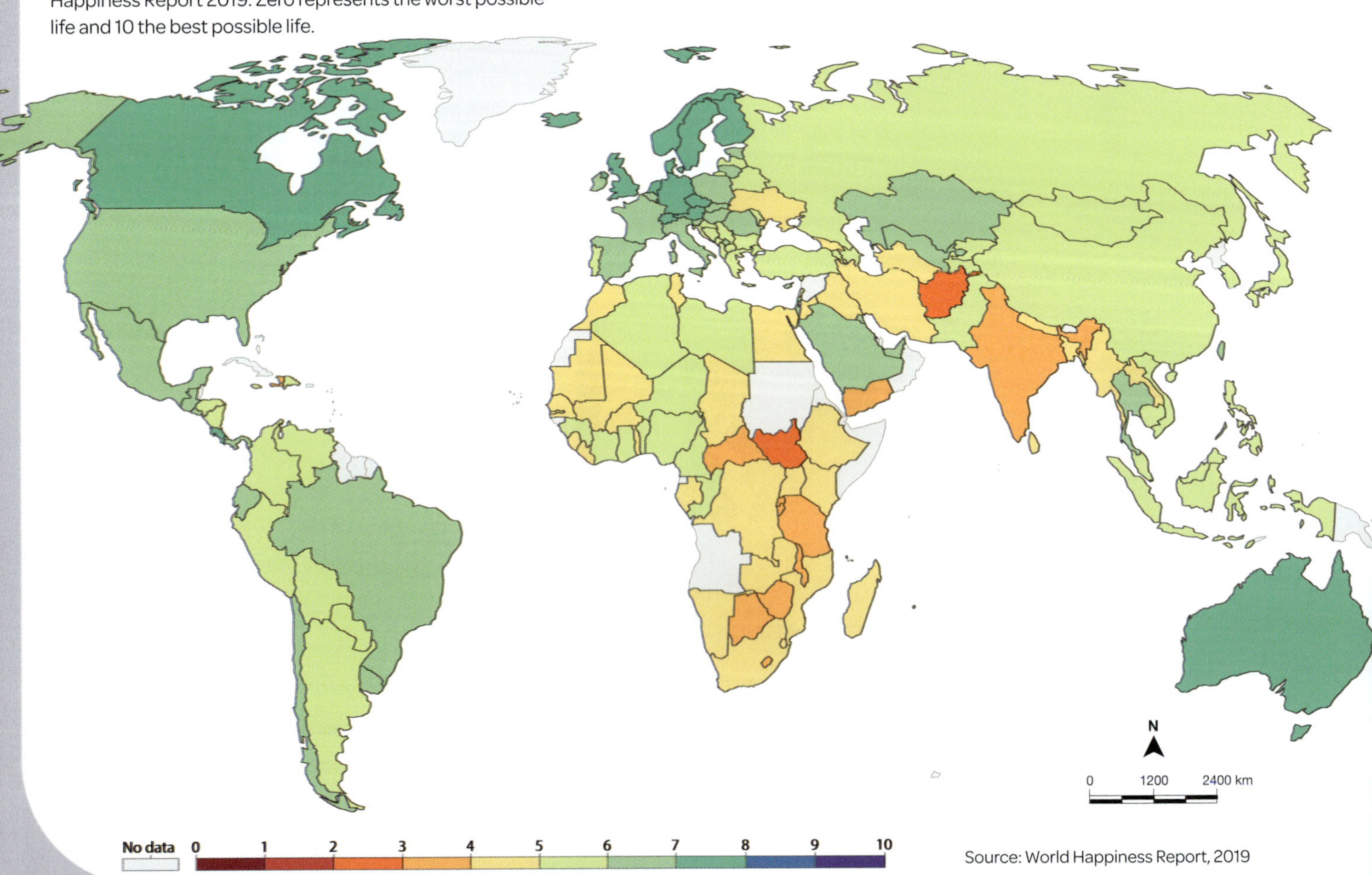

Source: World Happiness Report, 2019

Source 2

Finland has been named the Happiest Country in the World for 2018 in a poll run by the United Nations. Scandinavian countries filled five of the top seven rankings.

Happy countries

European countries feature highly in the list of the happiest countries in the world, holding 13 of the top 20 positions. Finland leapt from fifth last year to first this year to claim the title of the world's happiest country.

Finland is one of the most peaceful nations on Earth. Its residents enjoy high environmental standards, and access to more forest per square kilometre than any other European country. In 2019, Finland's Scandinavian neighbours – Denmark, Norway and Iceland – were second, third and fourth, with Sweden coming in seventh. Scandinavia just has to be the happiest region on Earth. On the other side of the globe, New Zealand and Australia are also happy places, too, coming in eighth and eleventh, respectively.

What makes these countries so happy? It's no surprise that the top-rated countries are all wealthy countries that enjoy a high quality of life. They also have a strong sense of community and respect for one another.

Unhappy countries

There is a strong link between unhappiness and the poorest and most dangerous nations on Earth. Countries in conflict, such as Syria (see pages 122–123), score poorly on the happiness index. Fifteen of the bottom 20 scores belong to countries in Africa that are suffering from conflict and poverty. The small African nation of Burundi is ranked the least happiest country in the world. Burundi has endured a conflict that has killed 1200 people, and forced 400 000 people from their homes.

Learning Ladder G4.7

Show what you know

1 What is the United Nations World Happiness Report?

2 Using SHEEPT, discuss why you think that Australia is rated as one of the happiest places on Earth.

Spatial characteristics

Step 1: I can identify and describe spatial characteristics

3 Why was Finland ranked the happiest country in 2018?

Step 2: I can explain spatial characteristics

4 Source 1: Use PQE to describe the distribution of happiness across the globe.

- a What patterns do you see? Are there particular areas in the world that are happier than others?
- b What data can you use to quantify the pattern you outlined?
- c What is an exception to the pattern you outlined?

Step 3: I can explain processes influencing places

4 What processes influence the world's unhappiest places?

Step 4: I can predict changes in the characteristics of places

5 Source 1: The two unhappiest countries shown are South Sudan and Afghanistan. Research why these countries score poorly on the World Happiness report and suggest what changes could be made to improve their ranking.

PQE, page 156
SHEEPT, page 158

How can we measure liveability?

Some of the key characteristics of a liveable city are that it is safe and comfortable, with good services available to the population. Australian cities perform well in liveability rankings.

Measuring liveability

Every person will have their own view about what makes a place more liveable or less liveable. A person's ideas about the liveability of a place will vary according to their age, background, income and lifestyle choices.

Every year the Economist Intelligence Unit (EIU) publishes a Global Liveability Index, where it ranks 140 major cities around the world, from most liveable to least liveable. In 2018 and 2019, the EIU rated Vienna in Austria as the most liveable city – this came after a string of seven years where Melbourne was in the number one position.

Damascus in Syria was assessed as the least liveable city from 2013 to 2019 because of the civil war that has raged for seven years (see pages 122–123).

The EIU uses these considerations (or criteria) to determine the liveability of places. Each consideration is given a percentage to show its importance.

1 *Stability* (25%): Extent of small and violent crime, the threat of terrorism and military and civil conflict.

Source 1

The 10 most liveable and least liveable cities, from the 2018 Global Liveability Index

Ten most liveable and least liveable cities, 2018

Source: Matilda Education Australia

Source 2

Vienna is the capital city of Austria. It scored 99.1 per cent to be judged the world's most liveable city for 2018. It attained perfect scores for security, health care, education and infrastructure in the Global Liveability Index. Improvements in security have seen Vienna replace Melbourne in top place.

2. *Healthcare* (20%): Health standards, the availability and quality of healthcare and the availability of medicine.
3. *Culture and environment* (25%): The comfort of the climate, level of goods and services, evidence of corruption and censorship, and availability of sporting and cultural activities.
4. *Education* (10%): Education standards and the availability and quality of private education.
5. *Infrastructure* (10%): Quality of roads, international and public transport, housing, water, energy and telecommunications.

Source 3

Damascus is the capital city of Syria, in the Middle East. It scored 30.7 per cent in the Global Liveability Index, and was judged the world's least liveable city for 2018. It scored poorly in all categories. The lowest ranking was 20 per cent for stability because Damascus is in the middle of a civil war between rebels and government forces that has forced millions to flee.

Learning Ladder G4.8

Show what you know

1. What are the characteristics of a liveable city? Hint: Use SHEEPT to help you answer the question.
2. Why does Damascus score so poorly in the 2018 Global Liveability Index?

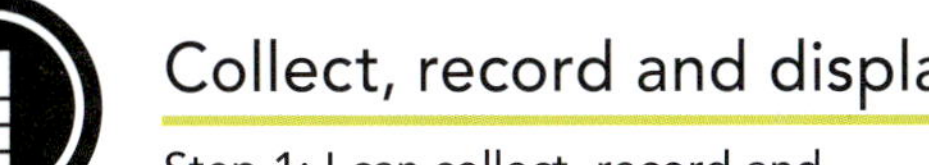

Collect, record and display data

Step 1: I can collect, record and display data in simple forms

3. How has the data been shown in Source 1? What is one advantage of showing data in this way?

Step 2: I can recognise and use different types of data

4. Is Source 1 a primary or secondary source of data? Justify your response.

Step 3: I can choose, collect and display appropriate data

5. Review the five criteria developed by Economist Intelligence Unit (EIU). Rank them according to the importance you would give each category. Justify your response.

Step 4: I can use data to support claims

6. Imagine you were a geographer who needed to collect data to identify whether a place could be classified as liveable. Create a list of appropriate primary and secondary methods that would help you come to a clear conclusion.

Data collection, page 144
SHEEPT, page 158

Why is Vienna the world's most liveable city?

Vienna was judged the world's most liveable city for 2018 and 2019 (see pages 130–131), after a comparison of 140 cities. It is one of the safest cities in Europe, and also offers great healthcare and education, as well as cheap and efficient public transport.

In addition, Vienna is one of the greenest large cities in the world. Half of the metropolitan area is made up of green spaces – with 620 square metres of green space for each of Vienna's 1.7 million inhabitants.

Vienna is surrounded by open space that was set aside back in 1905, and is one of the world's oldest **green belts** – a planned green space where no development is allowed. Since then, the green belt has been maintained and enlarged, and the latest urban development plan is looking to expand it even further.

Green spaces such as parks, sports fields, forests, meadows and wetlands give residents a chance for physical activity and relaxation. Trees produce oxygen, and help to filter out harmful air pollution that is usually a problem in large cities (see pages 124–125).

Barcelona
8.81 m^2

Los Angeles
90.19 m^2

Seoul
133.23 m^2

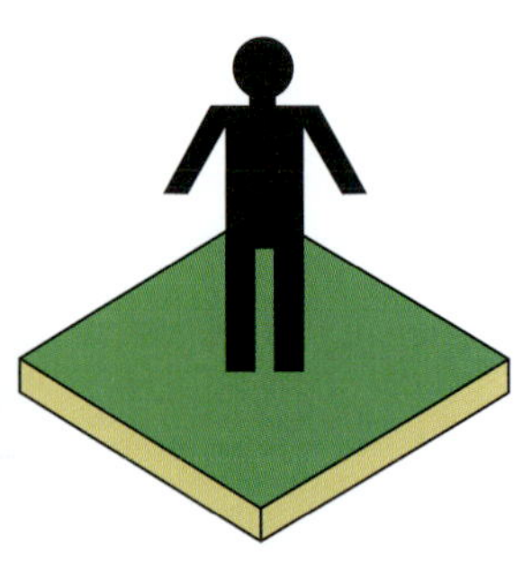
Tokyo
163.82 m^2

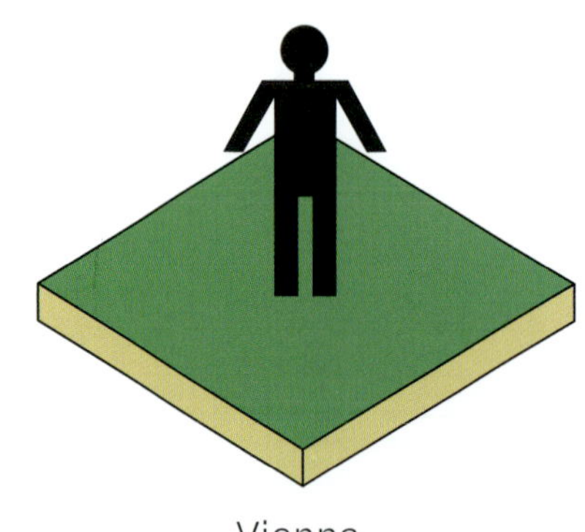
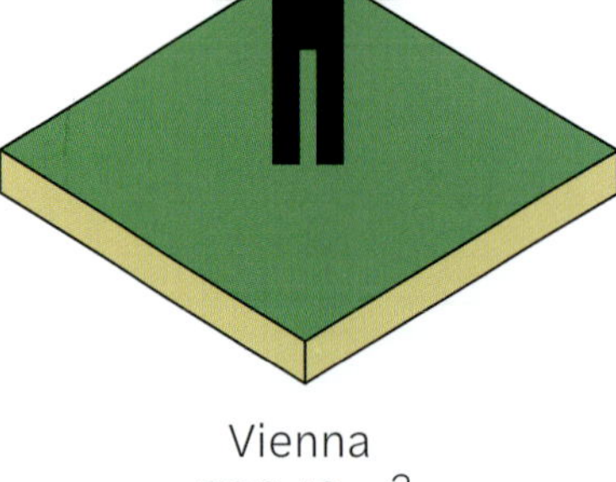
Vienna
620.16 m^2

Source 1

Green space in cities around the world per capita

Source: Humanitarian Data Exchange, 2018

Existing land uses 2016

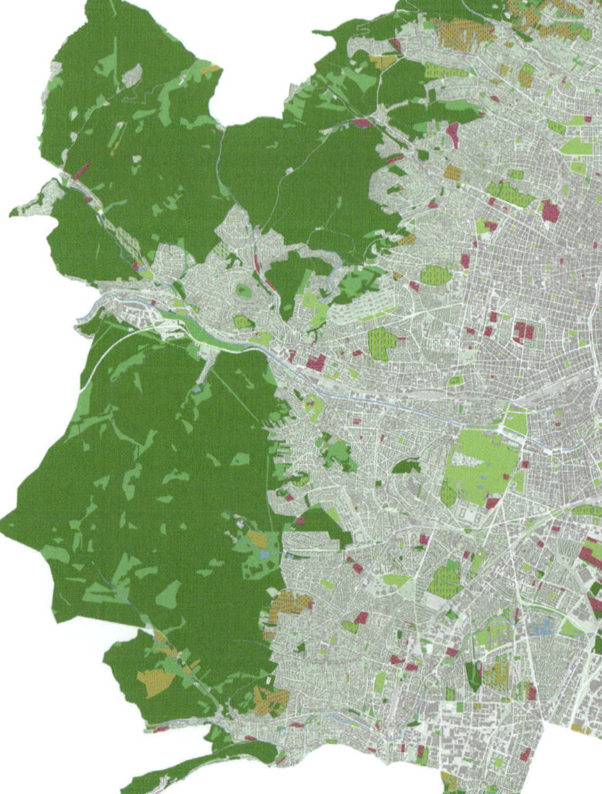

Source: Municipal Department 18 - Urban Planning and Development, 2016

Source 2

The large park at Schönbrunn Palace, Vienna, was opened to the public in 1779. The gardens and the palace were named a World Heritage Site in 1996. Urban parks and gardens provide safe routes for walking and cycling, and are good places for other recreation, such as picnics. Parklands play a critical role in cooling urban areas where buildings and roads heat up the environment.

Green space uses

- Forest
- Meadows
- Fields
- Vineyards
- Green houses and plantations
- Cemeteries
- Parks and gardens
- Outdoor sports facilities and public pools/campsites
- Bodies of water

N

0 0,75 1,5 3 km

Source 3

Green space in Vienna

Learning Ladder G4.9

Show what you know

1 Using SHEEPT, discuss why Vienna was voted the world's most liveable city in 2018 and 2019.

2 How do green spaces improve the liveability of a place?

Analyse data

Step 1: I can use geographic terminology to interpret data

3 Source 3: Where are the largest areas of green space found in Vienna?

Step 2: I can describe patterns and trends

4 Source 1: Look carefully at the pictograph and answer the following questions.

a How many square metres of green space per person does each resident of Vienna have?

b Which city in the pictograph has the least amount of green space per resident?

c How could you calculate the amount of green space per resident for your town or suburb?

Step 3: I can explain the reasons behind a trend or spatial distribution

5 Source 3: Which three types of green space dominate the areas on the fringe of the city of Vienna? Why do you think they are located here?

6 Source 3: Which type of green space is only located in the suburbs of Vienna? Why do you think they are located here?

Step 4: I can analyse relationships between different data

7 Use SHEEPT factors to explain why Vienna is the world's most liveable city.

HOW TO

SHEEPT, page 158

What challenges do low- and high-income countries face?

Countries around the world have different challenges to improve the liveability of places for their people. Those challenges depend very much on where you live.

Different challenges

In poorer **low-income countries**, some of the challenges to improve liveability are to provide a secure supply of food, give access to clean water (see page 64), provide somewhere for people to shelter, and supply education for all.

In richer **high-income countries**, governments are solving problems such as reducing travel times, improving technology, developing entertainment facilities and making places greener and more sustainable.

Low-income countries

Low-income countries have an income of $1045 or less per person, per year. Happily, the percentage of people living in low-income countries has fallen by more than 80 per cent.

In 1994, there were 3.1 billion people living in 64 low-income countries. But by 2014, there were 613 million people in 31 of the world's poorest countries. Most low-income countries are now in Africa – the only low-income countries outside Africa are Afghanistan, Cambodia, Haiti and Nepal.

Improving young lives

Ninety per cent of people aged between 10 and 24 live in low-income and **middle-income countries**. In general, young people in these countries have less access to education, healthcare and good work opportunities than young people in high-income countries. Children in low-income and middle-income countries also tend to have more family responsibilities than children in high-income countries.

Young women in low-income and middle-income countries face extra challenges. About 90 per cent of births to teenagers occur in low-income and middle-income countries – and the leading cause of death for young women ages 15 to 19 in low-income and middle-income countries is complications from pregnancy and childbirth.

Source 2

Students at a primary school in Kenya are having a lunch supplied by a school feeding program. Students receive a healthy meal, which is extra encouragement to come to school.

Source 1

In low-income and middle-income countries, children work in dangerous environments, such as these shipyards in Bangladesh. Child labour is cheap – and children living in poverty need to work to help support their families.

Education is the key to improving the liveability of people in low-income and middle-income countries, because:

- a child born to an educated mother is twice as likely to survive to the age of five
- educated mothers are 50 per cent more likely to immunise their children against disease
- individual earnings increase by 10 per cent for each year of school completed.

School feeding programs

Statistics from the United Nations show that 66 million primary school-age children in the world go hungry every day – with 23 million of these hungry children living in Africa. Further, 75 million school-age children do not attend school – of these, 55 per cent are girls and 47 per cent live in Africa.

To overcome this, feeding programs have been set up to supply meals for children who attend school. Meals are given to children at school during morning and afternoon meal and snack times. These meals might be a bowl of porridge or some crackers rich in nutrients. Families whose children regularly attend school might also receive packs of basic food items, such as rice and cooking oil.

Gross national income

Source: Matilda Education Australia

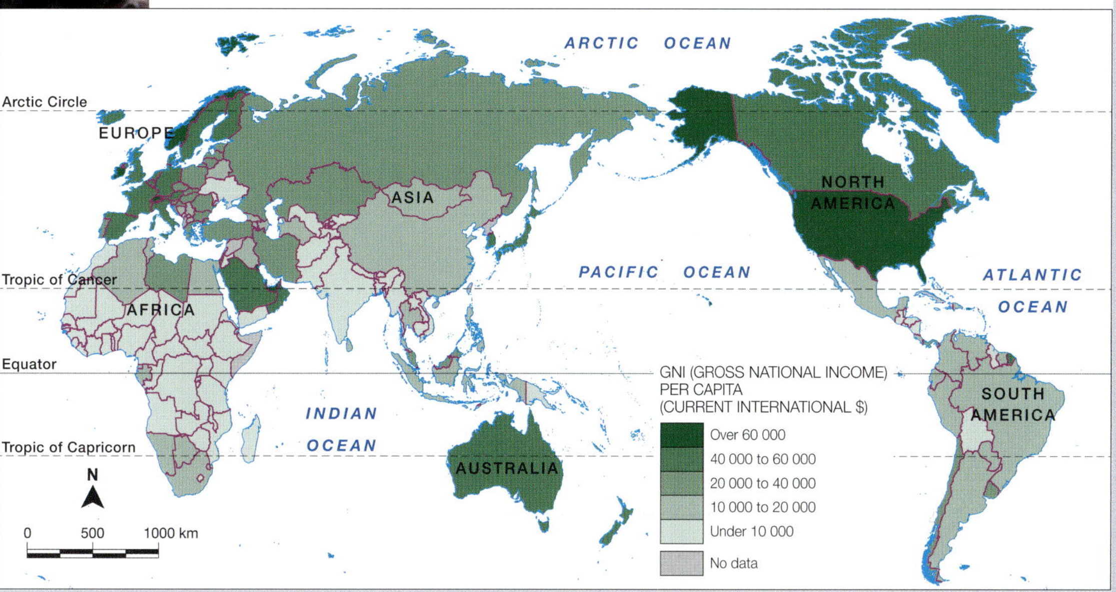

Source 3

Gross national Income per head of population. Gross national income is the total amount of money earned by a nation's people and businesses.

School feeding programs provide both educational and health benefits to the most vulnerable children in the world. They reduce hunger, improve the health of children, increase the numbers of children at school – and give parents a reason *not* to send their child to work.

School feeding programs in low-income countries often begin through funding by **humanitarian organisations** such as the United Nations World Food Programme. Schools have also been used to introduce other health programs, such as health education.

Improving liveability in high income countries

As cities grow and change, governments and urban planners are working to make cities more liveable. For people who live in cities in wealthy countries, this means:

- reducing transport times
- creating and protecting open space
- providing resources for recreation and entertainment
- providing affordable housing
- making sure there is work available
- providing a clean and sustainable environment.

Transport

As cities grow, traffic problems increase – as well as noise and air pollution. Australian cities (and many cities) have largely been designed for cars.

To manage the increasing volume of cars on the road, urban planners have introduced a number of measures to control city traffic. These include:

- introducing special traffic lanes for cars with passengers
- introducing 'park and ride' systems to encourage drivers to park their cars and travel into the city by bus, train or tram
- banning cars from the Central Business District and allowing only public transport, pedestrians and bicycles
- charging car drivers a toll when they enter the city centre
- restricting the times or days people can access the city by car.

Liveable cities promote public transport, walking and cycling instead of driving. Effective public transport systems are those within easy walking distance and provide a frequent and convenient service.

Source 4

The Marina Reserve in the inner-city Melbourne suburb of St Kilda is community space that has been created for young people. It includes a huge skatepark and an off-leash dog area.

Bike paths encourage people to commute to work or school by bicycle, as well as to ride for fun. Recently, over 500 cities in 49 countries had bike-sharing programs, where people could rent a bike for a period of time. A bike-sharing scheme was introduced in Paris in 2007 – and by 2012, bicycle trips in the city had grown by 41 per cent.

Open space

In liveable communities, most people live near an open space such as a park or playground. Parks provide green spaces in urban areas for plants and animals – they also cool urban areas and purify the air. For nearby residents they provide a space for activities such as exercise, ball sports, jogging and walking dogs.

The inner-city area of Prahran in Melbourne has among the least open space per head of population in Victoria. Stonnington Council is investing $60 million to build an urban oasis on top of a carpark. The community space, completed in 2019, will be 9000 square metres in size, and feature a forest, gardens, lawns and water features, as well as spaces for community events.

Source 5

An artist's impression of the urban park on top of a carpark in Prahran, Melbourne. In urban areas, parks provide healthy green spaces for residents to enjoy.

Resources for youth

In planning to make cities more liveable, the views of young people need to be considered. Public transport needs to be safe and reliable, and it needs to service local schools and shops, as well as sports and entertainment facilities.

Public spaces and facilities need to be designed to cater for the interests of young people. Young children need playgrounds and parks. Teenagers require sports grounds, stadiums, skateparks, mountain-bike tracks, cafes, cinemas and music venues.

Educational facilities need to be designed to cater for different types of learners, with flexibility in learning spaces and freely available technology, such as wi-fi and fast internet.

Learning Ladder G4.10

Economics and business

Step 1: I can recognise economic information

1 Source 3: What is meant by gross national income?

Step 2: I can describe economic issues

2 Source 3: What insights does gross national income give us when considering a country's liveability?

Step 3: I can explain issues in economics

3 If school feeding programs are often funded through organisations such as the United Nations, what potential economic issues could arise over time as communities develop a stronger need for these kinds of initiatives?

Step 4: I can integrate different economic topics

4 What are some of the challenges low-income countries face to improve liveability?

5 How did liveability improve for low-income countries between 1994 and 2014?

6 What are some of the challenges high-income countries face to improve liveability?

Step 5: I can evaluate alternatives

7 Find infomation online about a program run by a humanitarian organisation that is aimed at improving liveability for people in low-income countries.

How does a pandemic affect liveability?

The coronavirus pandemic swept across the world in 2020, affecting global liveability by affecting health, employment, and access to goods and services. Governments scrambled to slow the spread of the virus for which there was no cure.

The World Health Organization (WHO) declared the coronavirus COVID-19 a **pandemic** on 11 March 2020. A pandemic is a disease that spreads in multiple countries around the world at the same time, usually affecting a large number of people.

At this time the pandemic had reached every continent except Antarctica. The outbreak of COVID-19 is believed to have begun at a wildlife market in Wuhan in China.

In mid-March 2020, 80 000 cases and more than 3000 deaths from COVID-19 had occurred in China, where the outbreak began. However, because we live in a highly interconnected world, the virus quickly spread to the rest of the world.

More than 60 000 cases of COVID-19 and 2500 deaths were reported in countries other than China. Europe became the new **epicentre** for the virus as it spread quickly in Italy and Spain, along with large outbreaks in South Korea, Iran and the USA.

The COVID-19 pandemic sparked a public health emergency around the world, with severe restrictions placed on the movement of people to slow the spread of the virus to assist medical resources to deal with the crisis. In Australia, travel bans were put in place, gatherings of people were banned, and people arriving back from overseas or in contact with a person infected with COVID-19 were forced to isolate themselves for 14 days.

Source 1

Confirmed COVID-19 cases per million people on 14 March 2020

COVID-19 cases per million people by 14 March 2020

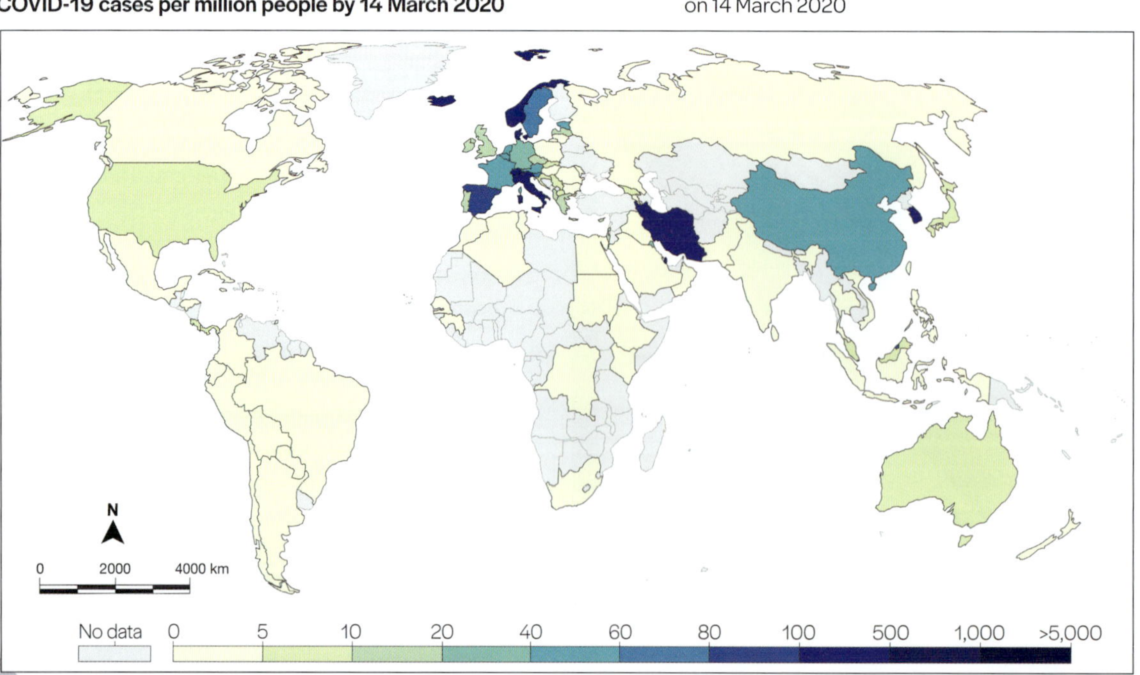

Source: European CDC

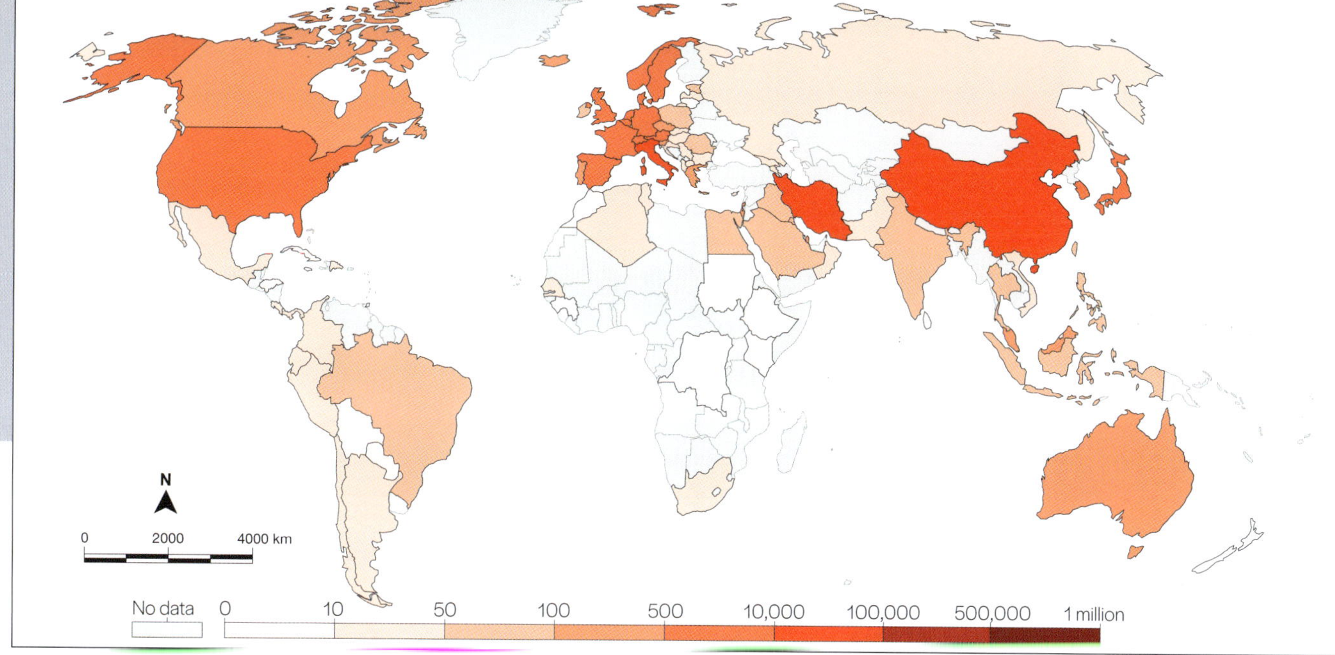

Number of cases of COVID-19 by 14 March 2020

Source: European CDC

Source 2

Confirmed COVID-19 cases worldwide on 14 March 2020

Early growth in COVID-19 cases

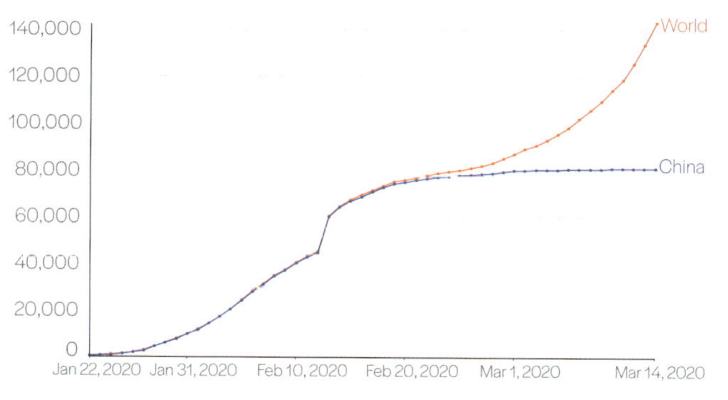

Source 3

Graph showing cases of COVID-19 virus by 13 April 2020

Source 4

Health personnel at Krakow Airport in Poland monitor the body temperatures of passengers arriving from Italy. They are detecting signs of COVID-19 in an attempt to slow the spread of the virus in Poland.

Learning Ladder 4.11

Show what you know

1 What is a pandemic?

2 How do you think the COVID-19 pandemic spread from Wuhan in China?

3 Source 4: What action is being taken here to slow the spread of the COVID-19 pandemic in Poland?

Analyse data

Step 1: I can use geographic terminology to interpret data

4 What is an epicentre? What was the original epicentre of the COVID-19 pandemic?

Step 2: I can describe patterns and trends

5 Source 2: USE PQE to describe the global spread of the COVID-19 virus by 14 March 2020.

Step 3: I can explain the reasons behind a trend or spatial distribution

6 Source 3: Explain the reasons behind the trends shown in this line graph.

Step 4: I can analyse relationships between different data

7 Source 1: Which countries were most greatly affected by the COVID-19 pandemic in March 2020? Why do you think this was the case?

HOW TO

PQE, page 156

Masterclass

Learning Ladder

States of peace on a global scale, 2018

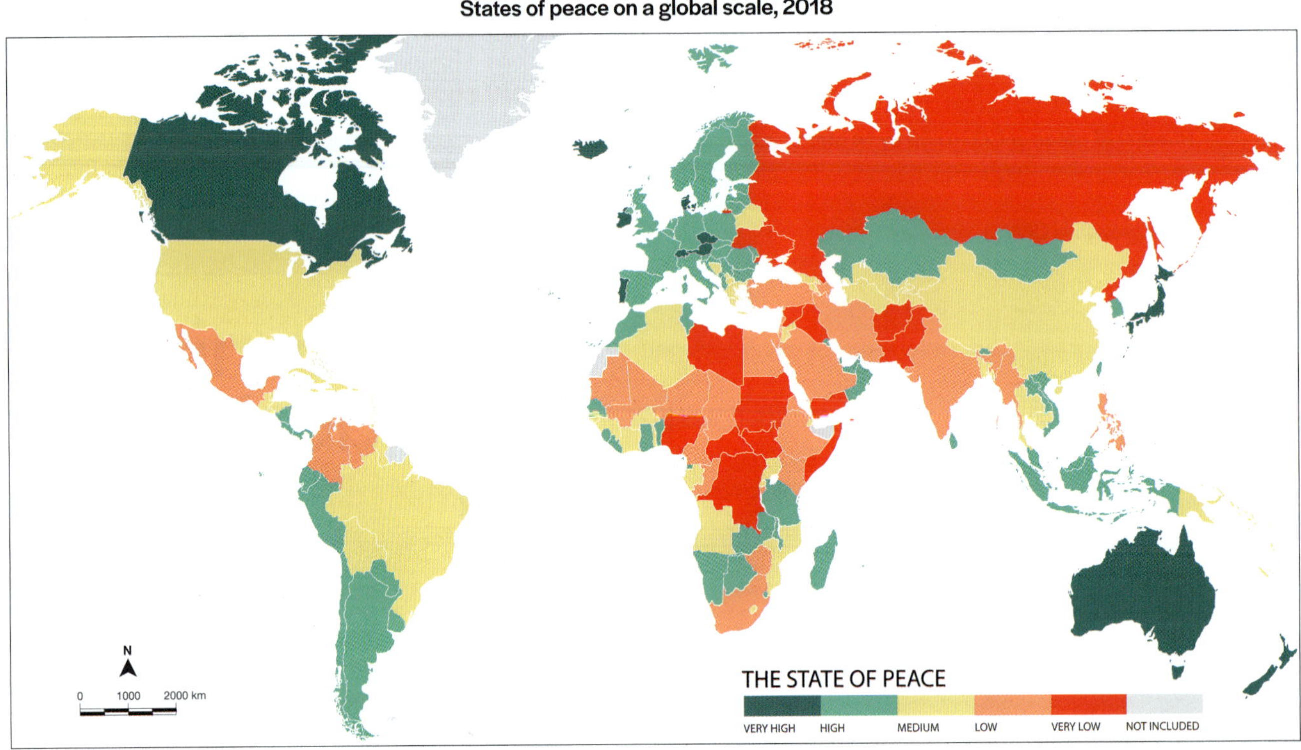

Source 1

Step 1

a I can identify and describe spatial characteristics

Identify and describe the different spaces you can see in Source 2.

b I can identify and describe interconnections

Give an example from Source 1 of a country where refugees are likely to come from. Give a reason for your selection.

c I can identify responses to a geographical challenge

Give examples of responses made by people living in war-torn countries.

d I can collect, record and display data in simple forms

Explain what the colour red in Source 1 represents.

e I can use geographic terminology to interpret data

Source 1: Is the northern or southern hemisphere the most peaceful?

Step 2

a I can explain spatial characteristics

Refer to Source 5 on page 137. Explain why planners are making this change to Prahran.

b I can explain interconnections

Refer to Source 3 on page 135. Explain why there is an interconnection between high-income countries and liveable cities.

Work at the level that is right for you or level-up for a learning challenge!

c I can compare responses to a geographical challenge

What two geographical challenges is the school feeding program trying to address (see page 135)?

d I can recognise and use different types of data

List two ways the data from Source 1 could have been collected.

e I can describe patterns and trends

Source 2: Use PQE to describe the distribution of peace on a global scale.

Step 3

a I can explain processes influencing places

Refer to Source 2. Why do you think this strip of houses has developed and where have people come from to move here?

b I can identify and explain the implications of interconnections

How has the interconnection between low-income migrants and open city land led to the development of the slum in Source 2?

c I can compare strategies for a geographical challenge

List three strategies to improve liveability for the people living in slums in Maharashtra in Source 2.

d I can choose, collect and display appropriate data

Look at Source 1 on page 130. For each city, select the criteria that they most need to improve to increase their liveability score.

e I can explain the reasons behind a trend or spatial distribution

Use SHEEPT to explain the pattern shown in Source 1.

Step 4

a I can predict changes in the characteristics of places

What changes might be introduced to a city centre to relieve traffic congestion (see page 136)?

b I can evaluate the implications of significant interconnections

Use Source 2 on page 115 to identify environments where people have had to adapt to extreme conditions.

c I can evaluate alternatives for a geographical challenge

What are key challenges for young people in low income countries (page 134)? What are the alternatives to improve their lives?

d I can use data to support claims

Refer to Source 2. Discuss whether this piece of data could be used to represent liveability in India. What other data would need to be collected to gain a clear understanding of this place?

e I can analyse relationships between different data

Which three countries have the most liveable cities in Source 3 (see next page)? Is there a relationship between these countries and their category in the Global Peace Index in Source 2?

Step 5

a I can analyse the impact of change on places

What impact did the outbreak of COVID-19 have on countries around the world in 2020 (see page 138)?

b I can explore spatial association and interconnections

Explain the spatial association between the patterns in Source 1 on page 128 and Source 3 on page 135.

Source 2

Aerial image of Maharashtra, India

Masterclass

c I can plan action to tackle a geographical challenge

Look at pages 132 and 133. How has Vienna taken action to improve the liveability of the city?

d I can evaluate data

Refer to Source 3. How relevant is data on healthcare when considering the liveability of a place? Use specific examples.

e I can draw conclusions from analysing collected data

Using data from the three sources, summarise the key elements that influence a place's liveability.

Source 3

The top 10 cities for liveability, 2018

Country	City	Rank	Overall Rating (100 = ideal)	Stability	Healthcare	Culture & Environment	Education	Infrastructure
Austria	Vienna	1	99.1	100	100	96.3	100	100
Australia	Melbourne	2	98.4	95	100	98.6	100	100
Japan	Osaka	3	97.7	100	100	93.5	100	96.4
Canada	Calgary	4	97.5	100	100	90	100	100
Australia	Sydney	5	97.4	95	100	94.4	100	100
Canada	Vancouver	6	97.3	95	100	100	100	92.9
Canada	Toronto	7	97.2	100	100	97.2	100	89.3
Japan	Tokyo	8	97.2	100	100	94.4	100	92.9
Denmark	Copenhagen	9	96.8	95	95.8	95.4	100	100
Australia	Adelaide	10	96.6	95	100	94.2	100	96.4

(100 = ideal; 0 = intolerable)

Capstone

How can I understand world liveability?

In this chapter, you have learnt a lot about world liveability. Now you can put your new knowledge and understanding together for the capstone project to show what you know and what you think.

In the world of building, a capstone is an element that finishes off an arch or tops off a building or wall. That is what the capstone project will offer you, too: a chance to top off and bring together your learning in interesting, critical and creative ways. You can complete this project yourself, or your teacher can make it a class task or a homework task.

Scan this QR code to find the capstone project online.

mea.digital/GHV7_G4

G5

Fieldwork

HOW DO WE USE DATA TO ANSWER GEOGRAPHIC QUESTIONS?

G5.1

How do we use data to answer geographic questions?

Fieldwork is an important part of Geography. It is one of the ways that geographers *learn by doing*, and it is a tool for discovering new things, learning about patterns and relationships, answering big questions and exploring the world around us.

Conducting fieldwork

When conducting fieldwork, geographers collect data. There are two main kinds of data:

- **quantitative data**
- **qualitative data**.

Quantitative data tends to be recorded in numbers; for example, the height of a building, the number of people in a population or the flow rate of a river.

Qualitative data tends to be more observational. You might write down descriptions of a place, conduct a field sketch or take photos as evidence to answer your research question.

Both aspects of data collection are very important, and you should try to collect both quantitative data and qualitative data in your fieldwork.

The steps you will follow to conduct your fieldwork study are:

- Step 1: Research question and hypothesis
- Step 2: Data collection
- Step 3: Data analysis.

Before you begin your own fieldwork task, let's explore each of these steps.

Research question and hypothesis

A **research question** is an overarching idea to be investigated. A good research question is one that is measurable and quantifiable. You need to be able to collect a clear range of data and answer the question within the limitations you are presented with. For example, will your fieldwork be conducted at school? Will you be able to survey members of the public?

After developing a research question, you will need to write a hypothesis. A **hypothesis** is an 'educated guess' about the answer to the research question. It's making a statement about what you think might happen.

"This desk job is killing me Jim. I need to be out in the field."

Source: CartoonStock.com

2 Data collection

Before you begin any fieldwork study, you need to develop a list of **primary methodologies** and **secondary methodologies**. You also need to agree, as a class, on your data collection methods.

Primary methodologies are things that you will do in the field in order to find evidence to answer your research question, and to prove or disprove your hypothesis.

Secondary methodologies are ways that you can collect data back in the classroom to help you answer the question; for example, research using websites, books or other publications.

You can record your methods in a table like the one on the right.

3 Data analysis

Once you have been out in the field and collected data using primary and secondary methods, you need to communicate your findings.

You can communicate your research via a report, or display it as a visual poster or presentation. When you communicate findings, you need to ensure you highlight any patterns, *interconnections* or significant data that allow you to answer the research question – and to prove or disprove your hypothesis.

In your study of Geography this year, you will explore water as a resource – its spiritual, economic, cultural and aesthetic value. You will also look at what makes a *place* liveable. Now you are going to conduct your own research in your local area to answer a research question and write a mini fieldwork report.

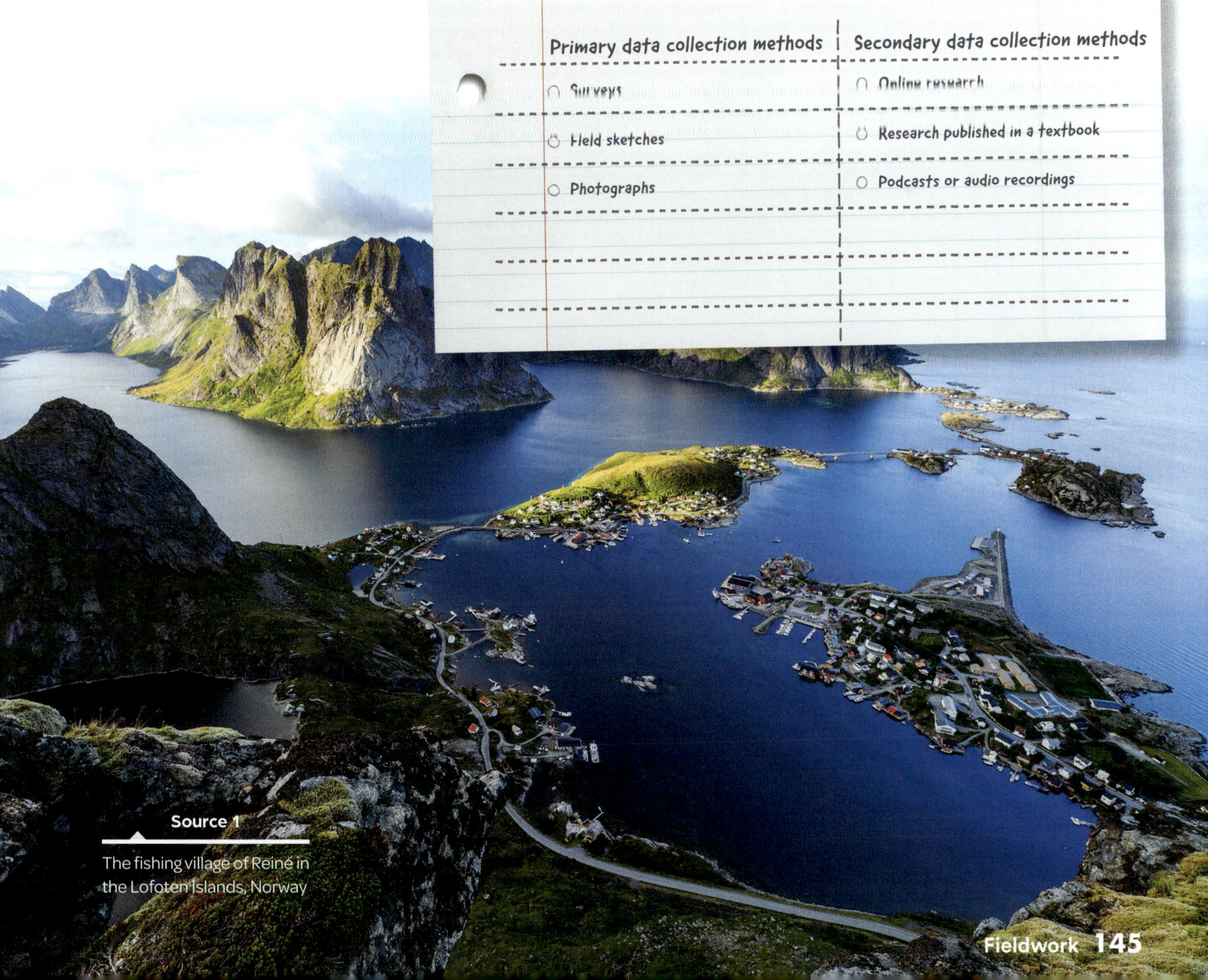

Primary data collection methods	Secondary data collection methods
○ Surveys	○ Online research
○ Field sketches	○ Research published in a textbook
○ Photographs	○ Podcasts or audio recordings

Source 1

The fishing village of Reine in the Lofoten Islands, Norway

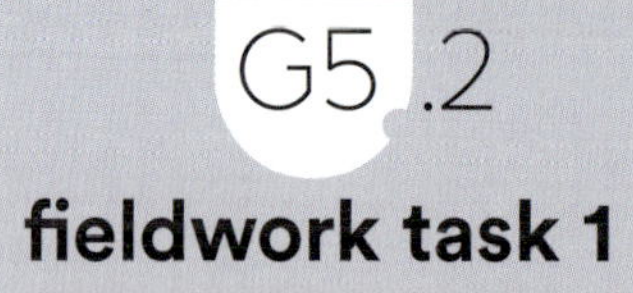

fieldwork task 1

How do you use water in your world?

Domestic water use accounts for only 12 per cent of Australia's total water use – yet Australia has one of the highest rates of water consumption per person in the world. The average daily water use is 340 litres per person, or 900 litres per household. Average water consumption ranges from 100 litres per person in some coastal areas, to over 800 litres per person in the **arid** inland areas.

1 Research question and hypothesis

Research question

How is water used in the local community?

Hypothesis

Water is used in a range of ways for both natural and human environments.

Source 1

Exploring the edge of the lake

Recreation in the local pool

2 Data collection

Some suggested primary methodologies you could use to collect data are listed below.

- Walk around your local environment and take photos as evidence of water use. Create a photo essay of these water uses, and add short labels to explain each use.
- Create a survey for your classmates or family to identify which type of domestic water use they think contributes most to our daily water consumption per person. Multiple-choice questions work best, as you can then create bar graphs to display your data. Questions you could ask include the following:

? Which domestic water use do you think contributes most to our daily water consumption per person?

☐ Toilets ☐ Sinks ☐ Hoses

☐ Shower ☐ Other:

? How is water used in the local community?

☐ Pools ☐ Dams ☐ Lakes

☐ Fountains

- Using appropriate techniques, complete a field sketch of water use, movement or value in your local community.

3 Data analysis

Once you have collected the data, you need to present and analyse your results. Use the steps listed in 'In the field' to help you complete your fieldwork analysis.

In the field

Research question

1 Describe the key differences between primary and secondary methodologies.

2 List the primary and secondary methods you will use to investigate the research question.

Data collection

3 Outline the characteristics and history of your local region in four or five sentences.

4 Name two primary fieldwork techniques you used to complete your fieldwork.

5 Name two secondary resources you used in the lead up to the fieldwork.

6 Answer this research question: 'How is water used in the local community?' Consider:
- How many different uses were present in your local area?
- Do you think that water is used wisely in these places?
- Which domestic water use do you think consumes the most water?

Data analysis

7 Which domestic water use was the most reported?

8 Why do you think this is?

9 Were you surprised by these results? Why or why not?

10 Was the hypothesis correct or supported by the evidence? Why or why not?

11 Record five key pieces of evidence that support your answer – these could be data, photos, sketches or surveys.

Evaluation of methods

12 Which methods were the most successful for collecting data to answer the research question? Why?

13 Which methods were the least successful for collecting data to answer the research question? Why?

14 Suggest three ways that water could be better conserved in your local region.

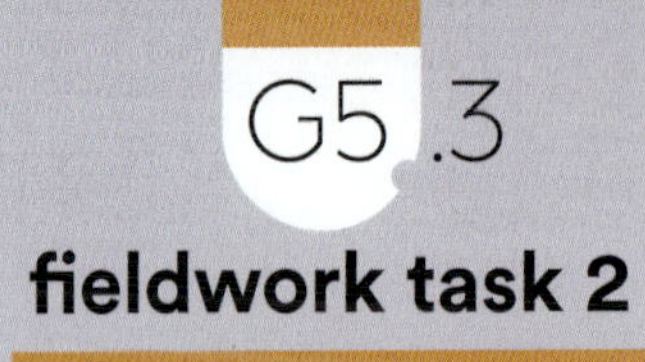

fieldwork task 2

How liveable is your place?

Within any community, there are things that can be improved to increase the overall liveability of that place. For example, Source 1 shows an empty housing development piled with bricks and overgrown with weeds. As Geographers, we can help town planners develop ideas about how to redesign *places* such as these – and make more useable *spaces* for the community.

Source 1

An empty lot, ready for building to begin

Source 2

Differing levels of liveability

Research question

How can liveability be improved in the local community?

Hypothesis

Different groups will value different facilities in the community.

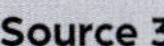

Source 3

Students checking survey and map materials

2 Data collection

Some suggested primary methodologies you could use to collect data are listed below.

- Walk around your local community and take photos of places where **infrastructure** could be improved to increase **liveability**. Create a photo essay of these examples, adding short annotations or labels to explain how they are being used or how they could be improved.
- Create a survey for your classmates or family to identify which areas in the community require improvement. Multiple-choice questions work best, as you can then create bar graphs to display your data. Questions you could ask include the following:

? Which aspects of the community do you value?

☐ Hospitals ☐ Schools ☐ Housing estates

☐ Shopping centres ☐ Parklands ☐ Other:

? Which aspects of the community need to be improved?

☐ Hospitals ☐ Schools ☐ Housing estates

☐ Shopping centres ☐ Parklands ☐ Other:

- Using appropriate techniques, complete a field sketch of a location that could be altered to improve the liveability of your local area. Annotate the changes that you suggest for this place.

3 Data analysis

Once you have collected the data, you need to present and analyse your results. Use the structure listed in 'In the field' to help you complete your fieldwork analysis.

In the field

Research question

1 Describe the key differences between primary and secondary methodologies.

2 List the primary and secondary methods you will use to investigate the research question.

Data collection

3 Outline the characteristics and history of your local region in four or five sentences.

4 Name two primary fieldwork techniques you used to complete this fieldwork.

5 Name two secondary resources you used in the lead up to the fieldwork.

6 Answer the research question: 'How can liveability be improved in the local community?' Consider:

- What is the definition of liveability?
- What facilities make your local region or area liveable?
- What facilities could be improved? Why?
- Brainstorm different groups of people within your local community, and suggest what facilities they might value. Give some reasons to back up your ideas.

Data analysis

7 Which facilities do people think make a place more liveable?

8 Are there particular areas in your community that require improvement? Describe these places.

9 Why do you think this is?

10 Were you surprised by these results? Why or why not?

11 Was the hypothesis correct or supported by the evidence? Why or why not?

12 Record five key pieces of evidence that support your answer – these could be data, photos, sketches or surveys.

Evaluation of methods

13 Which methods were the most successful for collecting data to answer the research question? Why?

14 Which methods were the least successful for collecting data to answer the research question? Why?

15 Based on your research, redesign an area in your local region to improve the liveability for the local community.

How do geographers use surveys?

In Geography, surveys are useful ways to collect data in the field. Surveys could be in the form of a questionnaire, where you ask questions about a topic to get an understanding of people's views.

You can also do a vehicle survey, where you tally the number of vehicles that pass in a particular amount of time. Surveys provide you with a sample of what is occurring in a natural or human *environment*.

Surveys can be written by hand or prepared on software such as Survey Monkey or Google Forms.

Some sample questions for a questionnaire about liveability follow.

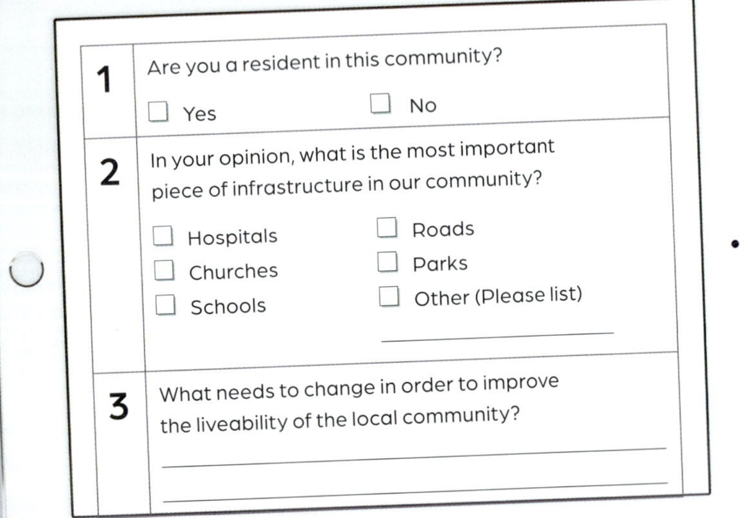

1	Are you a resident in this community? ☐ Yes ☐ No
2	In your opinion, what is the most important piece of infrastructure in our community? ☐ Hospitals ☐ Roads ☐ Churches ☐ Parks ☐ Schools ☐ Other (Please list) ______
3	What needs to change in order to improve the liveability of the local community? ______ ______

Source 1

A sample questionnaire

In the field

1. Are surveys a primary or secondary data collection method?
2. There are three main stages to fieldwork: defining the research question, data collection and data analysis. In which stage would a survey be a useful tool?
3. Write a survey question that would gather quantitative data about water use.
4. Write a survey question that would gather qualitative data about water use.
5. Why are multiple-choice questions often used in surveys?
6. Follow these steps to design a survey on a topic you are passionate about:
   a. Write an objective for your survey. Think about who will be reading the data and what decisions will be made based on the data you collect. Make sure your objective is short and clear.
   b. Write a title for your survey.
   c. Write two or three sentences to introduce and explain your survey to the people who will be completing it. Decide whether your survey will be anonymous or not.
   d. Choose how you will create your survey: on paper, on your computer using Word or Excel, or using an online platform such Survey Monkey or Google Forms.
   e. Using the method you chose in part d, write 10 survey questions. Keep your questions short and specific. Try to use multiple-choice questions to make data analysis easier.
   f. Swap your survey with someone else in your class. Proofread and evaluate each other's surveys. Give appropriate feedback to your classmate about how they can make their survey more effective.
   g. With your teacher's permission, conduct your survey in your class, school or community.

G6

Geo How-To

Knowing the locations of countries, their capitals – and even their flags – can be useful geographic knowledge. However, the key to success in Geography is understanding key skills and being able to apply them in different situations. This chapter will walk you through some of the key skills for Year 7 and provide you with examples of how to use them.

HOW TO

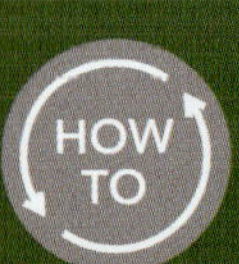

Mapping with BOLTSS (NA)

Every map needs BOLTSS! BOLTSS is an acronym – every letter of this word is the first letter of a key step to ensure that your map can be understood by its audience. When you create a map, use the BOLTSS checklist so you can be sure you have completed your mapping task correctly. The NA is to remind you to be neat and accurate.

Urban growth in Melbourne

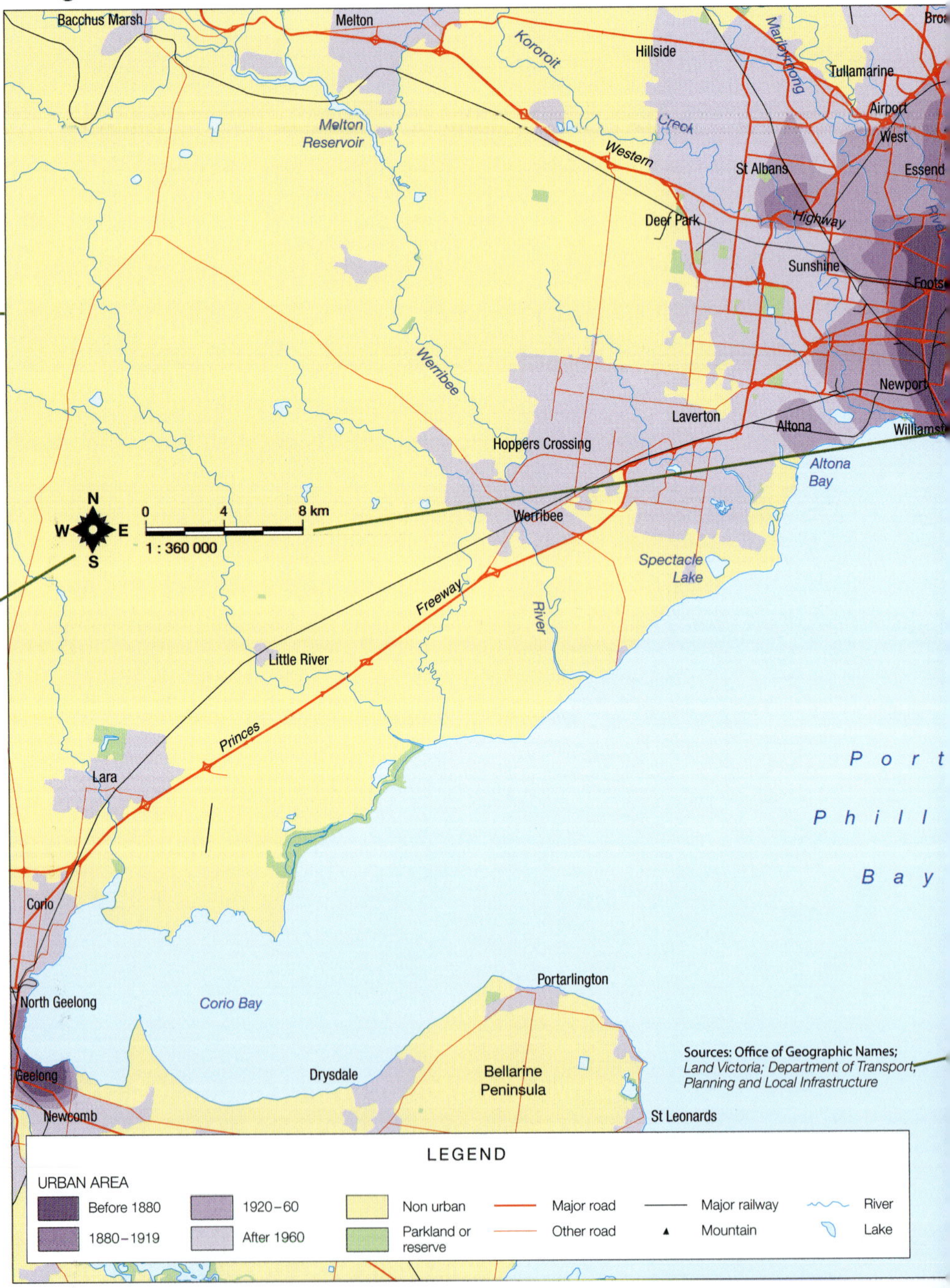

Bacchus Marsh
Melton
Kororoit
Creek
Hillside
Maribyrnong
Tullamarine
Airport
West
Essend
Melton Reservoir
Western
St Albans
Highway
Deer Park
Sunshine
Footsc
Werribee
Newport
Laverton
Altona
Williamst
Hoppers Crossing
Altona Bay
N
W
E
S
0
4
8 km
1 : 360 000
Werribee
Spectacle Lake
Freeway
River
Little River
Princes
Lara
Port Phillip Bay
Corio
North Geelong
Corio Bay
Portarlington
Geelong
Drysdale
Bellarine Peninsula
Newcomb
St Leonards
Sources: Office of Geographic Names; Land Victoria; Department of Transport, Planning and Local Infrastructure
LEGEND
URBAN AREA
Before 1880
1880–1919
1920–60
After 1960
Non urban
Parkland or reserve
Major road
Other road
Major railway
Mountain
River
Lake

Source: Matilda Education Australia

B Border

Borders are important, because they show the edges of the mapping field. A border provides a clear area for you to construct your map, and makes it look clear and neat.

O Orientation

An orientation (or compass) helps us understand direction when we read the map. Use all 16 points of the compass to give accurate directions. Orientations should be drawn as a 16-point compass.

N
nnw
nne
nw
ne
wnw
ene
W
E
wsw
ese
sw
se
ssw
sse
S

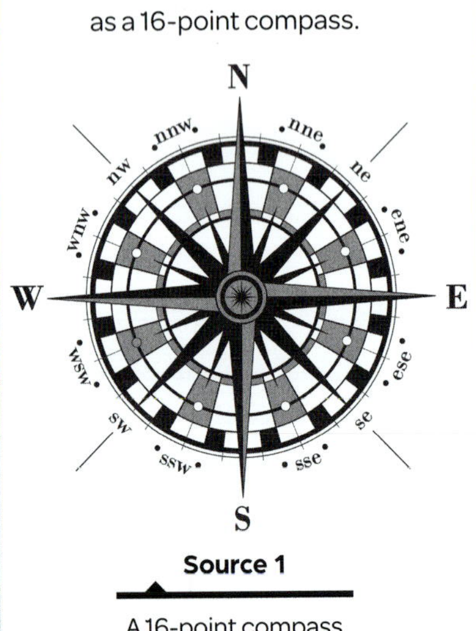

Source 1

A 16-point compass

Source 2

A sample legend

HIGHWAY
MAJOR ROAD
SECONDARY ROAD
RIVER
PROPOSED ROAD
BRIDGE
TOLL
TRAFFIC LIGHT
ONE WAY
HOUSING WITH NUMBER BESIDE ROAD
LIGHT RAIL & STATION
TRAIN RAIL & STATION
STATE BOUNDARY
CITY BOUNDARY
GRASS FIELD
SEA
FOREST
BUS STATION
TAXI STAND
LIGHT RAIL STATION
TRAIN STATION
AIRPORT
SEAPORT
APARTMENT
FACTORY
MUSEUM
CHURCH
CHINESE TEMPLE
MOSQUE
HINDU TEMPLE
SCHOOL
COMMUNITY HALL
POLICE STATION
FIRE STATION
HOSPITAL
LIBRARY
POST OFFICE
EMBASSY
SHOPPING MALL
HOTEL
PLACE OF INTEREST
INFORMATION KIOSK
HIGH COURT
PARKING AREA
SWIMMING POOL
FAST FOOD
FOOD COURT
CINEMA
PUBLIC TOILET
TELEPHONE
PETROL STATION
RESTING AREA
GOLF COURSE
PLAYGROUND
NUCLEAR REACTOR

L Legend

A legend – sometimes called a 'key' – is vital for understanding the map. Without a legend, we would not know what the colours and symbols stand for, and we would not be able to interpret patterns or the way something is spread out (which is called 'distribution').

T Title

A title tells us what the map shows, and gives us some understanding about it. Make sure you always add the date and the time to your sketch maps – this will allow you to monitor change over time in a location. The title of this map is 'Urban growth in Melbourne'.

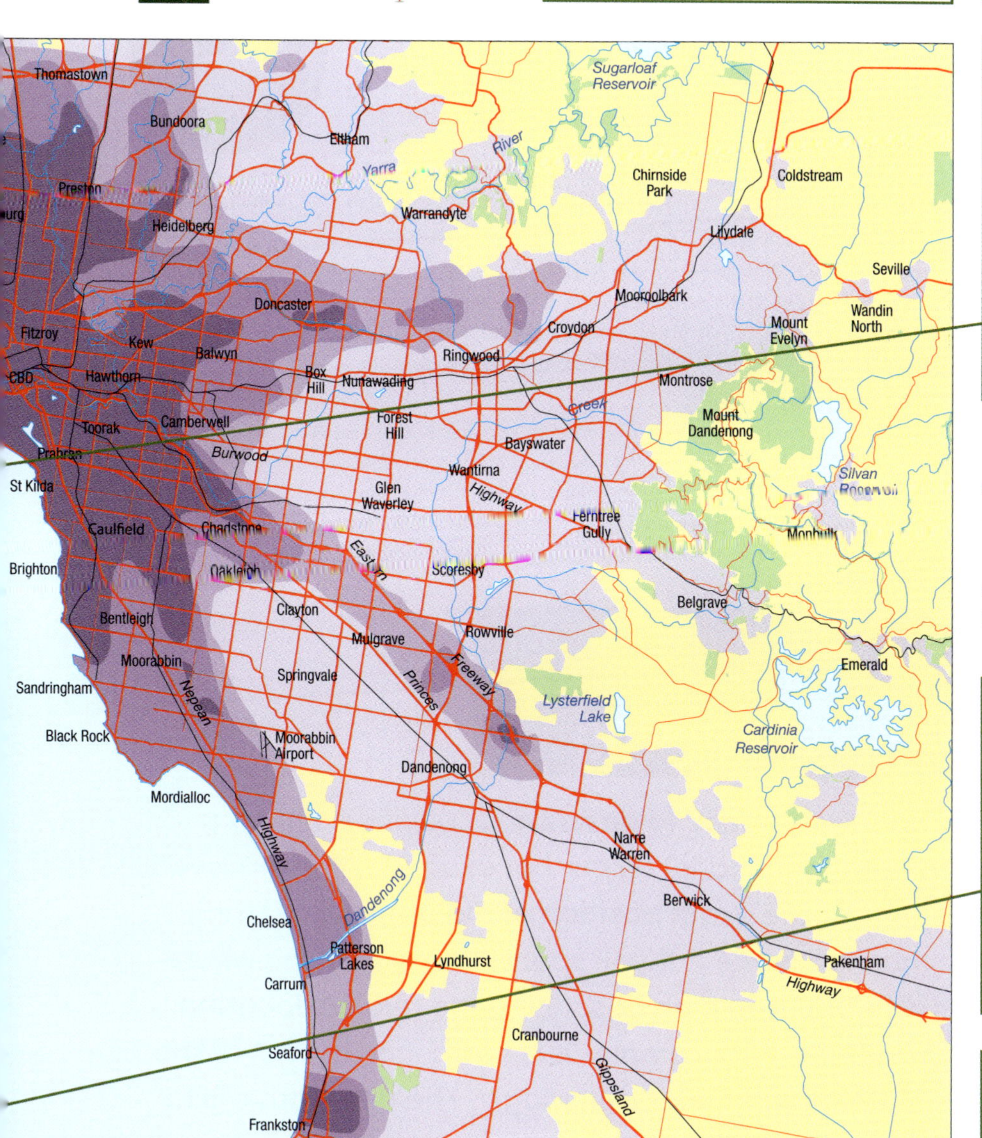

S Scale

A scale gives us information about how big something is in real life. So while a house on a map might be 2 cm across, it might actually represent a house that is 20 m wide in real life. There are many different types of map scales: linear, ratio and fraction.

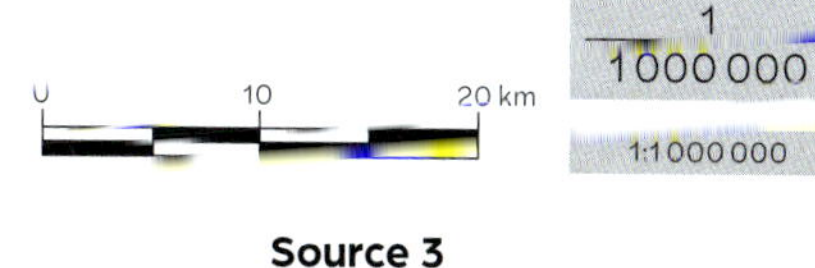

Source 3

Map scales: linear, fraction and ratio

S Source

When you create a map, it is important that you let the reader know where you got the information – which is known as your source. The source can also indicate whether or not the map is trustworthy.

(NA) Neatness and Accuracy

When we read a map, we rely on it being neat and easy to read – but we also expect that the data has been displayed accurately. So when you construct a map, it is important to correctly show the patterns and distributions you see in the data.

Source 4

Which sketch map is easier to read?

Mapping skills

Direction

Northern hemisphere

Equator

Southern hemisphere

Source 5

Earth

Imagine trying to find the location of a party without using any directional terms. How would you know where to go and where you needed to end up? Direction is a very important concept in Geography because without it we are all lost!

Up until now, direction may have been described to you as 'left' and 'right', 'above' or 'below'. While these are helpful, in Geography we also need to use compass points: north, south, east and west.

Source 6

A compass

Consider the world globe (Source 1). Around the centre there is a line of **latitude** called the **equator**. North of the equator is the northern hemisphere. Here we find continents such as North America, Europe, parts of Asia and Africa. South of the equator is the southern hemisphere – this is where we live!

Sometimes, we mistakenly use the word 'above' instead of 'north', or 'below' instead of 'south'. If you say something is above the equator, what you are really saying is that it is floating in the air over the top of it! If you say something is below the equator, you are really saying that it is buried beneath it! Be careful how you use directional terms in Geography to ensure you are sending people in the right direction.

Source 7

A map with grid references

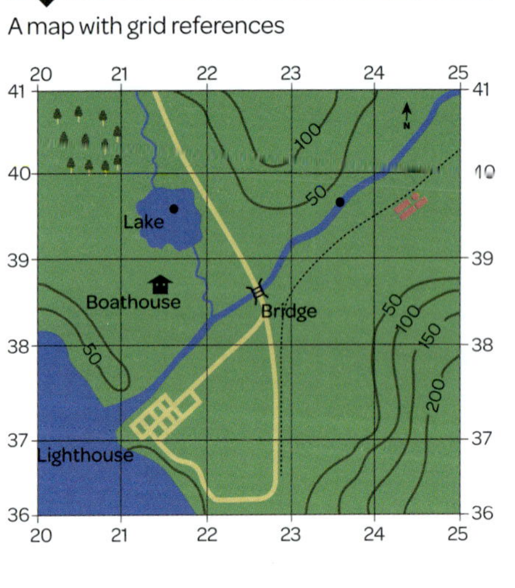

Source 8

An example of grid references

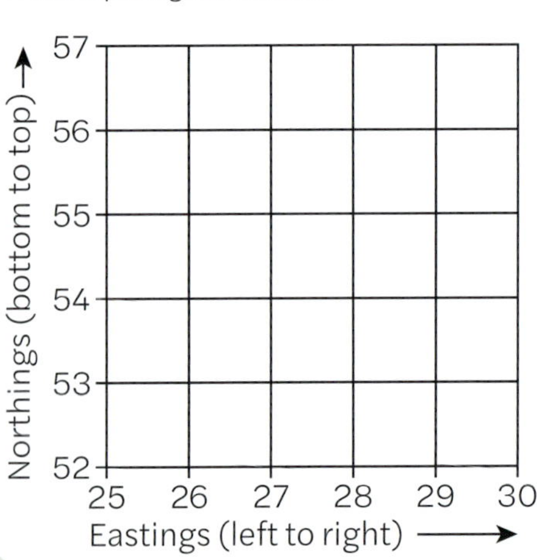

Grid referencing

It is not always easy to describe a specific point on a map using directional terms. For example, you may want to highlight the exact position of the boathouse using the map in Source 7. If you spent time explaining where this was, you would be describing the relative location of that place. For example, 'It is north-north-east of the lighthouse and west of the bridge'. The alternative is to use the grid references or area reference (AR) to show the exact or absolute location of that place.

To read an area reference, we first look at the numbers (or letters) associated with the vertical lines, reading from left to right – these are called eastings. We then look at the numbers associated with the horizontal lines, reading from the bottom of the map to the top – these are called northings. In this case, the boathouse is located at the four-figure grid reference AR2138. This reference takes us to the bottom left-hand corner of the grid square that the boathouse is located in. The lake is located in AR2139 and the bridge is located in AR2238.

Scale

Look at Source 5 again. Here we have an image of the Earth. Logically, we know that this image of the Earth is not to scale. In reality, the Earth is over 12 700 km in diameter and would not fit in this book!

Scale is important in Geography as it allows us to make maps and other images, and shrink them down to a size where we can see patterns, distributions and changes over time. Scale can be written in a number of ways. Typically, scale is displayed using a line that acts like a ruler, showing you how many centimetres on the map represent the real distance. Scale can be used to determine size and distance.

0 10 20 km

Source 9

An example of a linear scale

Scale 1:32,000,000
Mercator Projection

Source 10

Obviously this map is not to scale. It is a large area that has been shrunk down to fit on this page. The scale on this map helps us determine how big things are in real life.

PQE

In Geography, we use maps and graphs so that we can understand what is happening around us. In many cases, these maps and graphs provide us with information about patterns and the location (or distribution) of specific items.

The formula PQE helps us to describe these patterns and their distribution. P stands for pattern, Q stands for quantify (or 'how many') and E stands for exception.

P Pattern

A pattern is a trend in the data. When you are looking for a pattern, you need to read the legend and interpret what the colours or symbols mean.

On a graph or map, you might notice that all the data points tend to be clustered in one spot, or that the data points are distributed unevenly. You might need to use compass points or the names of places to describe where on the map these clusters appear. For example, when observing the map in Source 11, we notice that western USA prefers mint chocolate chip and OREO cookies and cream ice-cream, while the eastern side prefers vanilla and chocolate.

Descriptive words you can use include: *clustered, even, uneven, highly distributed, north, south, east, west, increase, decrease* and *fluctuate*.

Q Quantify

When we quantify our pattern, we need to use numerical data to provide evidence of what we see.

You could obtain data by using the legend, by using the scale to measure, or by conducting a count. You need to ensure that the data you provide relates directly to the pattern you recorded earlier.

For example, we noticed that western USA prefers mint chocolate chip and OREO cookies and cream ice cream, while the eastern side prefers vanilla and chocolate. To quantify, we could count the number of states that like these ice-cream flavours. Six states in the west prefer mint chocolate chip and four prefer OREO cookies and cream.

E Exception

An exception is a trend on the map or graph that doesn't 'fit in' with our original pattern statement.

When you spot an exception on a map or graph, it is good to quantify it to provide a comparison to our original operations. For example, we noticed that western USA prefers mint chocolate chip and OREO cookies and cream ice cream, while the eastern side prefers vanilla and chocolate. To quantify this, six states on the west coast prefer mint chocolate chip and four prefer OREO cookies and cream. However, Wisconsin, which is on the east side of the USA, has also voted mint chocolate chip as its favourite flavour – which is an exception to our original pattern.

What is the difference between qualifying data and quantifying data?

PQE helps us describe patterns and distributions. When we describe, we 'say what we see'. In a PQE analysis, we do not explain or give a reason why we see patterns, this is done in a SHEEPT analysis (page 158).

To quantify means to use percentages, counts, ratios or data from the legend to provide more details about the different patterns you are writing about. To qualify a statement means you use words such as 'large', 'many', 'small scale', 'enormous', 'broad' or 'tiny' to describe a pattern or change.

By using quantifiable data, we can more easily see key differences between locations or even monitor change over time. Imagine that your PQE analysis stated: 'There are a lot of people who like ice-cream in the USA'. Does this sentence allow you to create a detailed visual of what is happening around the country? Or does this quantified statement provide more detail: 'Six states in the west prefer mint chocolate chip and four prefer OREO cookies and cream'?

How do I start my PQE sentences?

When writing a PQE analysis, start sentences with the following key terms:

Pattern: Overall …

For example:

Overall, the west side of the USA prefers mint chocolate chip and OREO cookies and cream ice cream, while the east side prefers vanilla and chocolate.

Quantify: To quantify …

For example:

To quantify, six states on the west coast prefer mint chocolate chip and four western states prefer OREO cookies and cream.

Exception: However …

For example:

However, the state of Wisconsin, which is on the east side of the USA, has also voted mint chocolate as their favourite flavour, which is an exception to the pattern.

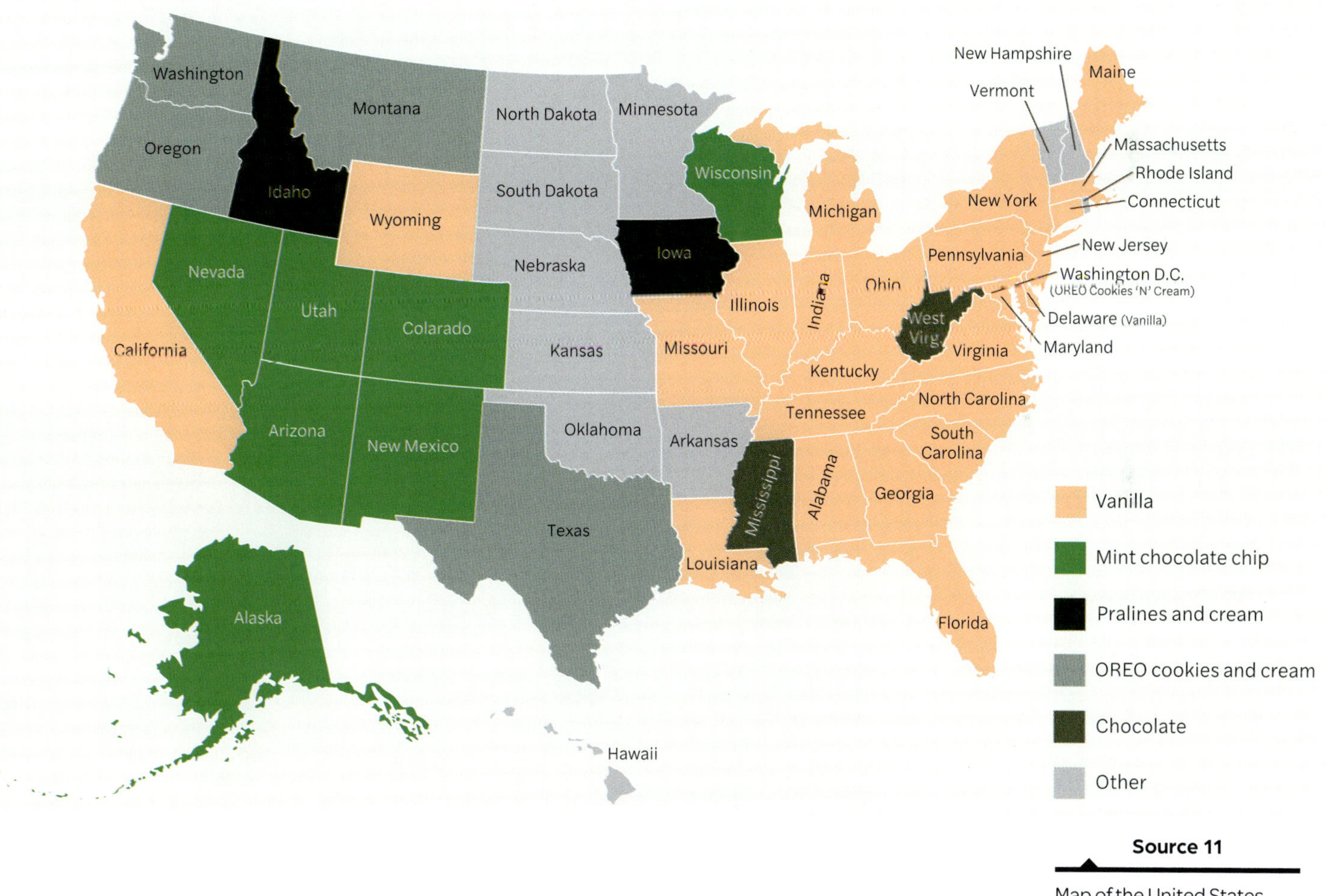

Source 11

Map of the United States showing favourite ice-cream flavours

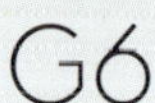

G6

HOW TO

SHEEPT

SHEEPT is an acronym that helps you remember the reasons why a spatial pattern occurs. This acronym stands for: Social, Historical, Economic, Environmental, Political and Technological.

S Social

Social factors are anything to do with people. Social factors include population, culture, language and religion.

H Historical

Historical factors are anything to do with our past. Historical events, buildings, people and changes to climate all influence what we see in our world today.

E Economic

Economic factors are those relating to money. In Geography, income, costs of things and how much money is spent can provide us with information about a *place*. Tourism is a huge economic factor for many regions.

E Environmental

Environmental factors are those relating to the natural or human environments on Earth. Humans can manipulate the environment to suit their needs.

How do I write a SHEEPT analysis?

SHEEPT is usually used to explain why patterns or distributions may be occurring in a particular region. It can also be used to expand our thinking when annotating images or considering new geographical content. When completing a SHEEPT analysis, you do not need to use each of the terms directly. The following is an example of how we can write a SHEEPT analysis for an image. The highlighted terms indicate the use of a SHEEPT term. Can you identify all the SHEEPT terms used?

Source 12 is an image of the Parthenon in Athens. This is a historical site that was once used as the city treasury in ancient Greece; however, today is it largely a tourist site. Tourism is important for Greece's economy and many locals are employed in the industry as tour guides or restaurant workers. Many tourists enjoy visiting Greece as it is well known for its warm, sunny weather; beautiful beaches; and rocky cliff faces. When tourists travel to historical or other sites in Greece, they are expected to follow local laws and customs. Due to the age of this historical site, it is constantly renovated by large machinery and monitored by officials to ensure tourists and locals are safe when exploring the grounds.

Source 12

The Parthenon in Athens attracts tourists wanting to find out about ancient Greek culture, history and democracy.

P Political

Political factors are those to do with the government or leading groups, and usually involve laws and policies.

T Technological

Technological factors relate to the different kinds of technology that we have access to. This could be in the form of gadgets, **spatial technology** – or even medical technology.

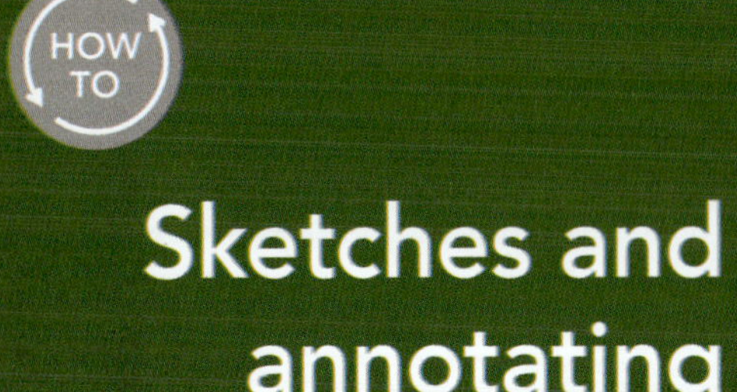

Sketches and annotating

Field sketches are an excellent way of recording data when you are investigating a research question.

Sketches allow you to annotate movement, patterns or any interconnections you see. Field sketching is not a test of your artistic skills – the idea is to record a simplified version of what you can see.

T Title 1

Provide a heading for your sketch that tells readers what it is and where it is located. You might wish to record both the absolute and relative locations.

O Orientation 

An orientation shows the direction that you were facing when you made your sketch. You need to use a compass so you can record a correct orientation.

A Annotations 

Annotations (or notes) are the most important thing to complete when drawing a field sketch.

Annotations allow you to record details about what you see, and explain how elements of your drawing relate back to the research question. Make sure that the lines to your annotations are made with a ruler – and that they do not overlap!

TROUPS CREEK WEST WETLAND
1 NARRE WARREN NORTH:
7.6.19 12.50pm 5

3 A path allows individuals to walk around the wetland and enjoy the outdoor environment

When you are making a field sketch, you need to remember TOASTIE!

TOASTIE will help you remember the key skills when making a field sketch.

It stands for Title, Orientation, Annotations, Scale, Time, Information and Edge.

An abundance of flora flourishes at the wetland

The retarding basin supports biodiversity and provides a scenic view.

6

A park bench allows individuals to sit and enjoy the outdoor environment.

S

2 E W

N

4

1m

7

Source 13

An annotated sketch

S Scale 4

Most sketches are not made to scale. However, all geographical maps and sketches require a scale to give the reader some indication of size.

To estimate a scale, use a ruler or pace out the area you have sketched. Then use a ruler to identify how large the same area is on your drawing.

For example, you might estimate that the path you are looking at is 1 m wide, and when you measure your drawing of the path it is 1 cm wide. So your rough estimated scale is 1 cm = 1 m.

T Time 5

By recording the time your sketch was completed, you can analyse how the environment changes over the course of a day or a month – or even years!

I Information

The annotations on your sketch need to be longer than one word. Annotations should be at least one sentence, and they should inform the reader and help them to identify patterns.

E Edge

Always include a border so it is clear where your sketch starts and ends.

Visual communication

Block diagrams

Block diagrams are a useful way of drawing landscapes in order to show what is happening above ground and below ground.

For instance, when we consider the movement of water, block diagrams can help us visualise (or picture) how water can move on the surface of Earth through rivers, oceans and streams – and also how it can move under the ground through water tables and pipes.

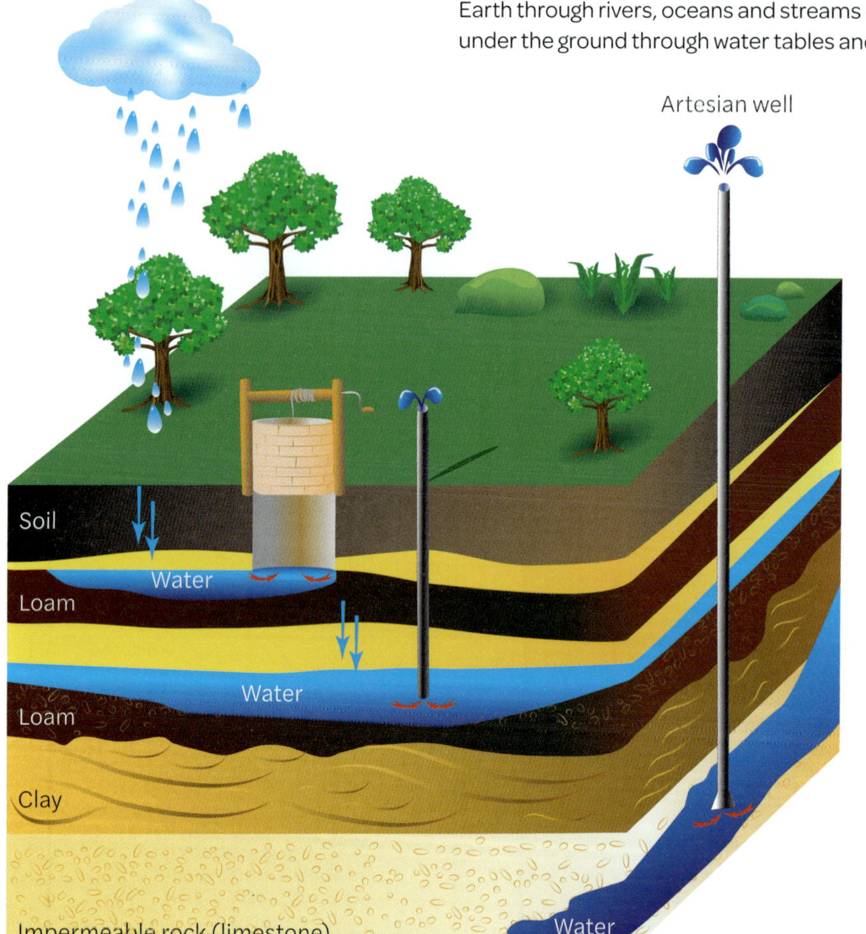

When you create a block diagram, make sure you draw a three-dimensional (3D) object that has an equal amount of drawing room both above and below the surface of Earth.

Annotations are important on a block diagram, as they help your audience to interpret your drawing.

Source 14

A block diagram

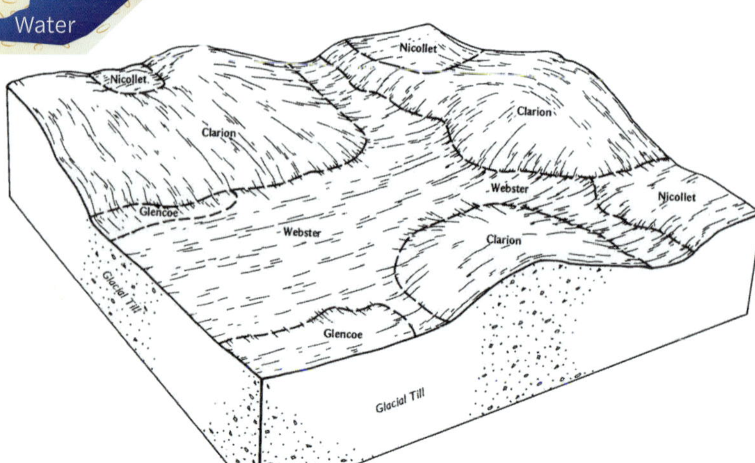

Source 15

Sketch of a block diagram

Flow diagrams

Flow diagrams are helpful to show processes, or to show how different parts of the environment are *interconnected*. Arrows show the movement between stages, or show how different items are connected.

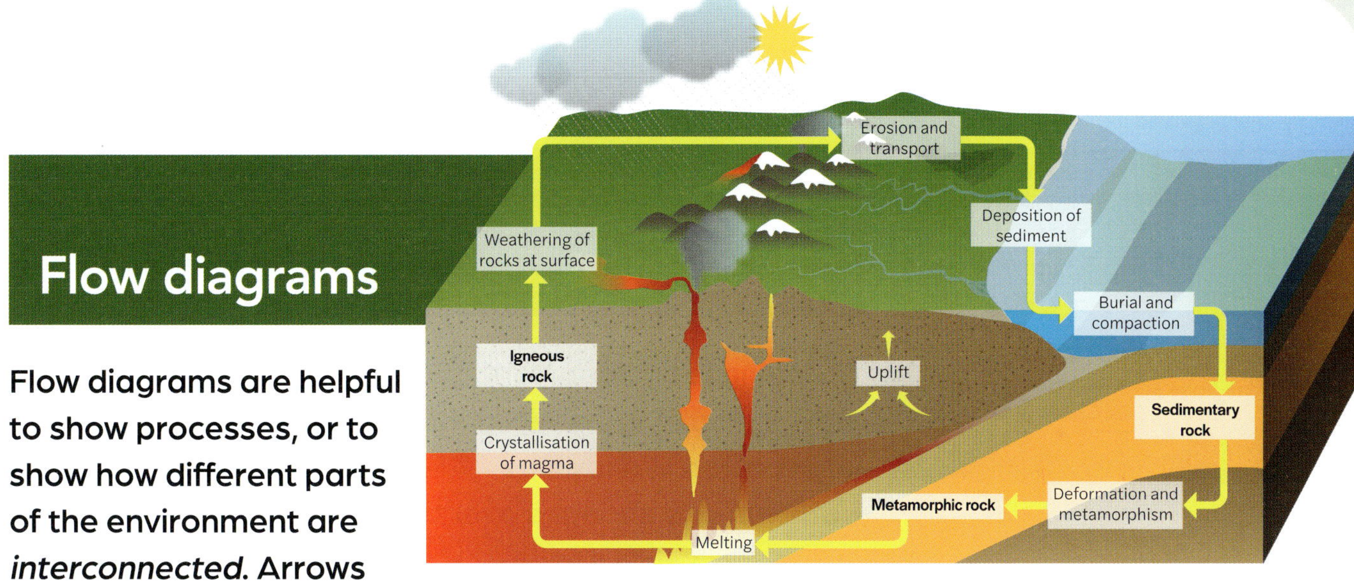

Source 16

A flow diagram

Photo essays

A photo essay is a way of presenting information visually. It is ideal for showing characteristics of a place or a process.

A photo essay usually includes a series of photos with specific annotations or captions. The captions provide brief background information about the key features of the image, or the meaning behind the selected image.

Source 17

A photo essay

Preparing for the flood with sandbags

Heavy rainfall could create a problem

Floodwaters are rising

The massive clean-up

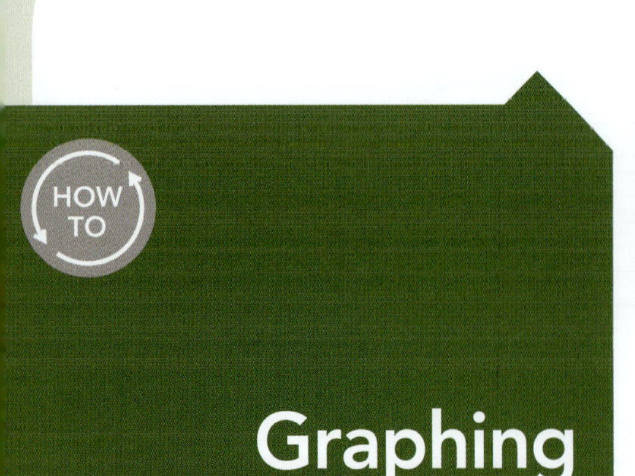

Graphing

Simple graphs

Graphing is an important way of displaying geographic information. It shows us patterns and changes over time.

When you are creating a graph, the word to remember is SALTS! SALTS is an acronym based on the first letters of four key things about graphs: Scale, Axis, Legend, Title and Source.

S Scale

The scale of your graph will depend on the data you are trying to visualise. To find out the axis scale you need for a graph, first find out the 'range' of the numbers, which means finding the lowest value and the highest value, and then fill in the numbers in between so you can mark your data points easily.

A Axis

Each graph has an *x*-axis and a *y*-axis. The *x*-axis is the horizontal axis and the *y*-axis is the vertical axis. Make sure you label each axis!

L Legend

A graph often uses colours to represent data. A legend shows the audience what these colours mean and how to read the data.

T Title

A title lets your audience know what your graph is showing.

S Source

When you graph information, it is important to acknowledge where you gained the data from. Maybe you collected it yourself or maybe you sourced it from a website. By stating the source of your information, the reader knows how reliable it is.

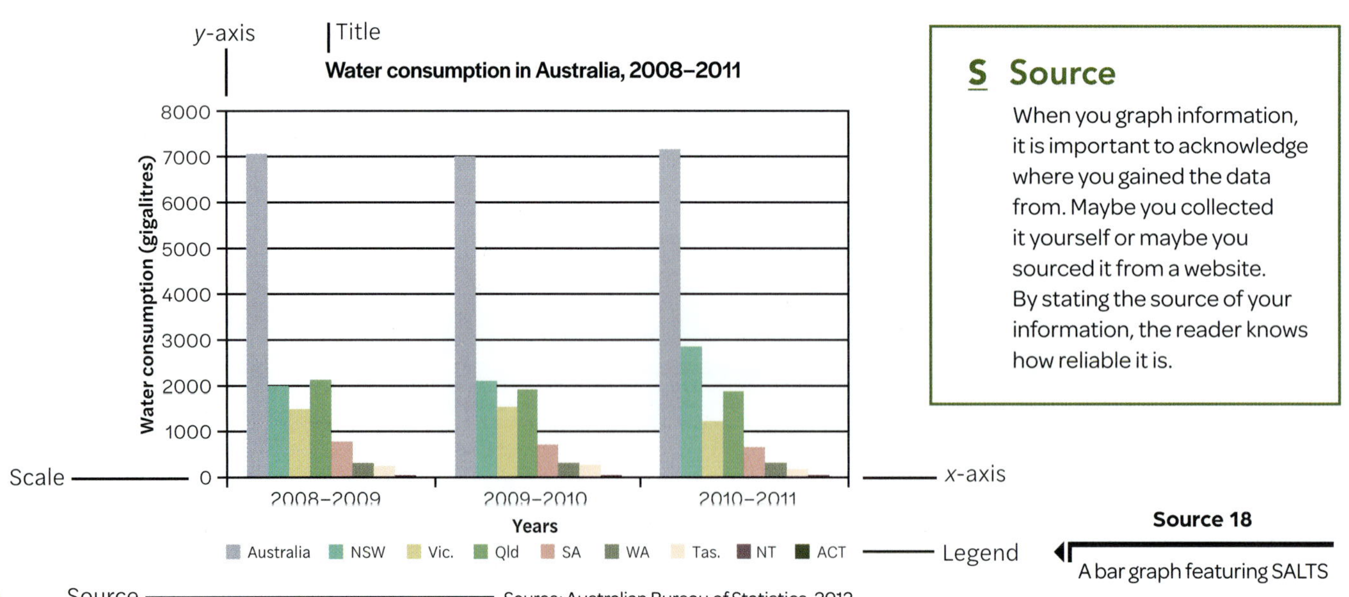

Source 18

A bar graph featuring SALTS

Climate graphs

Put very simply, a climate graph is two graphs in one. It shows two types of data on one graph. Climate is the average temperature and level of precipitation (rainfall, snow etc.) that a region usually receives over a year.

How do I read climate graphs?

This climate graph shows:

- the temperature (in degrees Celsius) on the left *y*-axis
- the amount of precipitation (in millimetres) on the right *y*-axis
- the months of the year on the horizontal axis, or *x*-axis
- the red line (or maximum temperature) varying throughout the year, peaking December to March and decreasing from April to July
- the amount of precipitation varying, shown with the blue bars. It tends to be highest in spring from August to November – with the exception of another peak in May.

The best way to describe the patterns we can see in a climate graph is by using two PQE analyses – one for temperature and one for precipitation.

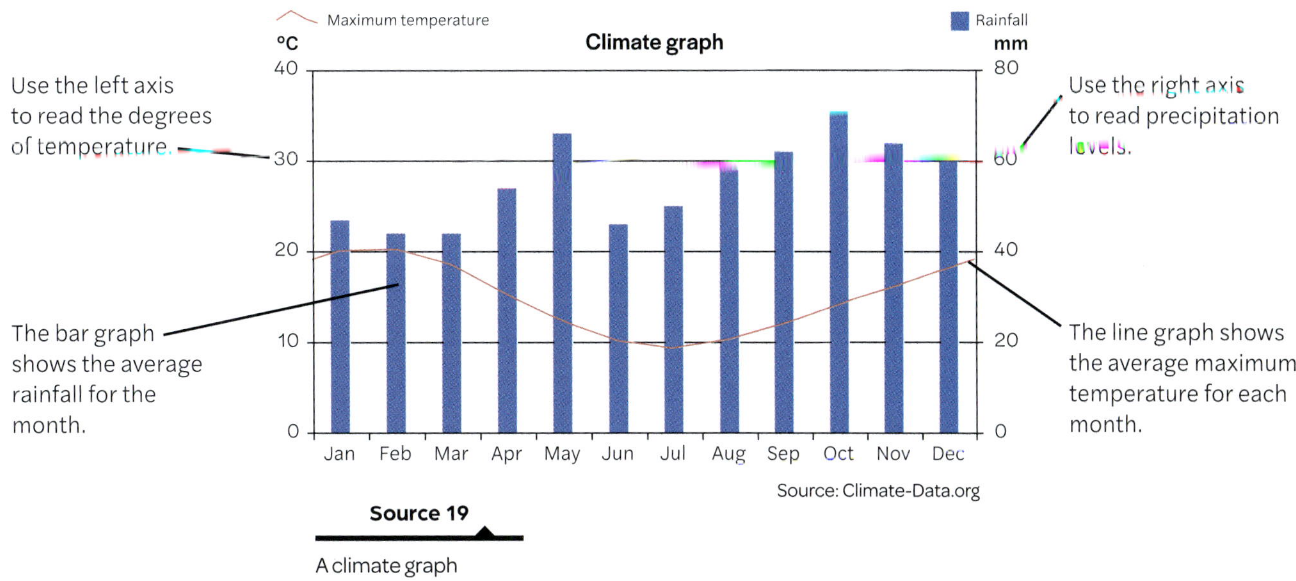

Source 19

A climate graph

What is the difference between weather and climate?

Look outside your window and describe today's temperature and the amount of rainfall there has been. Is it cloudy? Has there been a thunderstorm or is it beautiful sunshine? What you have just described is the weather. Weather changes daily and is usually predictable up to about 10 days in advance. Apps on your smartphone and the last five minutes of the evening news show you these predictions of daily weather.

Now close your eyes and describe the climate of Australia. Do you imagine Australia being hot and dry? Just because we can describe Australia as hot and dry, does not mean it is like this everywhere, all year round. Unlike the weather, climate helps us describe the yearly (annual) average temperature and level of precipitation (rainfall, snow etc.) in a region or country. Climate graphs help us visualise the climate of a region or country. Climate change describes how the average temperature and precipitation levels of a location have been altered over time.

Population pyramids

Population pyramids are graphs that show the number of females and males in particular age groups in a population, and they are like bar graphs turned on one side.

Population pyramids can be made on a local, national or global scale. On a population pyramid, female data is normally shown on the right side and male data on the left. The length of each bar represents the number of males or females within that age group.

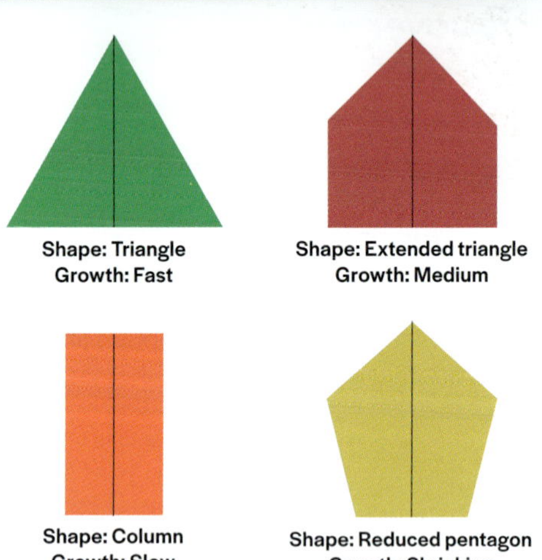

Source 20

Various population pyramids. The shape of the pyramid tells you about the population.

The shape of the pyramid tells us a lot about the population

Population pyramids come in different shapes: triangles, columns, pentagons and so on. The shape tells you about the population. If the shape is:

- a triangle: growing population, with more young people than old people
- a box: slow or stable growth, with equal numbers of old and young people
- an upside-down pyramid: aging or declining population, with more old people than young people.

How do I interpret a population pyramid?

The population pyramids in Source 21 show Australia's population in 1955 and 2018.

In 1955, there were lots of young people in the age range 0–14. Approximately 5.5 per cent of the population were males aged 0–4 years old, and 5.3 per cent were females aged 0–4 years old. We also had many working-age people. For example, 4.1 per cent of the population were males aged 30–34, and 3.8 per cent were females aged 30–34.

In 2018, there are significantly fewer young people in our population and more old people. The total population also grew to 25 million people.

Source 21

Population pyramids showing Australia's population in 1955 (left) and 2018 (right)

Australia's population in 1955 and 2018

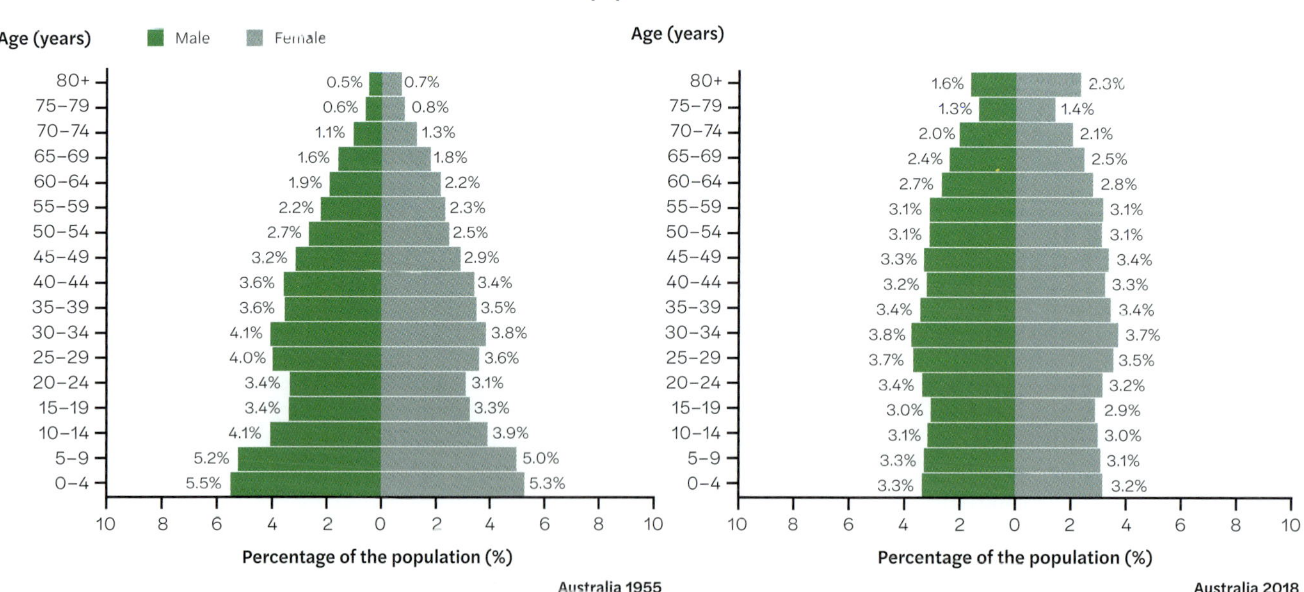

Source: PopulationPyramid.net

Satellite images

Satellite images are pictures of Earth taken from space.

Satellites orbit the Earth and take photos constantly. We can use this data to see a change in land cover or other spatial patterns over time. Satellite images taken during the day give us the most information.

Source 22

Satellite images taken at night provide different information than those taken during the day. This photo of Italy shows the extent of the urban areas.

How can I use colour to interpret satellite images?

When you are trying to interpret a satellite image, look for colours and shapes. Colour will normally give you some idea about land cover. For example, green usually indicates vegetation, blue is the colour of water and brown is usually desert or barren land. White can indicate snow or clouds, so here you need look at shapes. Clouds tend to look fluffy and, depending on the time of day, can cast a shadow onto the ground. Snow is usually on the tops of mountains, and you can see where it is melting or following the slopes. Looking for shapes is also helpful when you are trying to identify rivers or reservoirs, which can be seen as blue lines.

How can satellite images show change over time?

Seasonal variations can sometimes be very clear in satellite images – in spring and summer, green vegetation flourishes, while during winter, white snow may fall or vegetation may die, leaving brown, bare ground.

Source 23

This satellite image was taken at the same place over four seasons. Satellite images are useful to show change over time.

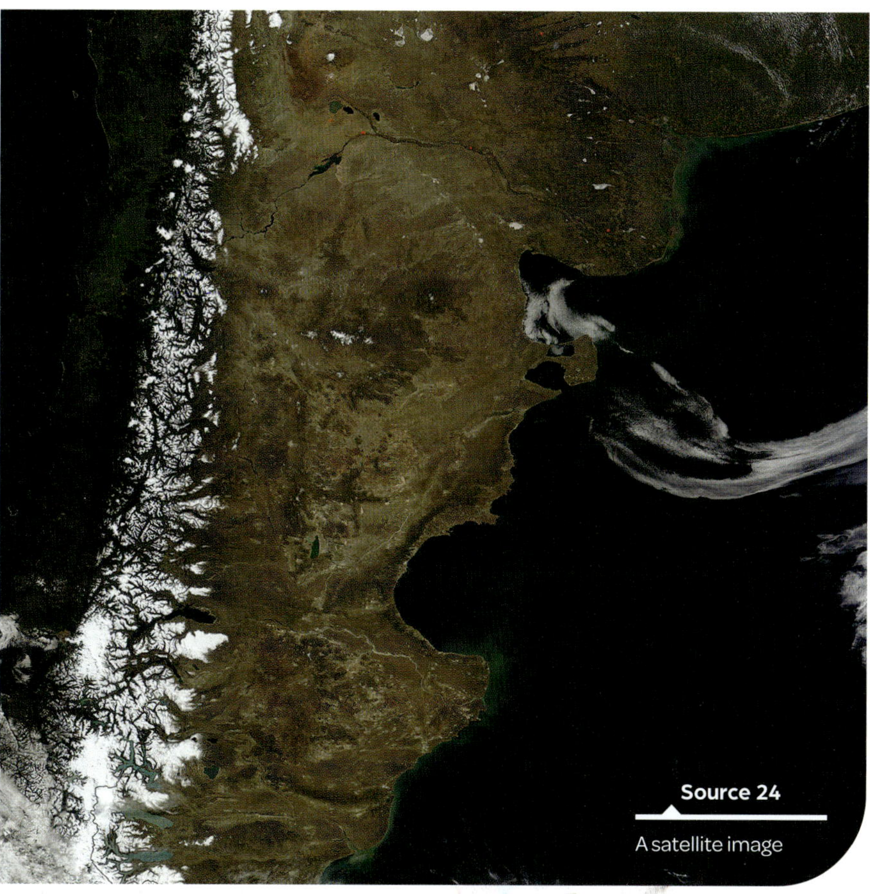

Source 24

A satellite image

Glossary

abrasion the process of scraping or wearing something away

aesthetics visual beauty

aid agency an organisation that provides assistance to communities and individuals who need it; e.g. World Vision

air pollution the release of harmful particles or materials in the atmosphere

air mass large body of air that is all the same temperature and humidity, and at the same level of pressure

aquifer underground layer of water-bearing rock where groundwater can be extracted

arid receiving little or no rain

atheist person who has no religion or belief in any god

benefit positive outcome

block diagram three-dimensional diagram that shows how a system or feature works

bore a deep well drilled into the ground to tap into underground water storages

carrying capacity the amount of material that can be held in a contained area such as a river

catchment an area where water is collected by the natural landscape

climate the average weather in a certain place, over a period of time

climate change a change in climate patterns, often refers to an increase in global temperatures apparent from the late 20th century onwards. Also referred to as global warming.

climate graph a graph showing average rainfall and temperature for an area over a period of time

community a group of people connected in some way and with a distinct identity

community of place a community based on a connection to a specific place or region

condensation a process by which a gas (for example, water vapour) is changed to a liquid (for example, water drops)

Connection to Country the deep spiritual, physical, social and cultural relationship between Indigenous Australians and the land

constitution a document stating the fundamental principles by which a nation, state or organisation is governed

continent one of Earth's seven main continuous expanses of land; e.g. Asia

contour lines lines on a map that join points of the same height above or below sea level

convectional rainfall rain that occurs when land heats the air above it, forcing the air to rise

cost resources that are consumed to achieve an outcome

cumulonimbus cloud dense, vertical cloud, formed from water vapour carried by powerful upward air currents; associated with heavy rain and storms

dam barrier constructed to hold back water

decentralisation movement of jobs, people and resources out of large cities and into regional areas

delta triangular region of sediment at the mouth of a river

density amount of material or number of objects in a specific area

deposit drop or place in an area, such as the deposition of sediment by a river

deposition formation over time of a layer of soil or rock

desalination process of removing salt from seawater

desert area that receives less than 250 millimetres of rain every year

developed wealthy country with advanced industrial and economic infrastructure

dew point temperature below which water begins to condense and dew (tiny droplets of water) is formed

drainage basin area of land where precipitation collects and drains into a river, bay, or other body of water

drought long period of very low rainfall, leading to a shortage of water

economic issue or problem problem involving the allocation or consumption of resources

ecosystem a community of plants, animals and other organisms in their physical environment

environmental resource any material from the environment that is useful or valuable to society; these can be renewable, non-renewable or perpetual resources

equator a line separating the Earth halfway between the the poles, dividing the Earth into northern and southern hemispheres

erosion removal and relocation of soil or rock by water, wind or ice

evaporation process by which a liquid (for example, water) is changed to a gas (for example, water vapour)

flood large amounts of water overflowing onto dry land

floodplain area of ground next to a river that floods when the river waters rise

flow-on effect indirect effect caused by the outcome of other actions

fossil fuel fuel made from the decomposed remains of things that lived millions of years ago (for example, coal, oil)

freedom power or right to act, speak, or think as one wants within the boundaries of the law

fresh water any naturally occurring water that is not salt water, including water in ice sheets, ice caps, glaciers, icebergs, ponds, lakes, rivers, streams, and groundwater

frontal rainfall rain that occurs when a mass of cold air pushes up warm air

frostbite damage to body tissues caused by exposure to extreme cold, typically affecting the nose, fingers, or toes

geographic characteristic naturally occurring feature of a place, such as its landforms and ecosystems

geothermal energy thermal energy generated and stored in the earth

glacier mass or river of ice that slowly moves down a mountain or valley, carving a u-shaped valley as it goes

gorge narrow valley between hills or mountains, usually with steep rocky walls and a stream running through it

GPS Global Positioning System – a satellite navigation system used to determine the position of an object or person on the ground

green belt undeveloped areas around cities set aside for open space

green wedge undeveloped areas within cities that have been set aside for open space

groundwater water that is located under the earth's surface in rock or soil

headwater source of a river or stream

high-income country country with an average annual income of US$12 376 or more per person

humanitarian organisation organisation that provides aid or assistance to people in need, for example, Red Cross

hydroelectricity electricity that is produced by water-powered generators

hypothermia life-threatening medical condition resulting from having an abnormally low body temperature

hypothesis proposed explanation used as the starting point for further investigation

ice cap covering of ice over the ground in a large area, especially in the polar regions

ice sheet layer of ice covering a large area of land for a long period of time

ice shelf floating platform of thick ice attached to a land mass

iceberg large floating mass of ice that has detached (or calved) from a glacier or ice sheet and is floating in open ocean

infant mortality death of children who are one year old or younger

infiltration process where water (rain, hail or snow) soaks into the soil

infrastructure physical and organisational requirement needed for the operation of a society, for examples, roads, water supplies

interconnections relationship between all living things, non-living things and processes; a key concept in geography

internally displaced person person forced to flee their home and shelter elsewhere within their country's borders

irrigation watering of crops and pastures from a water source other than precipitation (for example, with water from a river or lake)

latitude a measurement in terms of something's distance north or south of the Earth's equator

less economically developed country (LEDC) low-income country experiencing severe barriers to development; also referred to as developing countries

levee a wall built next to a river to stop it from overflowing

life expectancy the average number of years a person is predicted to live

liveability measure of how easy it is to live in a particular place based on factors such as climate, safety, healthcare, schools, employment and transport

liveable easy or pleasant to live in

lock a device used for raising and lowering boats between different levels on waterways, such as a weir

low-income country country with an average annual income of US$1045 or less per person

malnutrition lack of proper nutrition from one's diet, leading to health problems

meander bend or curve in a river or road

middle-income country country with an average annual income of US$1046–$12 375 per person

more economically developed country (MEDC) a country with a strong economy, in which most people have access to good education, health care, and employment opportunities

monsoon seasonal wind, occurring in the Asia and Pacific regions, that brings large amounts of rain

multi-faith embracing and respecting multiple religious faiths

multicultural embracing and respecting multiple cultures and backgrounds

natural hazard natural process, such as a landslide, which could cause damage, injury or loss of life

natural disaster natural event, such as a landslide, which causes widespread damage and loss of life

need items or services required for survival

nomadic choosing to move from place to place over time to find better conditions, rather than living in one location

non-renewable resource resource that cannot be regenerated or replenished within a human timescale once it is used up (for example, coal, oil)

oasis place in a desert where groundwater has come to the surface creating a green or fertile area

orographic rainfall rain that occurs when warm, moist air is lifted over a mountain

paddy field planted with rice that grows in water

pandemic a disease that spreads in multiple countries around the world at the same time, usually affecting a large number of people

per capita for each person

perpetual resource natural resource with an unlimited supply (for example, wind, sun)

phenomenon something that is observed to exist or happen

photo essay series of photographs used to present information visually

plunge pool deep basin at the foot of a waterfall created by the action of the falling water

polar the cold environments around the North and South Poles

population distribution pattern showing the number of individuals living in an area

population density measure of how many people live in a particular region or area

population pyramid graph showing the number of males and females living in age groups in a population

PQE acronym for Pattern, Quantify and Exceptions, used to describe spatial patterns or graphs

precipitation water that falls from the atmosphere to the ground as rain, snow, hail or sleet

primary methodology data-gathering activity undertaken in the field, such as field sketches

qualitative data non-numerical data based on qualities or characteristics

quantitative data numerical data based on measurements or counts

rain gauge device that measures the amount of rain that falls in an area

recycle convert into reusable materials

refugee a person threatened by conflict or a natural disaster who is forced to flee their home to seek safety, food and shelter in another country

renewable energy energy generated from a renewable resource such as sun, wind or water

renewable resource a natural resource that can regenerate or replenish within a human timescale (for example, trees)

research question idea to be investigated, or problem to be solved through research

reservoir human-made or natural water storage area that collects and retains water

reverse osmosis process in which a solvent is pushed through a membrane into another solution

run-off water that flows over the surface of the land and into rivers and lakes

rural–urban fringe region on the edges of cities that is close to rural areas or towns

safe water water that is safe to drink and free from contamination

salt pan shallow region where salt water has evaporated, leaving a layer of salt on the ground

SALTS acronym for Scale, Axis, Legend, Title and Source

satellite image image of Earth taken from space

saturated zone area deep below the soil where all the spaces between soil and rock particles are filled with groundwater

sea change decision to move to a coastal location for a better lifestyle

sea level average level of the sea, at a given point

secondary methodology data-gathering activity undertaken outside of field studies, such as research

secular something that is not connected to any religion or spiritual matter

sedimentary rock formed over a long period by sediment deposits

semi-desert region that is dry but receives more regular rainfall than a desert

SHEEPT acronym for Social, Historical, Environmental, Economic, Political and Technological factors

sketch simple drawing made to record data when in the field

social enterprise companies whose objectives are to benefit the community as well as to make a profit, where surplus income is are reinvested into social projects

spatial technology computer systems that interact with real-world locations in some way

spring groundwater from an aquifer that makes its way to the earth's surface

stream gauging station location where the speed, height or volume of a body of water can be measured

tectonic plate massive slab of solid rock that moves extremely slowly across the surface of the Earth

terrace flat area on the side of a hill, used for agriculture

TOASTIE acronym for Title, Orientation, Annotations, Scale, Time, Information and Edge

transpiration water vapour leaving plants through their leaves

tree change decision to move to a rural location for a better lifestyle

tropical used to describe the climate in the areas of the world between the Tropic of Cancer and the Tropic of Capricorn

turbine machine for producing power where a wheel or rotor, usually fitted with vanes or blades, is made to revolve by a fast-moving flow of water, steam or other fluid

urban based in or relating to a city or large town

urbanised living in or based in cities

valley low area of land between hills or mountains

virtual water total amount of water needed to produce a crop or product

want items or services that are useful but not required for survival

water balance the flow of water in and out of a system

water cycle the natural process by which water circulates between the earth's oceans, atmosphere, and land, involving precipitation, drainage in streams and rivers, and return to the atmosphere by evaporation and transpiration

water poor lacking in natural water resources

water resource natural body of water that can be used for different purposes

water rich having a large amount of natural water resources

water scarcity not having enough natural water to meet the needs of a community or region

water table the level below the earth's surface where water is found

water vapour water in its gas state; when liquid water evaporates, it becomes water vapour

waterfall cascade of water falling from a height

weir A small dam used as a water supply

Index

M

N

O

P

W

Y

Acknowledgements

The author and publisher are grateful to the following for permission to reproduce copyright material:

PHOTOGRAPHS: AAP/Reuters/Amit Dave, **14–15,** /Reuters Graphic, **55,** /Reuters/Siegfried Modola, **119**, /Thiess, **72**; agefotostock /Mint Frans Lanting, **68** (left), /Brett Price, **136**; Alamy /Andrew Aitchison, **135**, /Muhmmad Al-Najjar/SOPA/ ZUMA Wire, **131** (bottom), /Dinodia Photos, **141**, /Oskar Eyb/ imageBROKER, **147**, /Jean Paul Ferrero/AUSCAPE, **92** (left), /Khoon Lay Gan, **153** (top), /Tom Grundy, **54**, /ton koene, **117** (top), /Makhnach_S ,**154** (bottom), /Keith Morris, **149**, /Hugh Peterswald, **101** (bottom), /Planet Observer/UIG, **86**, /Suretha Rous, **58–9**, /81281/RooM the Agency, **63**, /Scattolin/Premium Stock Photography GmbH, **133**, /Alys Tomlinson, **101** (top), /Raul Touzon/National Geographic Image Collection, **26** (bottom), /Kristoffer Tripplaar, **148** (top), /Penny Tweedie, **i** (centre left), **83** (top), /Genevieve Vallee, **92** (right), /David Wall, **104**; Fairfax /Maps, *The Age*, **108**, crime statistics map, **120**, /Joe Armao, **102**, /Pat Scala, **99**; Getty Images /ZEIN AL-RIFAI/AFP, **123**, /Amer Almohibany/AFP, **111**, /Jon Arnold Images, **117** (bottom), /Auscape/UIG, **82**, /Andrew Beckett, **16,** /Doug Byrnes/Corbis Documentary, **97**, /Claver Carroll, **85** (top right), /MelindaChan, **4–5** /Russell Charters, **105** (top), /Steve Daggar Photography, **29**, /Sean Davey, **55**, /Mark Evans, **52–3**, /ALBERT GONZALEZ FARRAN/AFP, **127**, /Michael Fay/National Geographic Image Collection, **13**, /Christopher Groenhout/Lonely Planet Images, **96**, /Bartosz Hadyniak**, i** (centre right), **45**, /Mohammad Ponir Hossain/NurPhoto, **134–5**, /SAJJAD HUSSAIN/AFP, 124, /Ulet Ifansasti, **121**, /Dorling Kindersley, **16–17, 17, 27, 32–3, 36–7**, /Barry Kusuma, **103**, /Sharon Lapkin, **79**, /Natphotos, **1**, /NurPhoto, **138–9**, /Anton Petrus, **178**, /AFP/Pool, **40–1**, /Philip Quirk, **i** (background image), /Gokhan Sahin, **i** (centre), **122–3**, /sarawut**, i** (top right), **11**, /Chan Srithaweeporn, **i** (bottom right), **151**; /Puneet Vikram Singh, Nature and Concept photographer, **113**, /Dima Viunnyk, **i** (bottom left), **143**; iStock /agustavop, **171–6**, /Beboy_ltd, **145** (bottom), /BeyondImages ,**91** (top), /Michael Burrell, **145** (top), /JohnCarnemolla, **85** (top left), /chinaface, **85** (bottom), /calvindexter, **155** (top), /deimagine, **142**, /drogatnev, **12**, /Elvetica, **74**, /franckreporter, **155** (bottom), /georgeclerk, **62–3**, /Brett Ginsberg, **81**, /Bartosz Hadyniak, **15**, /Jaykayl, **163** (bottom left), /ZernLiew, **73**, /leonello, **154** (top), /LUHUANFENG, **152**, /lushik, **61**, /Magone, **61** (top right), /malerapaso, **30** (top), /ShaniMiller, **168–70**, /Milkos, **60**, /mrinalnag, **69**, /Minerva Studio, **23**, /pixdeluxe, **177**, /posteriori, **161** (top), /Phonix_a, **163** (top left), /Rakdee, **68** (right), /RapidEye, **4,** /Raylipscombe, **i** (top left), **2–3**, /SandraRose, **125**, /Saturated, **150**, /Scharfsinn86, **59**, /Sebalos, **164,** /sergiykelya, **163** (top right), /tsvibrav, **28–9**, **109**, /ttsz, **162** (top), /Vectorios2016, **157**, /WildLivingArts **60–1**, /AlexanderYershov, **10**, **44**, **78**, **110**, **142** (arch icon), /yuelan **158** (bottom); NASA, **34–5**, **39**,/Jeff Schmaltz LANCE/EOSDIS MODIS Rapid Response Team, GSFC, **167** (bottom right, left), /Reid Wiseman, **167** (top); National Geographic/Lynn Johnson, **64–5**, **66–67**; Nearmap, **93**; Newspix/Yuri Kouzmin, **100**; Danielle O'Leary, **153** (bottom); OzAerial/ Daryl Jones, **105** (bottom), **106**, **106–7**; Planet Labs Inc., **58** (top & bottom); Science Photo Library/Carlos Clarivan, **20–1**; Shutterstock/ AJP, **43**, /BlueOrange Studio, **128**, /Boris-B, **148** (bottom), /Amos Chapple, **24**, **24–5**, /creativemarc, **131** (top), /Antun Hirsman, **67**, /Eric Isselee, **158** (top), /kwest, **35** (top), /Rainer Lesniewski, **7**, /priyank05, **33** (top), /Samib123, **146**, /Anton Shahrai, **115**, /shopplaywood, **75**, /Fabio tomat, **33** (bottom), /UnknownLatitude Images, **110**, /Maythee Voran, **163** (bottom right); © The State of Queensland, Dept Natural Resources, Mines and Energy 2004 (CC BY 4.0) **90–1**; 2017-18 Mallee Region Photography © State of Victoria, Department of Land, Water and Planning, **94–5**; Thankyou Group, **30–1**, **30**, © UNICEF/UN055930/Gilbertson VII Photo, **48–9**; Wikimedia Commons/ SeanMack (CC BY 2.5), **84**, /Mattinbgn (CC BY 3.0), **95** (top).

OTHER MATERIAL: Statistics on map, Australian Bureau of Statistics, POPULATION DENSITY BY SA2, Australia - June 2018, map (CC BY 4.0), **84–5**; Extract, John 'Dudu' Nangkiriyn, quoted in *Climate change, water and Indigenous knowledge: A Community Guide to the Native Title Report 2008*, © Australian Human Rights Commission 2019 (CC BY), **68;** Bureau of Meteorology, © 2020 Commonwealth of Australia, /graphic, **52**, /map and graph (CC BY 3.0), **53**, /data on map (CC BY 3.0), **87** (top right); Graphic, © DiscoveringAntarctica BAS/RGS-IBG, www.discoveringantarctica.org.uk, **26** (top); Melbourne flood elevations map data, Floodmap.net, **42**; Cartoon, Mike Gruhn, CartoonStock.com, **144**; Illustration, Hydro Tasmania – Gordon Power Scheme, **28**; Maps, © Institute for Economics and Peace: Global Peace Index Report 2019**, 118–19**, **140**; Map data, liveaumap.com, 10 April 2019, **122**; Artist impression courtesy

Lyons Architects, **136–7**; Graph data, Melbourne Water Outlook 2018, **70**; Map, Murray–Darling Basin Authority (CC BY 4.0)**, 34**; Extract, Tina Rosenberg, 'Our Thirsty World', National Geographic, April 2010**, 64–5**; ourworldindata.org (CC BY 4.0)/ data from Gapminder, **127/**data from 'World Happiness Report 2019', **128**/ maps and graph, adapted from 'Coronavirus Disease (Covid-19) - Statistics and Research' by Max Roser, Hannah Ritchie and Esteban Ortiz-Ospina, 2019, ourworldindata.org and European CDC, **138**, **139;** Graphs adapted from, populationpyramid.net, **166**; Erin Schubert **160**, **160–1**; Map, SGS Economics & Planning, **98**; Graph data, © State of Victoria (Department of Environment, Land, Water and Planning) (CC BY 4.0), **73**; Daniel Flynn quote, Thankyou Group, **30**; Sketch, U.S. Department of Agriculture, **162** (bottom); Map, Vienna City Administration Municipal Department 18 Urban Development and Planning, **132–3**; Diagram, Timm Kekeritz, www.virtualwater.eu/, **60–1** (bottom); World hunger map, © World Food Programme 2019, **126**.

FLIP BOOK

good humanities

Flip your book over and begin your journey through **History**

Source 13

A painting depicting Egyptian workers harvesting wheat [Wall painting from tomb of Menna, Thebes.]

3 Notes not needed:

- Egypt copied military technology from others
- Egypt had a large army of charioteers
- Egypt was sometimes behind in military technology

4 You could then put your notes into paragraphs, either:

- Trade: for and against it being influential
- Conflict: for and against it being influential

Or ...

- What was influential: trade aspects and conflict aspects
- What was not that influential: trade aspects and conflict aspects

step 4 I can structure a piece of writing

History essays should have an introduction, several body paragraphs and a conclusion.

For those students who are starting out writing essays, a paragraph structure that is easy to learn is TEEL. TEEL is an acronym for:

- Topic sentence
- Evidence
- Explanation
- Link.

These words are explained below. Every paragraph should have TEEL.

Introduction

The introduction should:

- show you understand what the question is asking
- say your overall response to the question
- introduce the main points.

Paragraphs using TEEL

Paragraphs using **TEEL** should have the following:

- **T**opic sentence: one sentence that summaries the whole paragraph
- **E**vidence: use *specific* examples, not general examples
- **E**xplanation: how evidence supports your claim
- **L**ink: at the end of the whole paragraph, link the main point in the paragraph back to the main question.

Conclusion

In your conclusion, make sure to:

- summarise your main points
- restate your response to the question.

step 5 I can write a draft

Drafting tips

- Focus on answering the question; don't just write everything you know about the subject.
- Don't worry about making mistakes when drafting – you will fix this later.
- Don't worry too much about punctuation, grammar, spelling or vocabulary when drafting.
- Start with the paragraphs. Draft the introduction and conclusion last.
- If you can, use a computer, as it makes it easier to edit and proofread your work later.
- Write the first draft quickly. Then edit and proofread slowly.

Sentence starters

Here are some sentence starters for introductions:

- This essay will discuss ...
- In this essay ...
- The issue being focused on is ...
- The issue that will be focused on is ...

Other words you could use in sentence starters in place of *focused* are:

- described
- analysed
- evaluated
- explained
- explored
- justified
- outlined.

For conclusions:

- In conclusion, ...
- In summary, ...
- It has been shown / demonstrated that ...
- Therefore / Thus / Hence, ...
- To summarise, ...

For comparing: (when they are the same)

- By comparison, ...
- In the same way, ...
- Likewise, ...
- Similarly, ...

For comparing: (when they are different)

- However, ...
- In contrast, ...
- On the other hand, ...
- Then again, ...

For adding more:

- Additionally, ...
- Also, ...
- Firstly, ... Secondly, ... Thirdly, ... Finally, ...
- Furthermore, ...
- In addition, ...
- Moreover, ...

For giving examples:

- For example, ...
- For instance, ...
- An illustration of this is ...
- As an example, ...
- ... such as ...

For showing effects:

- As a result, ...
- For this reason ...
- It can be seen that ...
- The evidence suggests ...
- The result of this is that ...
- These factors contribute to ...
- Different ways to say 'caused':
  - resulted in
  - created
  - lead to
  - determined
  - is attributed to
  - meant that
  - is dependent on
  - forced
  - made.

I can edit and proofread

The point of writing is communication. This involves using words to pass ideas from your head into the head of another person, so keep your writing clear and simple.

Editing is checking for meaning, to make sure the text answers the question and meets the requirements of the task. For example, does your writing need a bibliography? Does it need labelled pictures? Proofreading means checking the grammar, spelling and punctuation of your work.

Edit first, then proofread.

Editing tips

Following are some editing tips:

- Always use headings, unless you have been told not to.
- Delete any words, sentences or paragraphs that do not help the piece of writing overall.
- Check what you have written against the requirements of the task. Ask yourself:
  - does my writing answer the question asked?
  - is it clear that I have answered the question?
  - is there an assessment schedule or rubric my writing will be marked against? Mark yourself against these criteria. Is there time to improve at least one aspect of what you've written?
- What is your worst paragraph? Why? What would it take to make it your best paragraph? Do that!

Proofreading tips

Proofreading is going back over your finished work looking for errors.

- Don't try to fix every different type of problem at once. Pick one thing to correct each time you proofread. For example, first look at spelling, then look at punctuation, then look at confused words, then look at making your vocabulary more interesting.
- Read your work aloud. Even better, record it, then play it back to yourself a bit later.
- Ask someone else to read your work aloud.
- Read the sentences backwards to check for spelling. Otherwise your brain will correct any spelling mistakes automatically as you read and you won't notice as many.
- If you know you are a bad speller, don't trust your instincts. Check words you are unsure about in the dictionary.
- Spellcheckers will not fully correct your spelling. A word can be spelt correctly but still be the wrong word, so don't rely on a spellchecker!
- Print out your essay and read a printed copy of it, rather than reading on screen.

Common errors

Following is a list of common errors:

- only use apostrophes for shortening words and ownership, not for plurals
- 'I done' should be 'I did'
- write shorter sentences, preferably less than 25 words
- only use capital letters at the start of sentences and for proper nouns
- confusing 'your' and 'you're'
- confusing 'there', 'their' and 'they're'
- writing formally: don't use 'I' (unless told to), '&', 'etc.', 'e.g.', 'i.e.', 'wanna', 'heaps', 'stuff'
- could of / would of / should of are incorrect; replace with could have / would have / should have
- confusing 'to/ too/ two'
- confusing 'much/many': much = for an item that can't be counted (e.g. water), many = individual items that can be counted
- confusing 'then' and 'than': then = something happening after something else; than = comparing
- subject–verb agreement. If the subject is a plural, the verb must be too, e.g. towels *are* in the closet
- too many commas. Instead of writing long sentences with lots of commas, write shorter sentences
- punctuation marks should go outside quotation marks
- spell out numbers nine and below
- avoid starting a sentence with 'and', 'but' or 'because'
- a full sentence should have a subject (doer), verb (action) and an object (the thing the verb is happening to)
- use the same tense (future, present, past) in the whole text
- avoid using boring words: very, good, bad, weird, interesting, crazy, mad, funny, strange
- 'alot' is not a word. It should be 'a lot'
- avoid 'passive' sentences. Instead of 'The army was led by Alexander', write 'Alexander led the army'.

Historical research

I can define the problem

To define the subject to be researched, get some background information and build up a list of keywords.

Start by reading a simple Wikipedia page about your subject.

Get keywords for your topic. Think of different ways of saying your topic, or google 'synonyms of ...' and insert your search term.

I can decide what information to find and where to find it

What type of evidence do you want?

Include these kinds of words in your search:

- facts, examples, definitions, quotes, artefacts, images, data, statistics
- primary and secondary sources
- databases, links, archives, collections, references, research, museums, journals, graphs, tables, letters.

Where is your evidence?

There are many different types of websites to look at: scholarly works, databases, archives, reference sources and information pages.

How credible is the evidence?

Ask yourself the following questions:

- Is the content *relevant*? Is it useful for my purpose? Does it contain links to other relevant sources? Is it at an appropriate reading level?
- Is the source *believable*? What type of source is it? (Published or official sources are better.) Who is the author? (Experts are better.) When was it published? (Newer is usually better.) Is the source biased?
- Is the source *true*? Is it backed up by other sources? Does it *sound* right? Does it fit in with other things you know?
- Does the source state where its information comes from? This means it is more likely to be credible (able to be believed).

I can find information

Online search strategies

Following are some search strategies:

- After you type in a search term, scan through the first page of results. If they are not relevant, change your search.
- Start with a wide search, then get more specific.
- Learn *from* your search. Change what you are searching for based on what you learn after you start searching.
- Be ready to stop a search if it is taking you in the wrong direction.

Tips for searching with Google

- Every word matters.
- The order of the words matters.
- Capitalisation doesn't matter.
- Punctuation doesn't matter.
- Specific search terms are better.
- Use these terms to narrow your search: AND, OR, NOT.
- A search with 'filetype:' will find specific files. e.g. 'Qin Shi Huang filetype:ppt' will find PowerPoint files about Qin Shi Huang.
- A search with 'site:' will find things *within* a website. For example: 'boomerang site: nma.gov.au' will find boomerang-related material from the National Museum of Australia website.
- Use the tabs along the top for different types of results such as images, news, videos, maps and books.
- Use a hyphen to exclude words and narrow your search. For example, 'Great Wall-takeaways' will find information about the Great Wall of China, not a fast food outlet.
- Search for a range of numbers using two full stops between speech marks: '..' For example:
  - '2001..2004' searches between 2001 and 2004
  - '..2004' searches before 2004
  - '2004..' searches after 2004.
- An asterisk acts as a wildcard. So, for example, 'teen*' will return results with any of the words *teen, teens, teenager* in them.
- Use exact phrase searching by putting speech marks around a search to find exact text.

I can extract information

This stage is note-taking and is the same as Step 2 in the Writing Process. Read that section on page 224.

I can organise and present information

This stage is very similar to Step 3 in the Writing Process. Read that section on page 224.

Research will be presented in a number of different ways, and will usually include some history writing. History writing is generally presented as text with perhaps some supporting pictures.

You should also edit and proofread your research, just as you do with your writing. Read Step 6 from the Writing Process on page 227.

I can evaluate information

You can improve every time you conduct research by asking yourself these questions after you finish:

- What worked? What didn't work?
- How could I work smarter next time?
- Can I apply what I've learnt to other situations?
- How could I have improved:
  - the project?
  - the way I worked on my project?
  - the way I managed my time?

Glossary

Acropolis a fortified part of a city that is built on a high hill. The name comes from the Greek words *akro*, meaning high and *polis*, meaning city. The most famous Acropolis was built in the 5th century BCE in Athens, Greece.

agoge military training camp at Spartan army barracks for boys as young as seven years of age to learn to be brave warriors

agora the agora was the centre of activity in ancient Greek towns and cities. It was where public meetings were held, where plays were performed and a busy marketplace.

afterlife the belief in life after death held by many ancient cultures, such as the ancient Egyptians who preserved bodies through mummification ready for the afterlife

agriculture the science and practice of farming, including the growing of crops and rearing of animals

ancestors a relation who lived a long time ago from who you descended

aqueducts built to channel water to settlements. They had to be built very accurately as the water had to run downhill over large distances.

aquila the silver eagle emblem on top of a standard that identified each legion as they marched into battle; the standard was carried by an *aqilifer*

archaeology the study of human history through the excavation of sites and the analysis of artefacts and physical remains found there. People undertaking archaeology are known as **archaeologists**.

artefact an object made by a person, such as a tool or implement, usually of historical interest

assassination the murder of a leader or important official for political reasons

Assemblies branch of government in the Roman Republic where plebeians democratically elected magistrates and passed laws

auxiliaries Extra soldiers who were recruited across the Roman Empire and used as specialist forces, such as archers

Bronze Age the period where people made and used tools made of bronze, from 3000 to 1200 BCE

Buddhism a philosophy that began in India and arrived in China from via the Silk Road in the first century CE. Buddhists believe that after people die, they are reincarnated – reborn into another life. Living a proper life would stop the cycle of rebirth, and this stage, called nirvana, could be reached by leading a good life and no longer wanting things.

bullying involves causing distress and helplessness to a person through the repeated and intentional use of words or actions by an individual or group of people

century these units of 100 Roman soldiers were led by a commander known as a centurion

Christianity a religion based on the teachings of Jesus Christ; Catholic and Protestant religions both preach Christianity

chronology the process of organising events into the order they happened to bring structure and order to events in time

citizen a legally recognised member of a country or society. In ancient Greece, citizens were men who had completed their military training. In ancient Rome, citizens were males born to parents who were citizens.

city–state because of ancient Greece's mountainous landscape, small farming villages developed independently into city–states. Each city–state, or *polis*, such as Athens or Sparta stayed independent rather than growing into a single Greek nation.

civilisation an advanced society with urban settlements, technologies, government, organised society and religion

civil case disputes between people and acts that cause loss to others

civil war war between citizens in the same country

clan Aboriginal nations are broken into clan groups. Clans are larger than a family and share a common language and kinship system

cohort each Roman legion was broken down into 10 cohorts of up to 600 men, the equivalent of a modern battalion

Common Era (CE) the name of the period of time from the birth of Jesus (year 1) until now

common law judge-made law in cases not clearly covered by statutory law. Common law decisions set precedents for similar cases to follow in the future.

concubine a woman who lived in the imperial palace in ancient China as part of the emperor's harem. The concubine's job was to keep the emperor entertained and amused. It was considered an honour to be an emperor's concubine, and wealthy men sent their daughters to the imperial palace, hoping they would catch the attention of the emperor. Concubines entered the palace at a young age – sometimes as young as 13 years old – and were not allowed to leave.

Confucianism a philosophy taught by Confucius (551–479 BCE) that taught that people should be honest, brave and knowledgeable and never violent or arrogant. The path to happiness lay in obeying the law, doing one's duty and respecting older people – especially parents and grandparents.

Connection to Country the deep spiritual, physical, social and cultural relationship between Indigenous Australians and the land

Constitution sets out the rules by which a country is run. The Australian Constitution established a system where the responsibility to make or change laws is shared between a federal parliament and six state parliaments.

consuls the elected leaders of the Roman Republic (509 to 27 BCE). Each year, the citizens of Rome elected two magistrates as consuls, who ruled together for a term of one year.

corroboree an event where Aboriginal Australians interact with the Dreaming through dance and music, with their bodies painted to indicate the clan and type of ceremony

criminal case involves an action that is considered to be harmful to society, such as murder or armed robbery

customary law the laws that regulate behaviour and punishments in Aboriginal society

democracy democracy: rule by the citizens, who elect officials and leaders

discrimination treating a person or group unfairly or as if they are inferior, especially if based on their race or gender

domus a wealthy patrician's house in ancient Roman towns, with multiple rooms and an outdoor atrium with an altar and statues of the household

Dreaming explains how life came to be and establishes the rules for relationships between Aboriginal people, the land and all living things

dynasty where one family maintains power over many years by handing on the throne to an heir, usually the oldest son

ecclesia political assembly of citizens in ancient Greece where citizens would come together to vote for new laws and to make important decisions.

Elder an Aboriginal or Torres Strait Elder is a person who earned respect and authority in a community. It is an Elder's duty to instil a respect for the land and culture in community members by teaching young people about their natural environment, and sharing knowledge about the Dreaming.

electorate the electoral divisions for the House of Representatives in the Australian Parliament and for lower houses of state parliaments. Each Member of Parliament (MP) represents one of the 151 electorates in the House of Representatives to provide a direct link between the people in their electorate and the parliament.

emperor a ruler or monarch of an empire, i.e. in 27 BCE Octavian became the Roman Empire's first emperor. Japan's emperor is the only current ruler who uses this title.

empire a number of nations or states ruled over by a single leader such as an emperor, king, queen or an oligarchy, i.e. The Roman Empire

eunuch a castrated male who could work as servants in the imperial palace because there was no fear that he may father a child to the emperor's wives or concubines. Eunuchs were usually from poor families, yet sometimes became trusted and powerful members of the imperial court.

executive the prime minister and senior ministers of the Australian government and the governor-general who have the power to administer or implement the law

export where goods or services produced in one country are shipped to another country for sale or trade. Exports are crucially important for a country's economy to add to its gross national income.

gladiator criminals or prisoners who were forced to fight, usually to the death, in a public arena for the entertainment of ancient Romans

Governor-General the royal representative in Australia who must approve all bills made by the Commonwealth Parliament before they become law

harassment aggressive pressure or intimidation that humiliates or embarrasses a person, such as sexual harassment (unwelcome behaviour of a sexual nature)

helots slaves in ancient Sparta who undertook manual labour

hieroglyphics a writing system using pictures and symbols such as the system used by ancient Egyptians

historian people who specialise in the study of history by using evidence to answer questions about the past

homo sapiens scientific name for the human species, meaning 'wise man'

hoplite an ancient Greek soldier

House of Representatives The Australian government is formed by the political party with the largest number of members in the House of Representatives. Each of the 151 members is elected for three years to represent one local area (or electorate) of Australia.

Ice Age the period of time between 1.8 million and 12 000 years ago when much of the Earth's water was frozen at the northern and southern tips of the planet and the sea level was over 100 metres lower than it is today

import a good or service brought into one country from another country. Along with exports, imports are the other key element of international trade.

Industrial Revolution when humans began to use machines to do work previously done by hand in the 18th and 19th centuries

initiation ceremony a ceremony to mark the passage of girls and boys into adulthood

insulae ancient Rome's crowded, multi-storey apartments lived in by plebeians. Often whole families lived in one room of an *insulae*, with up to 40 people in each housing block sharing bathing and cooking facilities in a central courtyard.

inundation summer flooding of the Nile River that deposited rich silt that fertilised the land

Iron Age the period when people used iron to make tools and weapons, between 1200 and 500 BCE

judiciary the Australian courts that have the power to interpret and apply the law

kinship extended family relationships in Aboriginal and Torres Strait Islander societies that are important to sharing culture and organising society. For example, a mother in the kinship system refers not only to a woman's birth mother, but also to her sisters.

kowtow to kneel and bow so low that your head touches the ground. In ancient China, everyone had to keep their head bowed if they were addressing the emperor.

legislature the Australian parliament that has the power to make the law

legion the main fighting units of the Roman army of about 6000 men. Legions were further broken up into six cohorts.

legionary ancient Roman soldiers

ma'at the order and balance in the Egyptian world and the heavens that also formed the basis of the principles of truth, order, balance and justice that underpinned law in ancient Egypt

magistrate elected official of the Roman Republic charged with running the city. Once a man had served as a magistrate he could join the Senate. Consuls were the two highest magistrates.

Mandate of Heaven the right to rule given to a king or emperor in ancient China by the gods. If the leader was selfish or cruel, it was a sign the gods had withdrawn his mandate to rule. Natural disasters such as famines and floods were also signs of heaven's displeasure with the king.

megafauna large animals over 40 kilograms, including extinct Australian megafuna such as the marsupial lion

midden waste deposits from meal preparation, often containing bones and shells, which provide evidence of what the people living in a certain area ate

Middle Kingdom the term used by ancient Chinese people to describe their country as they believed they were living in the centre of the world

moiety a group in society that coexists with only one other group in society. Marriage can only take place between members of opposite moieties.

monarchy rule by a single ruler who inherited the role

mummification preparation of a body for the afterlife in ancient Egypt. The major body organs were removed (except the heart) and the body was washed, embalmed and wrapped in bandages.

nations tribal groups of the Indigenous peoples of Australia

native title recognition that Aboriginal and Torres Strait Islander people have traditional rights to the land and water as set out in the *Native Title Act 1993*

Neolithic relating to the New Stone Age (between 10 000 years ago and 2000 BCE) when people used polished stone implements and farming commenced

nomads people with no permanent home who herd animals from place to place to find fresh pasture; those who follow this lifestyle are described as **nomadic**

occupational health and safety (OHS) laws, regulations and guidelines to ensure good standards of health and safety in the workplace

oligarchy a small group of people who have control over a country

oral history information about the past obtained through spoken interviews or in-depth conversations concerning experiences, recollections, reflections and lore

ostraka sherds of broken pots re-used that were as voting 'ballots' cast by Athenians at meetings of the *ecclesia*

Palaeolithic relating to the Old Stone Age (between 2.6 million and 14 000 years ago) when people used primitive stone implements

papyrus the first paper-like material used by a civilisation, made from the papyrus plant that grew by the Nile River in Egypt

parliament a group of elected politicians who makes laws for the country

paterfamilias the oldest male and head of the ancient Roman family who had complete control over all family matters, including the right decide if a baby should live or die decide if a baby should live or die

patrician wealthy landowners in ancient Rome whose families had held influential positions in society

perioikoi traders and craftsmen in ancient Spartan society who were not considered citizens

phalanx a battle formation used by ancient Greek hoplites to protect themselves in battle. They packed together tightly and overlapped their shields to provide a protective outer barrier to deflect arrows and spears. Spears in the front row were held forward to penetrate the enemy and those in the other rows were held skyward to deflect incoming missiles.

pharaoh the leader of ancient Egyptian society who was viewed as the link between the gods and the people, and therefore had both earthly and divine responsibilities

plebeians poorer farmers and craftspeople in ancient Rome who made up the bulk of the population

polis because of ancient Greece's mountainous landscape, small farming villages developed independently into a polis or poleis (plural). Each polis or city-state such as Athens or Sparta stayed independent rather than growing into a single Greek nation.

precedent common law decisions by a judge that are followed in similar cases

prehistoric the period of time before the invention of writing in ancient Sumer in about 3500 BCE

primary source a source that was created or existed at the time under study such as books, letters, artefacts and buildings

propaganda biased or misleading information used to promote a point of view

prosecutor the lawyer who conducts the case against the accused person in the court room. It is up to the prosecutor to produce evidence that will prove their claims against the accused person or disprove any disputed facts in court.

pyramid huge pyramid-shaped buildings built as tombs for royalty or to house the gods in civilizations such as ancient Egypt and the Olmec, Maya and Aztec civilizations in the Americas

quaestor an important public official in ancient Rome who looked after finances and oversaw important legal cases such as murder

radiocarbon dating the age of something that was once alive can be estimated by measuring the amount of radioactive carbon present in the sample. Carbon breaks down at a known rate, so the age of the sample can be estimated.

referendum changes to the Australian Constitution can only be made by a referendum, where a majority of Australian voters and a majority of states vote to approve the changes proposed by parliament

representative democracy a system of government such as Australia where citizens vote for government representatives to make laws and rule the country on their behalf

scribe well-educated and highly regarded person in ancient Egypt who could read and write, including complex hieroglyphics

seasonal migration where people move to another area for a period of time to ensure they do not exhaust their land of plants and animals

secondary source a source created after the time under study, such as a text books, websites and documentaries

Senate The Australian Senate is the house of that reviews legislation passed in the House of Representatives. The Senate has 76 members elected for a term of six years. Each of the six states has 12 senators, and the Australian Capital Territory and Northern Territory both have two senators. The Senate in Ancient Rome was an advisory body appointed by leaders to advise magistrates and leaders on proposed laws and actions, particularly involving finance and foreign policy.

Silk Road a 6500-kilometre trading network of routes over land and sea that connected China to Asia, Europe and the Mediterranean. Camel caravans carried goods along the route through the mountain ranges and deserts of southern Asia.

slave a person who is owned by another person, who is usually forced to work without payment. In ancient civilisations, slaves were at the lowest level of society.

socially responsible an obligation by an individual, organisation or business to improve society or the environment, such as reducing waste and recycling or working with charitable organisations

standard the emblem of a Roman legion carried into battle by an *aqilifer*. The Roman standard used to identify each legion was a silver eagle (or *aquila*) mounted on a pole.

Stone Age the period lasting from 2.6 million years ago until 2000 BCE where people made and used tools made out of stone. Historians refer to two different Stone Ages – Palaeolithic, before the development of farming (about 10 000 years ago) and Neolithic – after the development of farming.

stratigraphy artefacts are found in different strata (or layers) of rock and soil. By dating the stratas, the archaeologist can assume that the artefacts found there are the same age as the strata.

statutory law laws created and passed by parliament

Taoism a Chinese philosophy focused on living in harmony with oneself, with nature and with the universe

timeline a list or graph that displays events in chronological order

terra nullius land that the law says that nobody owns. British colonizers regarded the Australian continent as *terra nullius* as they saw no evidence of formal buildings and fences

timeline a list or graph that displays events in chronological order

totem a natural object, animal or plant that represents its people's spiritual link to the land. For example, totems for the Wurundjeri are Bunjil the wedge-tailed eagle and Waang the Crow and this totem is inherited from your father.

traditional owner an Indigenous Australian who is a member of a nation or tribe that has rights and responsibilities for an area of land or sea

trireme large and fast ancient Greek warships powered by 170 oarsmen. The bronze prow at the front of the trireme was used as a battering ram to smash into other ships.

tyranny rule by an individual who seized power illegally

universe all the things that exist in space and its contents such as stars, planets and gas clouds

villa in ancient Rome, villas were country estates owned by wealthy Romans

vizier chief advisor to the pharaoh in ancient Egypt

yin and ***yang*** concept from Taoism where all living things have two sides – *yin*, the female force: dark, cool and passive and *yang*, the male force: light, hot and active

workers compensation scheme ensures that employees who are injured while conducting their work can receive medical treatment and other assistance including rehabilitation to help them get back to work quickly

Index

B

Acknowledgements

The author and publisher are grateful to the following for permission to reproduce copyright material:

PHOTOGRAPHS: AAP Image, **22** (centre), /Vincent Lamberti/ Gundjeihmi Aboriginal Corporation, **25**, /University Of Melbourne/ AP, **26**, /Richard Wainwright, **105**; Alamy Stock Photo/Album, **55**, **138–9** (centre), /Atlaspix , **132**, /Alan Baker/Icon Images, **89**, /Bayar Balgantseren/imageBROKER, **174** (top), /Bjorn Svensson/age fotostock, **33**, /Chronicle, **131** (centre and top left), **148–9** (centre), /Classic Image, **108** (top), **171** (top), /Collection Christophel © Warner Bros/Helena Productions, **95**, /Dennis Cox, **194** (bottom), /Luis Dafos, **85** (bottom), /Dorling Kindersley ltd, **76–7**, /Adam Eastland , **90**, /Adam Eastland Art + Architecture, **144–5** (bottom), /Heritage Image Partnership Ltd, **67** (top), /Fine Art Images/Heritage Image Partnership Ltd , **9**, /Roman Garcia Mora/Stocktrek Images, **22** (top), /Alfredo Garcia Saz, **66–7**, /GL Archive, **221**, /Granger Historical Picture Archive, **74** (top), **82** (left), **207** (top), /Christian Goupi/age fotostock, **164** (bottom), /Heraldo Mussolini/Stocktrek Images, **44**, /The History Collection , **45** (top), **164** (centre left), /Rod Ho, **200** (centre left), /Peter Horree, **94** (top), /Morten Hübbe, **174** (bottom), /Ivy Close Images, **67** (bottom inset), /Karl Johaentges/Look, **194** (top), /Lanmas, **109**, **111** (top), /Erich Lessing/Album, **190**, /mazartemka/ Panther Media GmbH, **201** (bottom), /Moviestore collection Ltd, i (centre right), **134–35**, /National Geographic Image Collection, **71** (bottom), /North Wind Picture Archives , **120–1** (centre), /Natalia Okorokova, **85** (bottom), /Oldtime, **200** (centre right), /Hugh Olliff, **92** (top), /John Parrot/Stocktrek Images, **156–7** (bottom), /Alfred Pasieka/Science Photo Library, **14**, /Paul Seheult/Eye Ubiquitous , **131** (top right), /PaulPaladin, **72** (left), /Wiliam Perry, **124**, /The Picture Art Collection, **147**, **162–3**, **180**, /Pictures Now, **68**, /Photograph by ART Collection , **104** (centre), /Photograph by Peter Horree, **106–7** (centre), /Pocholo Calapre, **174** (centre), /Igor Prahin, **47** (top), /The Print Collector/Art Media/Heritage Images, **84**, /Prisma Archivo, **93**, /Science Photo Library, **79** (bottom), /jozef sedmak, **149** (top), /SIPLEY/ClassicStock, **80**, /TCD/Prod DB © Warner Bros , **98–9** (centre), /Travelscape Images, **23** (centre), **41** (top), /Universal Art Archive, **171** (bottom), /View Stock, **202**, /Warner Bros/A.F. ARCHIVE, **112–13**, /Warner Bros/Pictorial Press Ltd , **119**, /Werner Forman Archive/Heritage Image Partnership Ltd , **61** (top), /Jan Wlodarczyk, i (centre left), **56–7**, /World History Archive, **58**, /wyrdlight, **134** (inset); © Altair**4** Multimedia – Altair**4**.com, **59**; Auscape/Jean-Paul Ferrero, **2**; Australia National University/Stuart Hay, ANU, **28** (top); Illustrator: Peter Trusler © Australian Postal Coorporation, **29** (top); Bridgeman Images/The punishment of the Bastinado/ British Library, London, UK/© British Library Board. All Rights Reserved, **197**, /Figurine of a girl running, Sparta, 520–500 BC (bronze), Greek, (6th century BC)/British Museum, London, UK, **111** (bottom), /Roman Emperor, possibly Nero (37–68) c.1625–30 (oil on canvas), Grebber, Pieter Fransz. de (c.1600–53)/Musee des Beaux-Arts, Dunkirk, France, **146**, /A Stone Age man making a weapon, Jackson, Peter (1922–2003)/Private Collection/© Look and Learn, **6**, China: Wu Zetian (624–705), Empress Regnant of the Zhou Dynasty (690–705)/Pictures from History, **177**, /Hannibal crossing the Alps, Scarpelli, Tancredi (1866–1937)/ Private Collection/© Look and Learn, **137**; © The Trustees of the British Museum, **87**; DK Images/Russell Barnett, **152**, /Richard Bronson, **102–3**, /Paulo Donati, **158–9** (centre), /Guilano Fonari, **123** (bottom), /Sergio, **160–61** (bottom); Fairfax Images/Glenn Campbell, **22** (left), **35**, /Justin McManus, **28–9**; Getty Images/ Alinari Archives, **164** (centre right), /Auscape/Universal Images Group, i (top right), **21**, /Scott Barbour , **183** (top), /Torsten Blackwood/AFP, **31**, /Bloomberg, **150–1**, /Felix Cesare, **242**, /Matteo Colombo, i (centre), **107** (top), /De Agostini/Biblioteca Ambrosiana, **200–1** (top), /De Agostini/A. Dagli Orti, **82–3**, /De Agostini Picture Library, **92** (bottom), **117**, **156–7** (top), /DEA Picture Library, **77** (top), **83** (top), /DEA/G. Dagli Orti, **129**, **154**, /DEA/G. Nimatallah, **104** (top), /DEA/S. Vannini , **71** (top), /Dorling Kindersley, **78–9**, **96**, **99** (top right), **101** (top), **122**, **173** (top), **188–9**, /Cathy Finch/Lonely Planet Images, **37**, /Cyril Folliot/AFP, **53**, /Luigi Galante, **153**, /David Gifford/Science Photo Library, **15** (top), /Bill Heinsohn, **106** (left), /Heritage Images/Hulton Archive, **120** (bottom), /Hero Images, i (top left), **1**, /Hulton Archive, **155**, /Nigel Killeen, **23** (top), /Leemage/Universal Images Group, **75**, /Yifan Li, i (bottom right, background), **205**, /Andy Lyons, **121** (top), /MediaProduction, i (bottom left) **182–3** (bottom), /Mondadory Portfolio, **100–1** (centre), /Tracey Nearm, **65**, /Scott Olson, **76** (bottom), /Peter Parks/AFP, **42–3**, /Photos.com, **72** (right), /robertharding, **63**, /Marco Rubino/EyeEm, **205**, /ViewStock, **198**, /Paiwei Wei, **170**, /Lisa Maree Williams, **19**, /zorazhuang, **127**; Photographer John Gollings, **22** (right), **49**; iStockphoto, **208** (all), /4X-image, **200** (centre), /alessandro0770, **3** (left), /AlexanderYershov, **52**, **88**, **126**, **168**, **204** (arch icon), /aphotostory, **172–3** (bottom), /Arsty, **222**, /Bet_Noire, **167**, /boryak, **126**, /bubaone, **216** (economic), /Michael Burrell, **223** (paper), **224**, /DaddyBit, **100** (left), /dar woto, **216** (environmental), /dbencek, **98–9** (background), /DianaHirsch, **110**, /GlobalP, **85** (top), /gt29, **228**, /Hohenhaus, **52** (top), /JIWEI QU, **203**, /karayuschij, **230–3**, /Kemter, **161** (top), /KingWu, **234–5**, /loonger, **11**, /MaRabelo, **165**, /Nikiteev_Konstantin, **216** (cultural), /OliverChilds, **172** (top), /Page Light Studios, **91**, /PEDRE, **18** (bottom), /pixdeluxe, **236–40**, /PytyCzech, **193**, /ROMAOSLO,

139 (top), /sarra22, **140–1** (background), /Scarpio, **218–19**, /simonbradfield, **241**, /slalomp, **223**, **228** (pen icon), /SSSCCC, **54**, /stock_colors, **131** (bottom), **159** (top), /sx70, **216** (social), /t_kimura, **227**, /xavierarnau, **213**; Metropolitan Museum of Art/ Item 14.40.676, **145** (top), /Item 1975.1.1664, **185** (bottom), /Item 41.162.212, **209**, /Item 56.210.5, **220**, Item 08.200.9, **210–11**, Museums Victoria Item MM31472, Photographer: Lilian L Pitts, **46** (top); Newspix/Lyndon Mechielsen, **40–1**, /Gary Ramage, **32**; Shutterstock.com/The Art Archive, **181**, /artphotoclub, **191**, /Astor57, **4**, /AVIcon, **216** (political), /ChameleonsEye, **51**, /delcarmat, **98–9** (bottom), /Everett Historical, **225**, /Gianni Dagli Orti, **64**, /GoodStudio, **140–1** (bottom), /Gordine N, **73**, /Granger, **179**, /Peter Hermes Furian, **118**, **217**, /Constantinos Iliopoulos, **94** (bottom), /johnbraid, **16** (bottom), /Julinzy, **43** (top right), /Kamira, **108** (bottom), /Ava Peattie, **215**, /PlusONE, **185** (top), /PNIK, **125**, /Pyty, **212**, /Yu Zhang, **195**; Science Photo Library/ NOAA, **15** (bottom); Tourism Australia, **36**; Wikimedia Commons/ NordNordWest, **16** (top), /Fresco by Giulio Parigi (1571–1635), Stanza della Mattematica, Galleria degli Uffizi, **123** (top); Courtesy of Clare Wright, **3** (right); WorkSafe Victoria, **81**.

OTHER MATERIAL: Illustration, *Greece in spectacular cross-section*, Stephen Biesty, Oxford University Press. Illustrations © copyright Stephen Biesty 2006. Reproduced with permission of the Licensor through PLSclear, **97**; Graphic, 'Tunahamon's Tomb' by Wojciech Ostrycharz, Branzo, **74** (bottom); Map screenshot images are the intellectual property of Esri and is used herein with permission. Copyright © 2020 Esri and its licensors. All rights reserved, **175**; Global Oneness Project, **33**; From *Seeing the Light: Aboriginal Law, Learning and Sustainable Living in Country*, by Ambelin Kwaymullina, Indigenous Law Bulletin 2005, **32**; Painting, *Battle of the Brave*, by George Mamos, **114–15**; Calvert, Samuel & Gilfillan, J. A. (1865). *Captain Cook taking possession of the Australian continent on behalf of the British crown, AD 1770, under the name of New South Wales*, nla.obj-135699881, **34**; Joseph Lycett, *Aborigines Spearing Fish, Others Diving For Crayfish, A Party Seated Beside A Fire Cooking Fish*, 1817, National Library of Australia, nla.obj–138500727, **27**; Illustrations © Orpheus Books **60–1**, **78** (bottom); Tindale map ©Tony Tindale and Beryl George, 1974. South Australian Museum, **30**; *The Founding of Australia*. By Capt. Arthur Phillip R.N. Sydney Cove, Jan. 26th 17880/ Original [oil] sketch [1937] by Algernon Talmage R.A, Mitchell Library, State Library of New South Wales, Item FL3141725, **23** (bottom); Aboriginal Flag designed by Harold Thomas, **43** (top centre); Torres Strait Islander flag designed by Bernard Namok. Reproduced with permission by Torres Strait Island Regional Council, **43** (top left); Quote from personal communication with Beryl Timbery Beller, 14 March 2007, as published in *Bo-ra-ne Ya-goo-na Par-ry-boo-go, Yesterday Today Tomorrow, An Aboriginal History of Willoughby*, by Jessica Curry, 2008 Willoughby City Council, p 21, reprinted with permission, **34**.

While every care has been taken to trace and acknowledge copyright, the publisher tenders their apologies for any accidental infringement where copyright has proved untraceable. They would be pleased to come to a suitable arrangement with the rightful owner in each case.

FLIP BOOK

good humanities

Flip your book over and begin your journey through **Geography**

civics + citizenship | history | geography | economics + business

contents

Aboriginal + Torres Strait Islander peoples + cultures

The European + Mediterranean world

The Asia-Pacific world

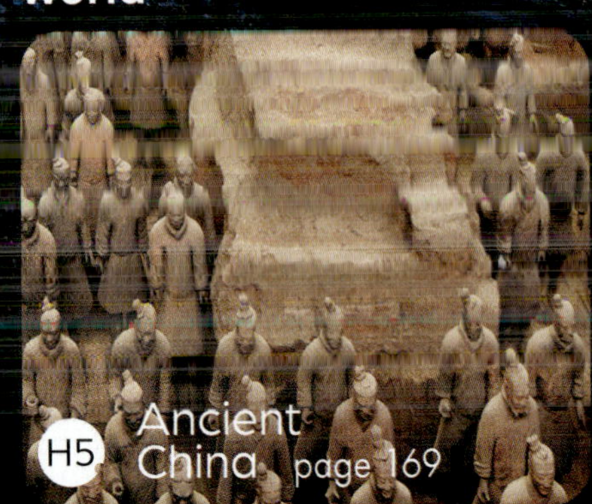

History concepts + skills

good humanities

Matilda Education Australia acknowledges the Aboriginal and Torres Strait Islander peoples of this nation. We acknowledge the traditional custodians on whose unceded lands we conduct the business of our company. We pay our respects to ancestors and Elders past, present and emerging. Matilda Education Australia is committed to honouring Aboriginal and Torres Strait Islander peoples' unique cultural and spiritual relationships to the land, waters and seas; and their rich contribution to society.

H0

Introduction to History

WHY DOES HISTORY MATTER?

page 2

historical skills

page 4

HOW DO I STUDY HISTORY?

learning ladder

page 10

HOW DO I USE THIS BOOK?

history overview

page 14

HOW ARE OUR HISTORIES CONNECTED?

H0.1

Why does History matter?

History is the study of our past. It is easy in our busy modern world to forget that every invention, every country and every idea has a history – a long chain of events that lead to right now. Studying History helps us to understand our world.

Thinking like a historian

People who specialise in the study of History are called historians. So, as a History student, you are a trainee **historian**! Your goal when learning about history is not to memorise facts or dates; your goal is to think like a historian.

History is both a *process* and *a way of thinking*. A historian's role has three parts to it:

1. to ask questions
2. to examine sources
3. to use the evidence to answer questions, or to tell a story about the past.

Three reasons why History matters

Exploring History will help you to develop powerful skills

Many people think that History is just a set of facts about events that happened in the past. However, History is more than just facts. Studying History gives you access to a new set of skills. Think of these skills as your 'superpowers' that nobody can take away from you. This year you will practise skills such as asking interesting questions, expressing opinions, thinking critically, processing information, analysing, researching and communicating. These skills will help you to succeed inside and outside your History classroom.

Source 1

Look closely at the tree in this photo. What can it tell you about events in the past? Think like a historian: ask questions, examine sources and use evidence. Ask yourself, 'Why did humans remove a large panel of bark from this tree a long time ago? What might they have used it for?'.

Source 2

Ancient Greek historian Thucydides. [Theophil von Hansen, *Statue of Thucydides* (1884), Austrian Parliament Building, Vienna.]

Source 3

Modern Australian historian and author Dr Clare Wright

2 Exploring History is the closest thing you'll ever get to time travel

Do you want to know how the Egyptian pyramids were built? Do you wonder why the Great Wall of China was constructed? Are you interested in lions fighting gladiators? Do you want to know how democracy started? Have you ever wondered what Australia looked like 60 000 years ago? The study of History allows you to take a trip back in time to discover some truly fascinating people and events – and to understand the *how* and *why* of our world.

3 Exploring History helps form your identity

Perhaps most importantly, studying History can help us to understand ourselves better. Human history is a long story about everything that people have ever done. There is a lot to learn from this story. We can learn what we should and should not do, and we can also learn about who we are.

Learning Ladder H0.1

1 List three reasons why we study History.

2 What does a historian do?

3 Historical facts are important and give us context, but why are skills so important in studying History?

4 What is *your* history? Write down at least five words or phrases about your own history.

5 Source 2 and Source 3 are separated by more than 2000 years. Imagine you could interview both of these historians. Write down the most important question you would ask each of them about their roles.

6 As a class, discuss the problems with this statement: 'In today's world, History does not matter anymore'.

How do I study History?

In Year 7 History, you will focus on five historical skills:

- chronology
- using historical sources as evidence
- continuity and change
- cause and effect
- historical significance.

You will also practise the important skills of writing and research. You can read an introduction to each of these skills over the following pages before practising them in your first historical study. The History How-To section on pages 205–229 will also support you step by step when you begin to apply these skills.

Chronology

The word **chronology** refers to the process of organising events into the order they happened. This brings structure and order to events in time. A **timeline** is a way to show or represent chronology.

Historians use a set of common terms to help them organise dates chronologically and to create timelines. These terms are also important in historical writing. The most commonly used terms are listed in the table in Source 1.

Term	Stands for ...	Meaning
BCE	Before Common Era	The period of time before the birth of Christ (before 1 CE)
CE	Common Era	The period of time after the birth of Christ (after 1 CE)
BP	Before Present	Years before the present day (often standardised to mean 'years before 1950')
c.	circa	The Latin word for 'approximately'
era		A span of time that can be described using a significant event or individual. For example, the Roman era.
decade	10 years	
century	100 years	
millennium	1000 years	

Source 1

The language of chronology

Counting time

A century is a period of 100 years. We live in the twenty-first century, which spans from 1 January 2001 to 31 December 2100. We are also living in the **Common Era (CE)**. In the Common Era calendar there is no Year 0, so new centuries always begin in a Year 1. For example, the first century CE spanned the years 1–100 CE. The second century CE spanned the years 101–200 CE.

Some Common Era dates are listed in the table in Source 2.

Span of years	Century CE
1–100 CE	first century CE
101–200 CE	second century CE
1001–1100 CE	eleventh century CE
1801–1900 CE	nineteenth century CE
1901–2000 CE	twentieth century CE
2001–2100 CE	twenty-first century CE

Source 2

Examples of the Common Era (CE) calendar, divided into centuries. We are currently in the twenty-first century CE.

BCE dates are a bit different because they 'run backwards', so the larger the number, the further away in time it is. For example, 4 BCE is further away from the Common Era than 1 BCE, as you can see on this timeline.

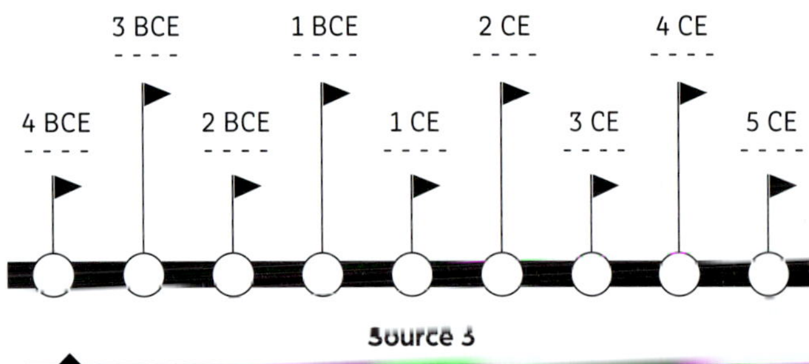

Source 3

BCE dates 'run backwards', so 4 BCE is further away in time than 1 BCE.

The years in BCE centuries also 'run backwards'. So where the first century CE spans the years 1–100 CE, the first century BCE also consists of 100 years but in reverse order: 100–1 BCE.

The table in Source 4 shows the span of years for some BCE centuries. Have a close look at the order of the years in each BCE century.

Span of years	Century BCE
2100–2001 BCE	twenty-first century BCE
2000–1901 BCE	twentieth century BCE
1900–1801 BCE	nineteenth century BCE
1100–1001 BCE	eleventh century BCE
200–101 BCE	second century BCE
100–1 BCE	first century BCE

Source 4

Examples of the Before Common Era (BCE) calendar, divided into centuries.

Divisions of time

Instead of always using the BCE and CE calendars, historians sometimes divide time up into larger periods to make events easier to understand. Often these periods will have a theme or special feature that connects them. There are many ways historians divide up time. Two of the most common are time periods based on the materials people used to make tools, and whether they were able to record events in their daily lives in some way.

Prehistory and History

Historians often refer to 'history' and 'prehistory' (which is where we get our word **prehistoric**). The invention of writing in ancient Sumer (modern-day Iraq) in about 3500 BCE is used by historians to separate the past into history and prehistory.

Prehistory refers to the time before recorded history. Once people could write, they left behind many more sources of information about the past for historians to study. The periods are summarised in this table.

Prehistory	Any time before 3500 BCE
History	Any time after 3500 BCE

Source 5

Prehistory and history

Stone Age, Bronze Age and Iron Age

This classification is based on the most advanced material people used at that time. During the **Stone Age** people created and used tools made out of stone. Historians refer to two different Stone Ages – before and after the development of farming. During the **Bronze Age** people made and used tools made out of bronze, which is a mixture of metals. The **Iron Age** was when people used iron to make tools and weapons.

Age	When was it?
Old Stone Age (Palaeolithic)	2.6 million years ago to 14 000 BCE (before development of farming)
New Stone Age (Neolithic)	10 000 BCE to 2000 BCE (after development of farming)
Bronze Age	3000–1200 BCE
Iron Age	1200–500 BCE

Source 6

The Stone Age, Bronze Age and Iron Age took place at different times in different parts of the world, so the dates vary. This table provides a rough guide.

History in the Australian education system

In Australian schools, History is divided into four periods. This year you are studying the ancient period, from 60000 years BP–650 CE. You will study a different era each year of your lower secondary schooling.

Year 7	Ancient History 60 000 years ago–650 CE
Year 8	Medieval History 650 CE–1750 CE
Year 9	Pre-modern History 1750 CE–1918 CE
Year 10	Modern History 1918 CE–Present

Source 7

You will study a different era each year of your lower secondary schooling.

Source analysis

Using sources to answer historical questions and build narratives is central to what historians do. Sources come in two main types: primary and secondary, as outlined in the table in Source 9.

This book is a **secondary source** – it was created by teachers and authors. There are many **primary sources** inside this book, such as ancient texts and photos of ruins and artefacts. Your challenge is to use the primary sources in this book, and the book itself, as sources of information to answer historical questions.

Source 8

Making stone tools in the Stone Age . [Peter Jackson, *A stone age man making a weapon* (1963), gouache on paper, Treasure no. 1]

Type of source	Definition	Examples
Primary sources	Created at the time. Primary sources show the perspectives of the people who experienced the event itself. They might have unique information about an event because they were actually there.	books, diaries, photographs, archives, letters, artefacts, buildings, ruins
Secondary sources	Created afterwards. Secondary sources are created by historians who combine primary sources to interpret the past, or to tell a narrative about it.	textbooks, websites, documentaries

Source 9

Primary and secondary sources

Uncovering bias

Just because primary sources were created at the time being studied doesn't mean they are always reliable. Primary sources can be biased (or unfair). Perhaps an ancient scribe hated a leader, so they wrote nasty and untrue things about them. Historians might then find this ancient text centuries later and read it. Should they believe what it says? The way to minimise bias is to look at many different sources. Secondary sources can be helpful for this, as they often bring together many primary sources.

The most common secondary sources you will use are websites. Judging how reliable websites are is very important. There are many websites that don't state where their information is from, making it hard to work out how reliable they are.

Source 10

Textbooks are secondary sources.

H3.3

How did gods influence daily life?

The people of ancient Greece believed that the gods of Mount Olympus controlled everything in the lives. To honour the gods, they would pray daily at a shrine in their home, or they would visit the temple of a specific god if help were required in a particular area of their lives.

Gods and goddesses

Ancient Greeks believed in many different gods and goddesses that controlled everything in their lives and their environment. People thought it was important to please the gods, as happy gods helped you and unhappy gods punished you.

Ancient Greeks believed that a family of the most important gods and goddesses lived at the top of Mount Olympus in northern Greece.

Honouring the gods

To honour the gods, ancient Greeks built temples in every town, each one dedicated to a specific god or goddess.

[illegible] an acropolis (see pages xx). Temples were elaborately decorated, inside and out. A statue of the god for whom the temple was built was erected inside.

Temples were cared for by priests who had the power to talk to the gods. People would visit different temples according to what they were praying for. For example, those wanting help in [illegible]

Continuity and change

History is a story of continuity and change: some things stay the same while others change. Some things, such as technology, have changed a lot over the course of history. Other things, such as the bond between parents and children, have changed very little.

Narratives (historical 'stories') and timelines will help you to understand what has changed and what has remained the same. You will learn how to recognise continuity and change, and how to describe how quickly – and to what degree – change happened.

Cause and effect

Historians often try to figure out *why* things happened. You will learn about different kinds of causes:

- individuals and groups that take part in some action (called *actors*)
- conditions: social, political, economic, cultural or environmental
- short-term triggers
- long-term trends.

Causes and effects can be organised using timelines or by writing historical narratives. You will learn that most causes themselves were actually caused by something that happened even earlier, and most effects have effects further into the future.

Historical significance

History is everything that has ever happened. So how do we decide what is worth studying?

Determining how historically significant something is requires a judgement call, or *evaluation*. Historians have come up with models to help guide us when we are trying to decide what is important. In *Good Humanities*, we use the model of historian Geoffrey Partington. To help work out how important something is, ask these five questions:

1. **How important was it to people at the time?**
2. **How many people were affected?**
3. **How deeply were people's lives affected?**
4. **For how long did these effects last?**
5. **How relevant is it to modern life?**

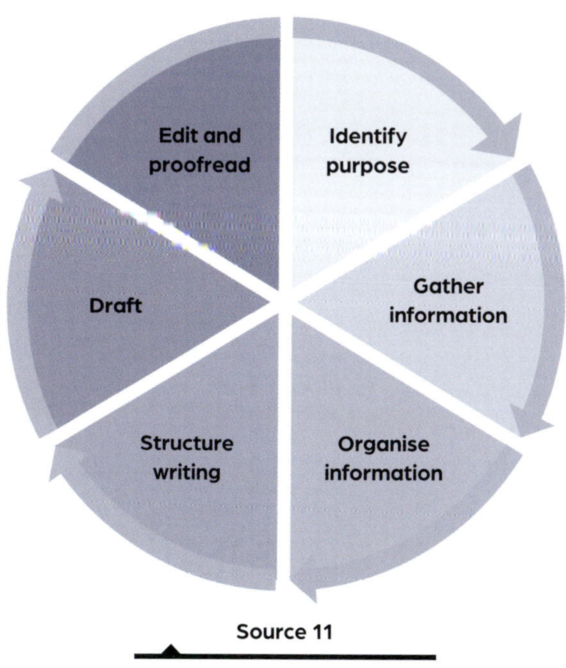

Source 11

The writing process has six steps.

Skills for historians

Historical writing

One of the most important skills you will learn is being able to communicate. Writing is a central communication skill for a historian, and you will use this skill throughout your life.

The writing process can be split into these six stages:

- **Step 1:** Identify the writing purpose
- **Step 2:** Gather information
- **Step 3:** Organise information
- **Step 4:** Structure writing
- **Step 5:** Draft
- **Step 6:** Edit and proofread.

Historical research

A core historical skill is researching. In this book, we split the research process into six parts:

- **Step 1:** Define the problem
- **Step 2:** Decide what information to find and where to find it
- **Step 3:** Find the information
- **Step 4:** Extract the information
- **Step 5:** Organise and present the information
- **Step 6:** Evaluate the information.

The History How-To section on pages 205–229 discusses these stages in depth.

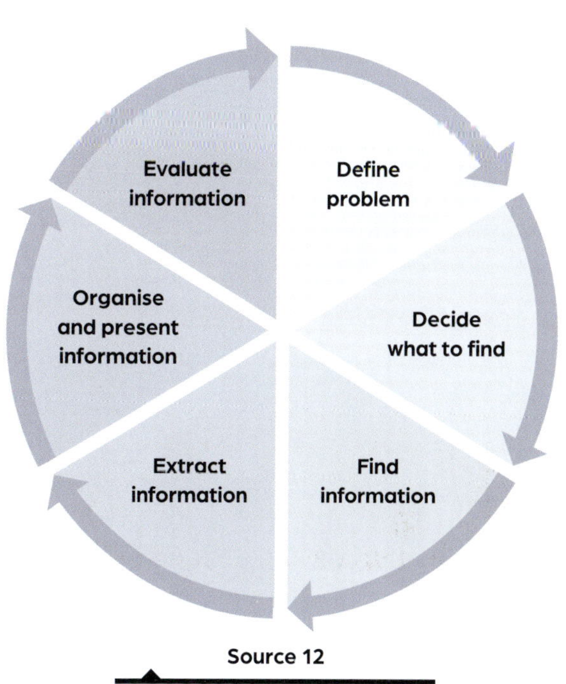

Source 12

The research process has six steps.

Source 13

Alexander the Great is considered to be historically significant. [Charles Le Brun, *Alexander Entering Babylon (The Triumph of Alexander the Great)* (1665), oil on canvas, 450 × 707 cm, Department of Paintings of the Louvre, Paris, France.]

Avoiding plagiarism

Plagiarism is when you use someone else's words or ideas and present them as your own work. It is forbidden. Plagiarism is when you:

- copy and paste from a website, or copy but change the words around
- use information from the internet but leave the web source out of your list of sources
- copy work from another student.

Plagiarism is a problem because it can be dishonest – you are pretending that someone else's writing or research is your own. But if you just plagiarise others it also means that you don't gain any new understanding, knowledge or skills.

You can avoid plagiarism by taking your own notes from research sources *in your own words*. When writing or researching, write using *your own notes*.

Learning Ladder H0.2

1 What important invention is used to divide time into prehistory and history?
2 Source 7: What historical period will you be studying in Year 7?
3 In which century is the year 1550 CE?
4 Source 6: What are the three ages that historians use to divide up time based on the most advanced material used?
5 Source 9: Describe the difference between a primary and a secondary source.
6 List one thing in human life that has been continuous since the Stone Ages.
7 What types of causes are there?
8 How can we decide what is important in history?
9 Why is learning how to write important?
10 List two reasons why plagiarism is bad.

How do I use this book?

Good Humanities has been built to help you thrive as you move through the Level 7 Humanities curriculum and to enable you to demonstrate your progress in every single lesson. The History section of this book includes five chapters of ancient history and a History How-To skills section. The History How-To section is vital – and you should refer to it often.

Climb the Learning Ladder

Each chapter begins with a Learning Ladder. The Learning Ladder is your 'plan of attack' for the skills you will practise in each chapter. It lists the five historical skills you will be learning, and has five levels of progression for each of those skills.

Each skill described in the Learning Ladder is of a higher difficulty than the one below it.

To be able to achieve the higher-level skills, you need to be able to master the lower ones. Practising doing activities at all the levels will help you to do more difficult skills, such as evaluating. This approach is called 'developmental learning' – and it puts you in charge of your own learning progression!

Read the ladder from the bottom to the top. As you progress through the chapter, you will climb up the Learning Ladder.

Source 1

The Learning Ladder helps you to take charge of your own learning!

learning ladder

Step	Chronology	Source analysis	Continuity and change	Cause and effect	Historical significance
step 5	I can describe patterns of change	I can use the origin of a source to explain its creator's purpose	I can evaluate patterns of continuity and change	I can evaluate cause and effect	I can evaluate historical significance
step 4	I can distinguish causes, effects, continuity and change from looking at timelines	I can use my outside knowledge to help explain a source	I can analyse patterns of continuity and change	I can analyse cause and effect	I can analyse historical significance
step 3	I can create a timeline using historical conventions	I can find themes in a source	I can explain why something did or did not change	I can explain why something is a cause or an effect	I can apply a theory of significance
step 2	I can place events on a timeline	I can list specific features of a source	I can describe continuity and change	I can determine causes and effects	I can explain historical significance
step 1	I can read a timeline	I can determine the origin of a source	I can recognise continuity and change	I can recognise a cause and an effect	I can recognise historical significance

Check your progress

Each chapter is divided into 13 or 14 sections. Each section is designed to cover one lesson, but sometimes your teacher might decide to spend more time (or less time) on a particular section. A section is two or four pages long.

At the end of every section, you will find a block of questions called 'Learning Ladder', which has two different types of questions or activities:

1. **Show what you know:** These questions ask you to look back at the content you have read and viewed and to show your understanding of it by listing, describing and explaining.

2. **Learning Ladder:** These activities are linked to the Learning Ladder. You can complete one of the questions or all of them. In each chapter you will complete several activities for each level of the Learning Ladder, as well as for each writing and research stage. This will sharpen your historical skills.

Source 2

Check your progress regularly. You can attempt one or all of the Learning Ladder questions.

key individuals key events

Throughout every chapter, you will discover a variety of features that focus on key individuals and events that have changed the course of history. Many of these features link to interactive sources, so you can explore further using your eBook.

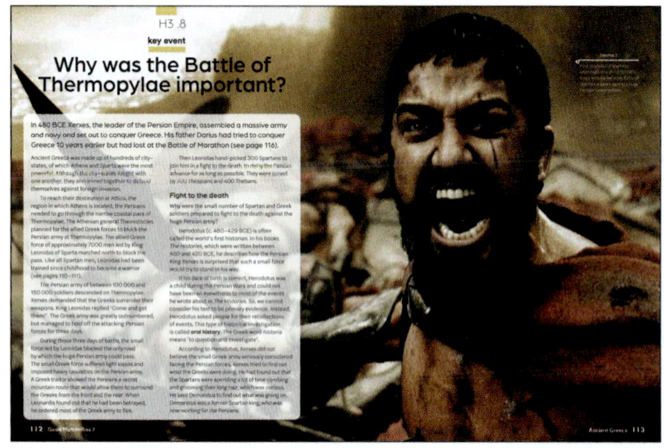

Source 3

Discover the people and events that have shaped our world.

civics+ citizenship

economics +business

The study of Humanities is more than just History and Geography – it is also the study of Civics and Citizenship, and Economics and Business. In every chapter of this book you will discover either a Civics and Citizenship lesson or an Economics and Business lesson. School is busy and you have a lot to cover, so designing a textbook where the important Civics and Citizenship and Economics and Business content is placed meaningfully next to relevant History or Geography lessons makes good sense, and will help you to connect your learning.

As you work through the Civics and Citizenship and Economics and Business sections in this book, you will be working your way up a Learning Ladder for these subjects too!

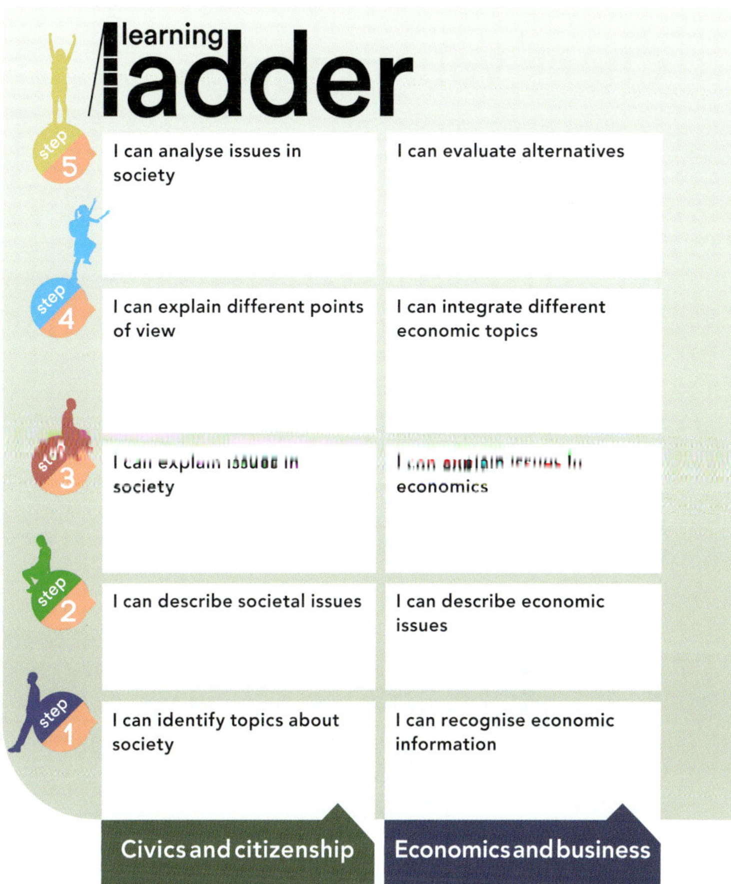

Source 4

Explore Civics and Citizenship, and Economics and Business, alongside your History course.

History How-To

In the middle of the book, you will find a skills section called 'History How-To'. In this section, there are explanations about how to perform each skill, including writing and research skills. There are *lots* of worked examples. Refer to it often, especially when answering the Learning Ladder questions and Masterclass activities at the end of each chapter.

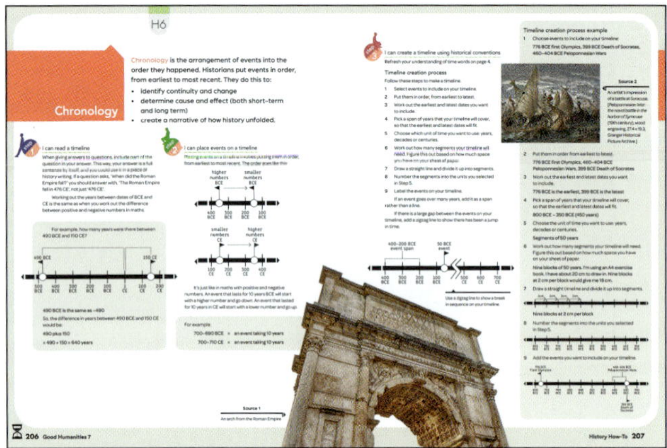

Source 5

The History How-To section is your key to success – refer to it often!

Masterclass

At the end of each chapter is a review section. The questions here are organised by the steps on the Learning Ladder. You can complete all of the questions *or* your teacher might direct you to complete just some of them, depending on your progress.

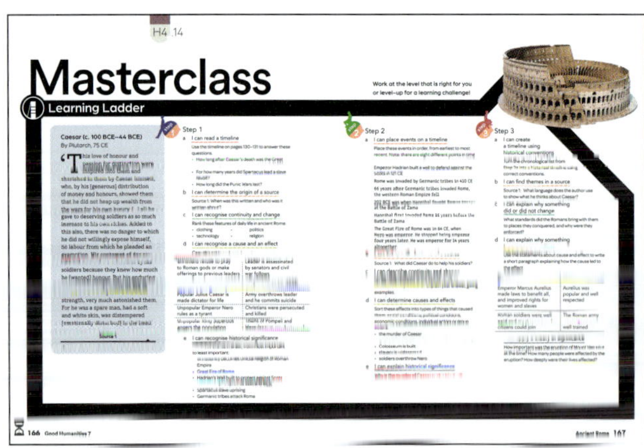

Source 6

You did it! We knew you could! The Masterclass is your opportunity to show your progress. Take charge of your own learning and see if you can extend yourself.

Capstone

After you complete a chapter, it's time to put your new knowledge and understanding together for the capstone project to show what you *know* and what you *think*. In the world of building, a capstone is an element that finishes off an arch, or tops off a building or wall. That is what the capstone project will offer you, too: a chance to top off and bring together your learning in interesting and creative ways. It will ask you to think critically, to use key concepts and to answer 'big picture' questions. The capstone project is accessible online; scan the QR code to find it quickly.

Source 7

The capstone project brings together the learning and understanding of each chapter. It provides an opportunity to engage in creative and critical thinking.

Learning Ladder H0.3

1 What are the different types of questions in this textbook? Describe them in your own words.

2 How can you use the Learning Ladder to monitor your progress in Year 7 History?

3 As a class, discuss the idea of 'monitoring your own progress'. Why is this important?

4 Read through the steps of the History Learning Ladder and consider where you might already be up to for each skill, based on your prior learning.

How are our histories connected?

History helps us connect things through time. Over the next four pages you will read a brief history of our world, from the beginning of the universe to the end of the ancient world, which is the period you will be studying this year.

The early Earth

The **universe** is approximately 13.8 billion years old. Scientists believe that the universe began with an explosion referred to as the 'big bang'. Earth itself came into being about 4.5 billion years ago, most likely from a swirling mass of debris left over from the creation of the Sun.

Life first emerged on our planet about 3.6 billion years ago. The first simple animals appeared about 600 million years ago and the first mammals 160 million years ago. The first primates appeared around 60 million years ago.

Source 1

An artist's impression of the big bang

Source 2

The human lineage

Ancient humans

About 250000 years ago the first humans evolved in Africa. We would later refer to this species as ***homo sapiens*** (meaning 'wise man'). Over the next 200000 years, humans migrated out of Africa into all the major landmasses. This happened in a few different waves: first to the Middle East, then India, Asia and Australia, then to Europe and the Americas.

Humans are the most widespread animal we know of in the history of the planet. Scientists believe that humans may have left Africa because of overpopulation in some areas or because they were looking for more food and resources. We don't know for sure why our distant **ancestors** left Africa – it may have been out of pure curiosity.

Early humans lived during the period known as the '**Ice Age**' (or Pleistocene era). During the Ice Age, much of Earth's water was frozen at the northern and southern tips of the planet – which meant that the seas were more than 100 metres lower than they are today. When the Ice Age ended, the water melted and the seas rose to their current levels.

The Stone Age can be divided into two ages: the **Palaeolithic** ('Old Stone Age') and the **Neolithic** ('New Stone Age'). The Neolithic Age began with the invention of farming, one of the two most important developments in history. (The other development was the **Industrial Revolution**, when humans began using machines to do work previously done by hand.)

Source 3

A world map showing the Ice Age, when sea levels were lower and land bridges joined the Australian mainland to Tasmania and Papua New Guinea.

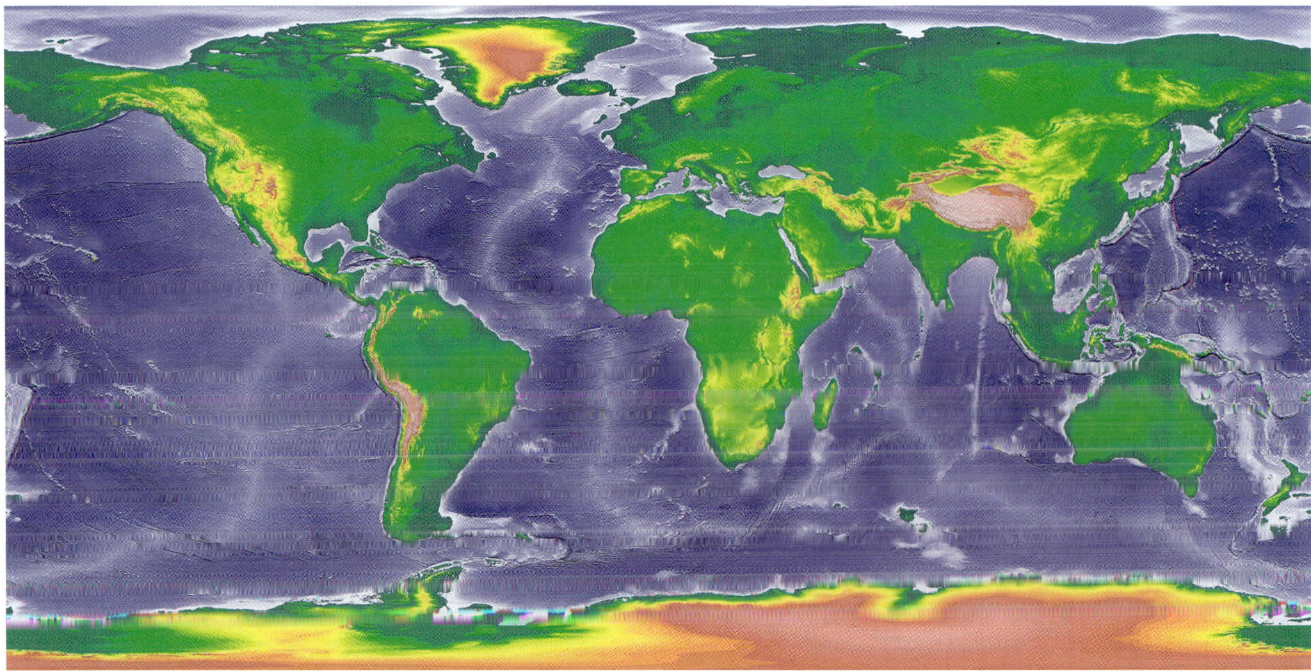

Source: NOAA/NCEI, 2008

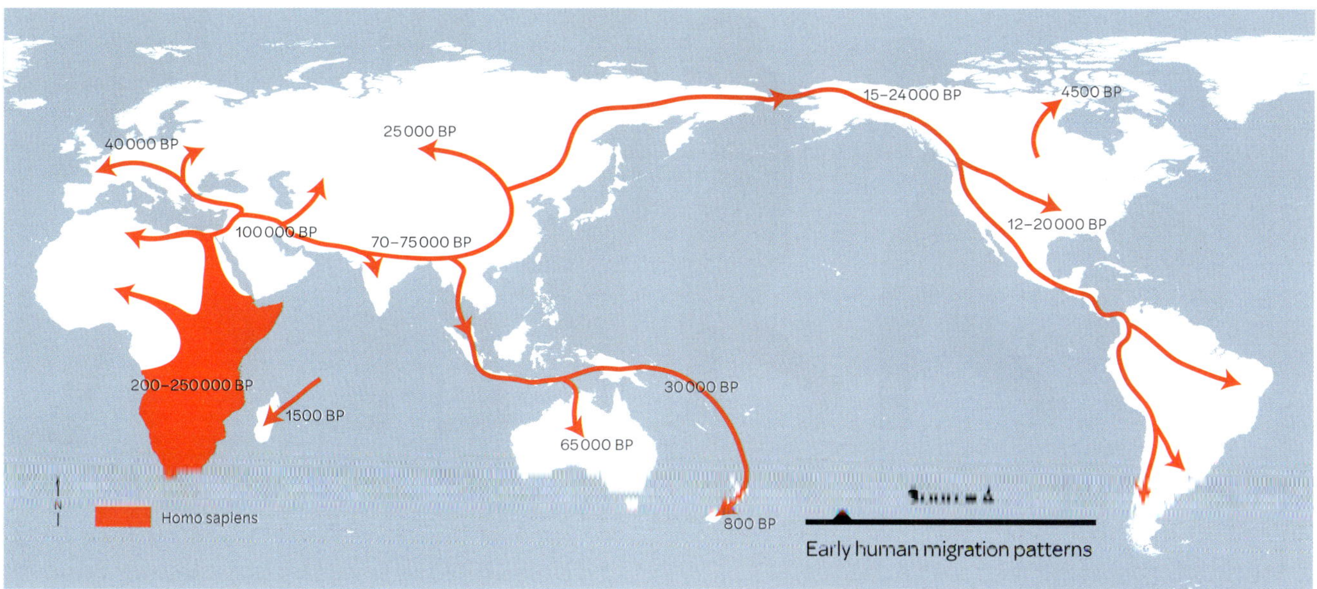

The Neolithic revolution

The Neolithic 'revolution' saw humans gradually change from hunting animals and gathering wild plant foods to **agriculture**. Instead of roaming around like **nomads**, people grew plants from seeds they dropped into the soil. Humans also began keeping animals with them in settled areas, rather than following herds around or hunting for animals.

The Neolithic revolution led to huge changes in lifestyle. People lived in permanent settlements and didn't move around all the time.

Source 5

One of Europe's best-preserved Neolithic sites is Skara Brae on the Orkney Islands, off the coast of Scotland. [Historic Environment Scotland, *Skara Brae* (1999). grooved ware and earth sheltering, Orkney Islands, Scotland.]

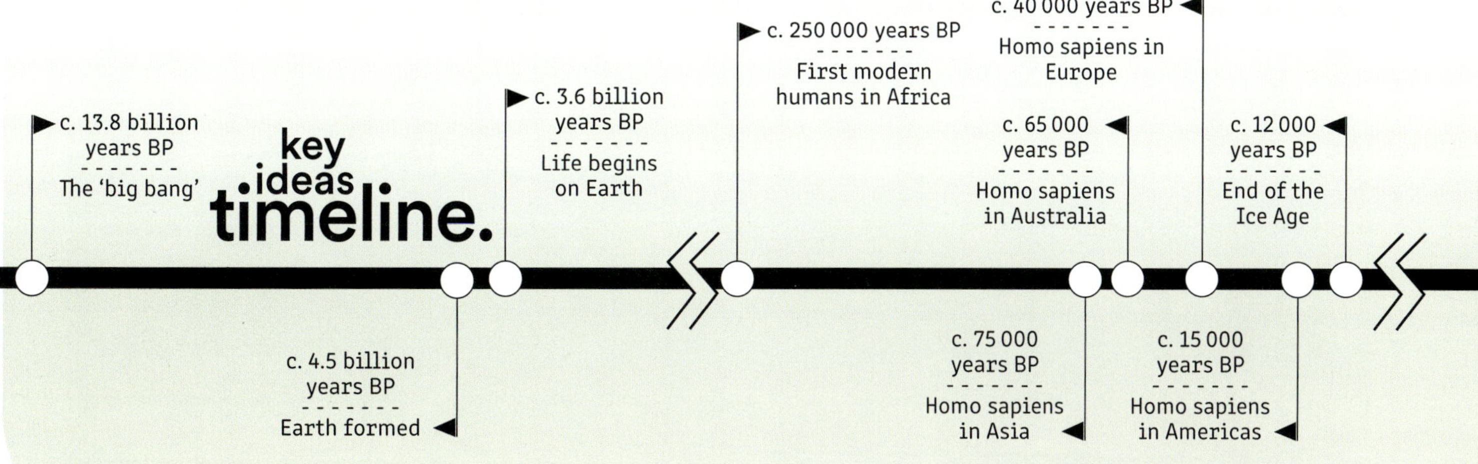

However, being farmers rather than nomads meant that people had to work longer hours, and their diet became more limited. The amount of work involved in farming was also unequal, especially between men and women. But because farming was more efficient than hunting and gathering, Neolithic people produced more food, so the farmer populations increased more quickly. This might explain why Neolithic ways became widespread.

Civilisation, culture and religion

Because farmers produced more food than they needed, it allowed some people to do things other than hunt for food, eventually leading to the complex human society we call **civilisation.**

The main features of civilisations are:

- production of readable texts
- consistent living patterns, such as living in towns or cities, or moving to certain areas at certain times of the year
- specialised occupations
- oral or written laws
- shared religious faith
- shared cultural values
- stable food supply
- well defined social structures
- technology and tool use
- trade.

Civilisation, culture and religion made groups of humans different from each other.

Religious leaders were possible in civilisations because better food production freed up time and allowed some people to focus on religious matters. Certain religions are especially important in world history because they spread widely and are followed by millions of people.

The river civilisations

The first civilisations sprung up along the valleys of major rivers. These areas had fresh water and fertile soil because the rivers flooded regularly. Not everyone was needed to grow food, which allowed some people to specialise in craft, writing, administration and religious practice. Rivers allowed people to travel easily, and to protect their territory from invaders.

Mesopotamia

One of the earliest groups to be called a 'civilisation' by European archaeologists lived in Sumer in Mesopotamia in 3500 BCE. Mesopotamia lies between the Tigris and Euphrates rivers in modern-day Iraq. These rivers flooded unpredictably and violently. The area contained independent cities that didn't generally unite, except perhaps to build **irrigation** canals. The area was central, allowing trade between faraway places such as Egypt and India. The first writing was developed in Sumer, in about 3500 BCE.

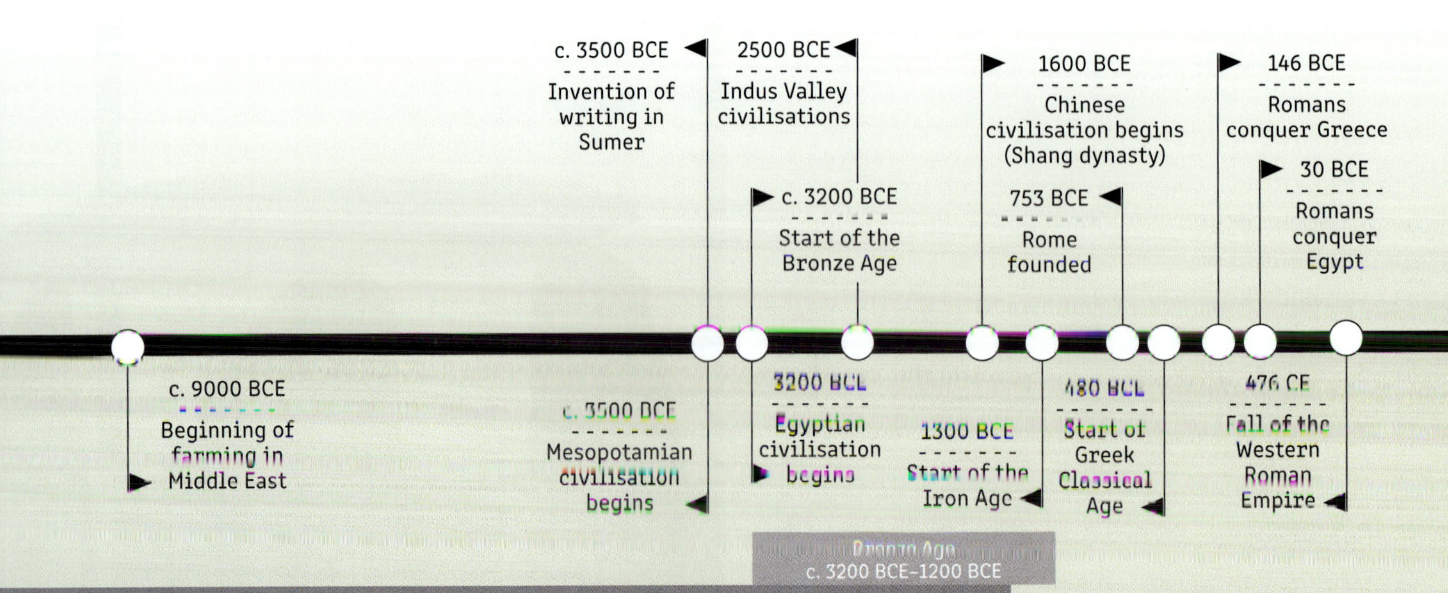

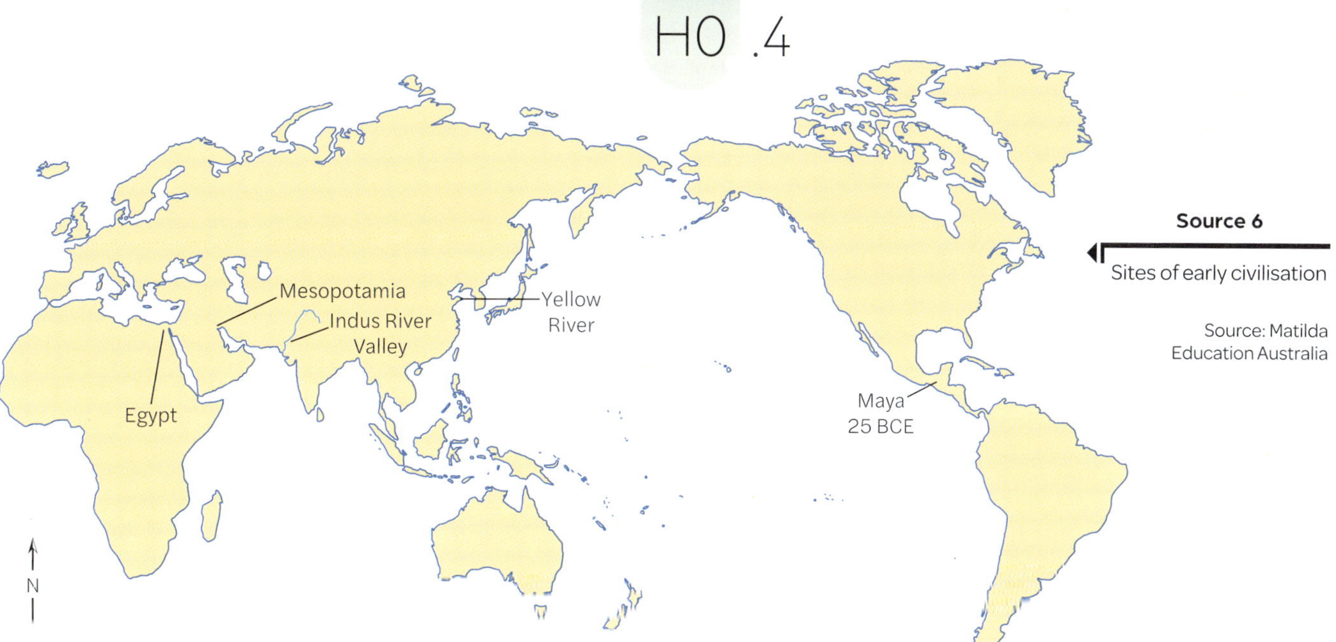

Source 6

Sites of early civilisation

Source: Matilda Education Australia

Egypt

In about 6000 BCE in northern Africa, groups of hunter-gatherers began settling along the Nile River and growing crops. The Nile River allowed for easy long-distance travel between the north and the south, which led to separate kingdoms in the Upper and Lower Nile. These kingdoms were unified by about 3200 BCE, which led to the development of a 'civilisation'.

Indus River Valley

A number of civilisations developed along the Indus River Valley, in north-west India and Pakistan. There is evidence of farming from about 4000 BCE. By about 2600 BCE, many towns and cities were established. Indus River Valley civilisation was at its peak from about 2500 BCE to 2000 BCE. Civilisation in this area came to an end in about 1500 BCE, for reasons that aren't known for sure.

Yellow River (Huang He)

The Yellow River civilisation is considered to be the cradle of Chinese culture. Some estimates suggest the area was settled by 9500 BCE. It developed into a fully-fledged Bronze Age civilisation by about 2500 BCE, when early tribes united into kingdoms to find solutions to common problems: flooding and irrigation.

Learning Ladder H0.4

1. Source 4: List the order in which humans migrated to the six continents.
2. What is the major difference between the Old Stone Age and the New Stone Age?
3. Rank the features of civilisation in order of importance, based on your opinion.
4. Source 2: Describe the physical changes you can see as humans evolved from earlier forms.
5. Source 4: Why do you think humans migrated in this pattern?
6. Source 3: What would have been some benefits and disadvantages of living at this Stone Age site?
7. Would you have preferred to have lived in the Old Stone Age or the New Stone Age? Justify your response.

Aboriginal and Torres Strait Islander peoples and cultures

H1

WHO WERE THE FIRST AUSTRALIANS?

page 24

source analysis

page 26

HOW DO WE KNOW ABOUT ANCIENT AUSTRALIANS?

economics + business

page 34

WHO OWNS THE LAND?

civics + citizenship

page 42

HOW IS IDENTITY FORMED?

Aboriginal and Torres Strait Islander readers are warned that the following chapter contains images of deceased persons.

How can I understand Aboriginal and Torres Strait Islander peoples and cultures?

While Neanderthals roamed Europe, and long before civilisations were formed in ancient Egypt, India, China, Greece or Rome, Aboriginal and Torres Strait Islander peoples were trading, burying their dead, engaging in elaborate rituals and producing stunning art here in Australia. Ancient Australia was home to the beginnings of the oldest continuous living culture on earth – a culture that continues to this day. By studying ancient Australia, we learn about a people who are at the heart of modern Australian identity.

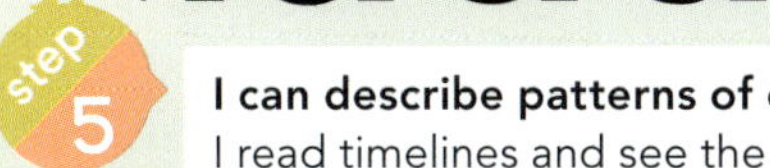

Step	Chronology	Source analysis	Continuity and change
step 5	**I can describe patterns of change** I read timelines and see the 'big picture'. I group timeline events and see if they show patterns of change. I know typical historical patterns to look for.	**I can use the origin of a source to explain its creator's purpose** I combine knowledge of when and where a source was created to answer the question, 'Why was it created?'	**I can evaluate patterns of continuity and change** I answer the question, 'So what?' about patterns of continuity and change. I weigh up different things and debate the importance of a continuity or a change.
step 4	**I can distinguish causes, effects, continuity and change from looking at timelines** I read timelines and find events that are linked by cause and effect. I find things that are the same or different from then until later times.	**I can use my outside knowledge to help explain a source** I have enough outside knowledge about Indigenous Australian people and cultures to help me explain a source.	**I can analyse patterns of continuity and change** I see beyond individual examples of continuity and change in ancient Australia and identify broader patterns, and I explain why they exist.
step 3	**I can create a timeline using historical conventions** When given a set of events, I construct a historical timeline, making sure I use correct terminology, spacing and layout.	**I can find themes in a source** I look a bit closer into a source and find more than just features. I find themes or patterns in the source.	**I can explain why something did or did not change** I answer the question 'why?' something changed or stayed the same between historical periods.
step 2	**I can place events on a timeline** When given a list of Indigenous Australian events, I put them in order from earliest to latest, the simplest kind of timeline.	**I can list specific features of a source** I look at an Indigenous Australian source and list detailed things I can see in it.	**I can describe continuity and change** I have enough content knowledge about ancient Australia and modern Australia to recognise and describe what is similar or different.
step 1	**I can read a timeline** I read timelines with Indigenous Australian events on them and answer questions about them.	**I can determine the origin of a source** I can work out when and where an Indigenous Australian source was made by looking for clues.	**I can recognise continuity and change** I recognise things that have stayed the same and things that have changed from ancient Australia until now.

Source 1

Gwion Gwion (formerly called Bradshaw) rock art, depicting a fish hunt. This artwork was produced more than 20 000 years ago. [Raft Point Gallery, Kimberley Region, Western Australia.]

Warm up

Cause and effect	Historical significance
I can evaluate cause and effect I answer the question 'So what?' about cause and effect. I weigh up different things and debate the importance of a cause or effect.	**I can evaluate historical significance** I answer the question 'So what?' about things that are supposedly historically important. I weigh up events against each other and cast doubt on how important things are.
I can analyse cause and effect I don't just see a cause or effect as one thing. I determine the factors that make up causes and effects.	**I can analyse historical significance** I separate out the various factors that make something historically important in ancient Australian history.
I can explain why something is a cause or an effect I can answer 'how?' or 'why?' a cause led to an effect in ancient Australia.	**I can apply a theory of significance** I know a theory of significance. I use it to rank the importance of ancient Australian events.
I can determine causes and effects Applying what I have learnt about ancient Australia, I can decide what the cause or effect of something was.	**I can explain historical significance** I answer the question 'why?' about things that were important in ancient Australia.
I can recognise a cause and an effect From a supplied list, I recognise things that are causes or effects of each other in Indigenous Australian cultures.	**I can recognise historical significance** When shown a list of things from ancient Australian history, I can work out which are important.

Chronology

1 The rock paintings in Source 1 were created more than 20 000 years ago. What year BCE would 20 000 years ago be?

Source analysis

2 Source 1: Describe what you can see in the rock painting, including human and animal figures. What type of activity is being recorded in the artwork?

Continuity and change

3 Aboriginal and Torres Strait Islander peoples are part of the oldest living continuous culture in the world. Discuss with a partner what you think this means.

Cause and effect

4 Source 1: The rock painting shows a fish hunt. Why might the painter have created this artwork?

Historical significance

5 Source 1: Is this rock painting historically important? Explain how you came to your answer.

How ancient is Australia?

Confirmed evidence shows that the first humans arrived in northern Australia more than 65 000 years ago. Emerging information suggests that the first humans could have arrived 120 000 years ago. Living in a new land, the first Australians developed innovative land management techniques and rich cultures tied to the land they lived on.

Source 1

Excavation by archaeologists at the Warratyi rock shelter in the Flinders Ranges found a bone from a diprotodon that suggests humans were hunting and eating megafauna (giant animals) about 49 000 years ago. The megafauna became extinct as a result of climate change during the last Ice Age.

Source 2

An axe-grinding stone found in Kakadu National Park in 2017 is the oldest confirmed evidence of the occupation of Australia 65 000 years ago. The tool-making technology developed by early Australians was the most advanced in the world.

Source 3

The discovery of Australia's oldest human at Lake Mungo, along with other artefacts, provided evidence about Australia's ancient past and how people lived. Mungo Man is thought to have died 43 000 years ago.

Source 4

A charcoal drawing found at Narwala Gabarnmang in the Northern Territory is Australia's oldest known rock art, thought to be painted 28 000 years ago. Rock paintings give us clues about the everyday life and beliefs of the first Australians.

key ideas timeline.

- c. 65 000 years BP — Estimated time of human arrival in Australia
- c. 49 000 years BP — Evidence that humans were hunting megafauna in Australia
- c. 43 000 years BP — Mungo Man dies – his skeleton later becomes the earliest evidence of humans in Australia
- c. 40 000 years BP — Tasmania occupied by this time
- c. 30 000 years BP — First Australians dispersed across whole continent
- c. 28 000 years BP — Australia's oldest known rock art created

Source 5

The Kulin Tanderrum (ceremony) is the annual meeting of the five clans of the Kulin Nation in Melbourne: Wurundjeri, Boon Wurrung, Taungurung, Dja Dja Wurrung and Wadawurrung. Through ceremonies, stories and art, Australian Aboriginal peoples connect to the Dreaming (see page 36) and the ancestral spirits that created the world.

Source 6

The Torres Strait Islands formed when the land bridge between Australia and Papua New Guinea was submerged at the end of the Ice Age, 8000 years ago. Torres Strait Islander peoples are sea travellers and traders.

Source 7

In 1770, James Cook landed in Australia and claimed the east coast of the continent for Great Britain. In 1788, the First Fleet arrived to establish a colony at Sydney Harbour. [Algernon Talmage, *The Founding of Australia. By Capt. Arthur Phillip R.N. Sydney Cove, Jan. 26th 1788* (1937 CE), oil on canvas, 106.5 x 77 cm, State Library of New South Wales, New South Wales, Australia.]

c. 12 000 years BP
The land bridge between mainland Australia and Tasmania submerged, isolating the Tasmanian population

c. 8000 BCE
Land bridge between Australia and Papua New Guinea submerged by rising seas

c. 4000 BCE
Dingo introduced to Australia

Learning Ladder H1.1

Show what you know

1 When did the first Australians arrive on our continent?

2 Source 5: How do Aboriginal people connect to the Dreaming?

Chronology

Step 1: I can read a timeline

3 True or False:

a There is evidence of Aboriginal rock paintings before the land bridge separating Tasmania and the mainland flooded.

b Australia's first known rock art was created about 30 000 years BP (before present).

c Humans were hunting megafauna in Australia earlier than 50 000 years BP (before present).

Step 2: I can place events on a timeline

4 List these events in order, from oldest to most recent.

3000 BCE Backed blades were being used as tools

47 000 years BP Bone tools were used in South Australia

4000 years after the bone tools were used in South Australia, rock engravings were created in South Australia.

63 000 years BP Axe-grinding stones were used in the Northern Territory.

Step 3: I can create a timeline using historical conventions

5 Convert all the dates in the list of events in question 4 into BCE and use them to create your own timeline.

Step 4: I can distinguish causes, effects, continuity and change from looking at timelines

6 Which events isolated the first Australians on the continent of Australia? How do you think this has impacted the development of their cultures?

HOW TO

Chronology, page 206

Who were the first Australians?

Current evidence confirms that modern humans arrived in Australia from Africa at least 65 000 years ago. Indigenous Australians are the Aboriginal and Torres Strait Islander peoples of Australia, who are descended from these first peoples.

Arrival in Australia

Aboriginal and Torres Strait Islander people living in Australia today are the descendants of the first people to come to Australia at least 65 000 years ago. An important **archaeological** find in Kakadu in 2017 confirmed the arrival date. More than 10 000 **artefacts** were uncovered in a layer of soil that has been dated to at least 65 000 years BP.

The archaeological dig took place on the Mirarr people's land. It was led by Associate Professor Chris Clarkson from the University of Queensland. In the top layer of soil, archaeologists found shells, bones and spear heads. Further down, they found axes and different kinds of grindstones.

'We have technologies developing here that we don't see developed anywhere else in the world for another 30 000 years – even 50 000 years ago in the case of Europe. They didn't have axes until 10 000 years ago,' Clarkson said.

To reach Australia, the first peoples had to cross close to 100 kilometres of ocean, which suggests they also had reliable seagoing craft and astronomical navigation skills.

Creation stories

Some Indigenous Australians believe that the world was created during the **Dreaming**, and that their ancestors have been here since the beginning of time.

Aboriginal stories, songs, dance and art explain that the Earth's surface was once nothing but mud. The landscape changed when ancestor spirits formed the mud into mountains, rivers, islands and other landforms. They gave life to humans, plants and animals.

World: Ancient human migration routes

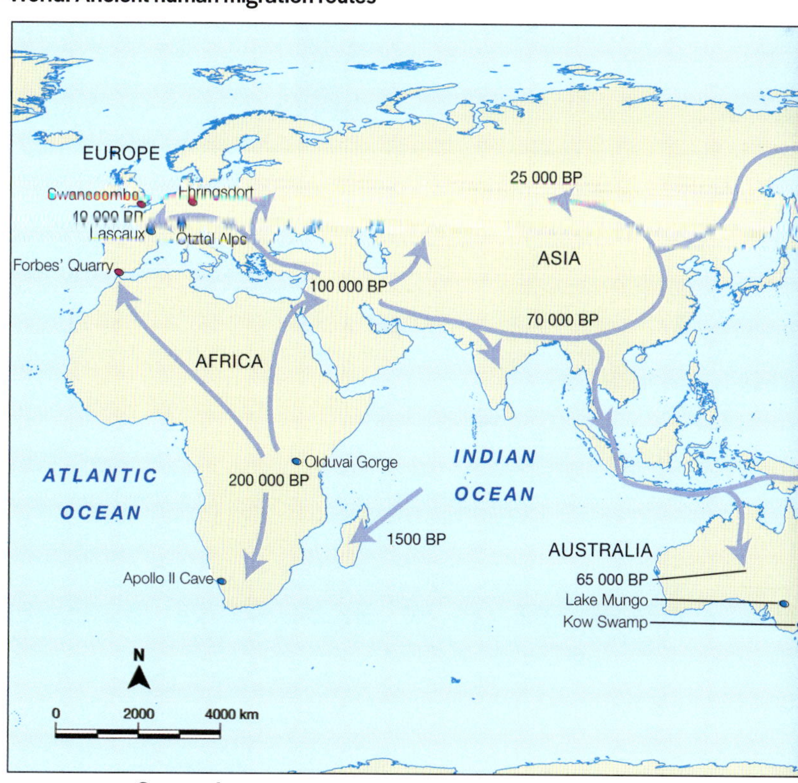

Source 1

World map showing migration routes

Source 2

Scientist Dr Elspeth Hayes (left) with May Nango (right) and Mark Djandjomerr (centre). May is a traditional owner and custodian of the archaeological site that is on Mirarr land, located in the Kakadu National Park in the Northern Territory.

Out of Africa

DNA evidence shows that the Indigenous Australians are linked to the first people who left Africa between 50 000 and 70 000 years ago in search of new hunting grounds and places to live.

As descendants of the first people to leave Africa, First Nations people in Australia have the oldest continuous culture on Earth. Migration of early peoples to Australia stopped about 50 000 years ago.

Arriving in Australia

Indigenous Australians were among the first modern humans to travel through Asia and Australia, which were then unknown territories. They are thought to have crossed from Asia into Australia during the last Ice Age (about 110 000 to 12 000 years ago) when the sea level was about 120 metres below its current level.

During the journey from South-East Asia in the Ice Age, it would have been possible to walk most of the way, although they would still have had to make sea crossings of over 90 kilometres. It is not known what kind of boat was used for these journeys.

Source: Matilda Education Australia

Learning Ladder H1.2

Show what you know

1 What was found at Kakadu from 65 000 years ago?

2 How did the first peoples get to Australia from Africa?

3 What difficulties would people with ancient technology have faced making a 90-kilometre sea crossing?

4 For how long do creation stories suggest Aboriginal and Torres Strait Islander people have been in Australia?

Chronology

Step 1: I can read a timeline

5 Look at the timeline on page 22. When is the first evidence of humans on the Australian mainland and in Tasmania?

Step 2: I can place events on a timeline

6 Find the dates of these events in the text and list them in order, from oldest to most recent.

Ancient migration to Australian continent stops

First people arrive on Australian continent

First people leave Africa

End of Ice Age

Step 3: I can create a timeline using historical conventions

7 Use the events in question 6 to create a timeline using correct conventions. Make sure you use equal spacing for equal years.

Step 4: I can distinguish causes, effects, continuity and change from looking at timelines

8 Look at the timeline on page 22. Which events show stability over time?

Chronology, page 206

How do we know about ancient Australians?

The discovery of Australia's oldest human remains at Lake Mungo, along with other artefacts, provide evidence of Australia's ancient past and how people lived.

Mungo Man

Ancient Australian history was rewritten when the bones of Australia's oldest and most complete human were uncovered at Lake Mungo in 1974.

The 43 000-year-old skeleton was named 'Mungo Man' after the dried-up lake basin where he was found. The skeleton was taken to the Australian National University, Canberra. Further research revealed that two of Mungo Man's lower teeth had been removed during adolescence, possibly as an initiation into adulthood.

Archaeologists began to understand that Australia's first peoples were far more advanced than other early humans in Europe at the same time, such as the Neanderthals. Ancient Aboriginal people had a sophisticated culture that included a complex language, innovative tools and a belief system with ceremonies that are still practised today.

Burial ceremonies

Archaeologists were amazed at the ritual of Mungo Man's burial. He had been laid out with his hands in his lap, and his body covered in a red ochre (paint) that had been transported from hundreds of kilometres away. The remains of a small fire were also found nearby.

'To find on the shores of Lake Mungo the extraordinary ritual of ochre and fire was a moment of sheer wonder,' said Dr Jim Bowler, the geologist who found the skeleton.

Funeral ceremonies are different for each Indigenous group. Some believe the spirit of the dead person remains where they died. Others believe that the spirit of the dead person joins their spiritual ancestors in the Dreaming.

Source 1

The 43 000-year-old skeleton of Mungo Man is a primary source of evidence uncovered by archaeologists in 1974. The skeleton was taken to the Australian National University in Canberra for further study. In 2017, the skeleton was returned to his original resting place after a campaign by traditional Mutti Mutti, Paakantyi and Ngyampaa owners to have him returned to his Country.

Source 2

This painting is a primary source created by convict artist Joseph Lycett. His paintings recorded Indigenous Australians in the early colonial period.

[Lycett, Joseph, *Aborigines spearing fish, others diving for crayfish, a party seated beside a fire cooking fish* (c. 1817), watercolour, 28 × 17.7 cm, National Library of Australia, Australian Capital Territory, Australia.]

In some Aboriginal communities, people are not allowed to say the name of a deceased person. This is to make sure the spirit is not called back to this world. Many Indigenous Australians respect this tradition today.

Further evidence

The ancient people who lived in the Lake Mungo region left behind a variety of materials that help us to understand how they lived, and how they used resources from their environment.

Aboriginal people used stone tools, such as large flat grindstones, to grind seeds into flour and to grind points for spear tips. They used stone knives and scrapers to butcher animals, to clean skins and to make wooden tools. Stone axes were ground to a sharp edge and used to cut wood.

We have also learnt about early human life at Lake Mungo from studying middens. A **midden** is a place where people leave the waste from their meal preparation. The middens at Lake Mungo provide evidence that the people living there ate fish and shellfish, as well as small mammals such as native rats, bandicoots and wallabies.

Learning Ladder H1.3

Show what you know

1 What did ancient Australians eat?

2 How did finding Mungo Man change the way people thought about ancient Aboriginal people?

3 What different beliefs about the afterlife did some Aboriginal Australians have?

4 Source 2: What primary source evidence would the artist have needed to paint this picture accurately?

Source analysis

Step 1: I can determine the origin of a source

5 Source 1: From when and where is this source?

Step 2: I can list specific features of a source

6 Describe the various activities people are doing in Source 2.

Step 3: I can find themes in a source

7 What kinds of technology did ancient Australians use? Use Source 2 and the text in your answer.

Step 4: I can use my outside knowledge to help explain a source

8 Source 1: Was it right that Mungo Man was taken away to a university for study? Was it right for local peoples to ask to have the skeleton back? Justify your answers.

Source analysis, page 209

Source 1

These ground-edge axes with handles are similar to the world's oldest ground-edge axe, which was found in the Kimberley region of Western Australia.

What was life like in ancient Australia?

The first Australians adapted to the challenges of their new home. They invented new tools and devised new ways of managing the land.

Surviving in a new land

When early humans arrived in Australia more than 65 000 years ago, they lived alongside giant animals, called **megafauna**.

An excavation by archaeologists at the Warratyi rock shelter in South Australia found a bone from a giant wombat, and fragments of an eggshell from a giant flightless bird. These finds suggest that humans were hunting and eating megafauna 49 000 years ago.

The occupants of the shelter may have also used megafauna skins. A 40 000-year-old sharpened bone point found at the site could have been used for needlework to make garments from the skins.

There is documented evidence that Aboriginal peoples maintained large pastures of murnong (yam daisy) around Melbourne. This required deliberate cultivation to ensure the yams were edible.

Seasonal migration

To be sustainable, early Australians practiced **seasonal migration** – moving to another area for a period of time to ensure they did not exhaust their land of plants and animals.

They used a range of spears, clubs and boomerangs to hunt kangaroos and wallabies, and they used boomerangs and nets to help in hunting birds. They used rocks to make fish traps in rivers and creeks.

In 2014, a team of archaeologists discovered an edge-ground axe dated to 49 000–44 000 years BP in Winjana Gorge in Western Australia. It is the oldest example of an edge-ground axe with a handle found anywhere in the world.

Managing the land

The first Australians developed great knowledge of land and resource management. The movement of their camps was coordinated with patterns of climate, plant growth and animal movement. Wildlife was observed to find water holes.

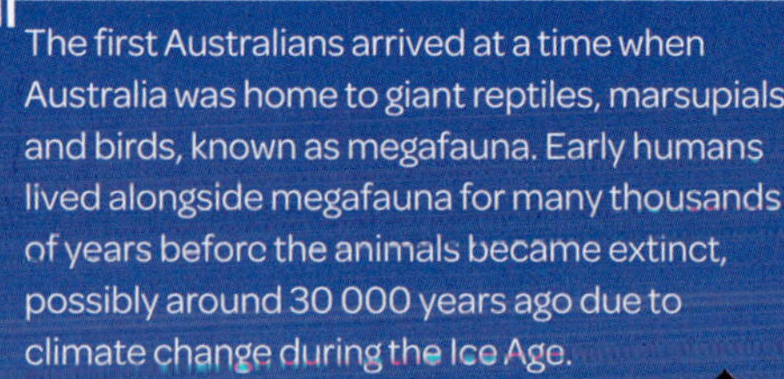

Source 3

The first Australians arrived at a time when Australia was home to giant reptiles, marsupials and birds, known as megafauna. Early humans lived alongside megafauna for many thousands of years before the animals became extinct, possibly around 30 000 years ago due to climate change during the Ice Age.

Trading

Indigenous Australians created extensive trade routes across the Australian continent, facilitating trade and exchange over many thousands of kilometres. If a person trading items wished to enter the Country of another First Nations group, they would need to seek permission and this would be granted after the required waiting time, and usually with a ceremony. In some cases, they would be given a message stick and this would be a symbol of having permission to be on Country that was not their own. This is similar to a travel visa today.

Source 2

Fire was important in land management for first Australians and remains so today. Firesticks were used to burn grasslands to promote regrowth of lush new grass, which attracted species such as kangaroos. Fire was also used as a tool to drive game towards hunters. Modern firesticks are still used for control-burning today.

Learning Ladder H1.4

Show what you know

1 List all the things Indigenous Australians used megafauna for.

2 What evidence is there that the first Australians were farmers?

3 How did the first Australians manage the land sustainably?

Cause and effect

Step 1: I can recognise a cause and an effect

4 Copy this table and match the cause with the effect.

Cause	Effect
Sea levels are lower	Megafauna become extinct
Climate change	More kangaroos come to eat new grass
Sea levels rise	Sea voyage from South-East Asia to Australia is made shorter
Firesticks used to burn grassland	Tasmanian Indigenous people stranded

Step 2: I can determine causes and effects

5 Why did Indigenous people burn grassland?

Step 3: I can explain why something is a cause or effect

6 What effects might there be from having edge-ground axes?

Step 4: I can analyse cause and effect

7 Megafauna became extinct due to climate change during the last Ice Age. Is this similar or different to modern animals becoming extinct? Justify your response.

HOW TO

Cause and effect, page 215

Who are Australia's Indigenous peoples?

Australian Aboriginal and Torres Strait Islander peoples are direct descendants of the first humans that emigrated from Africa. Aboriginal and Torres Strait Islander peoples developed distinct cultures in a range of diverse environments.

Source 1 is a reproduction of N.B. Tindale's 1974 map of Indigenous group boundaries existing at the time of first European settlement in Australia. It is not intended to represent contemporary relationships to land.

Tribal boundaries in Aboriginal Australia, Norman B. Tindale, 1974

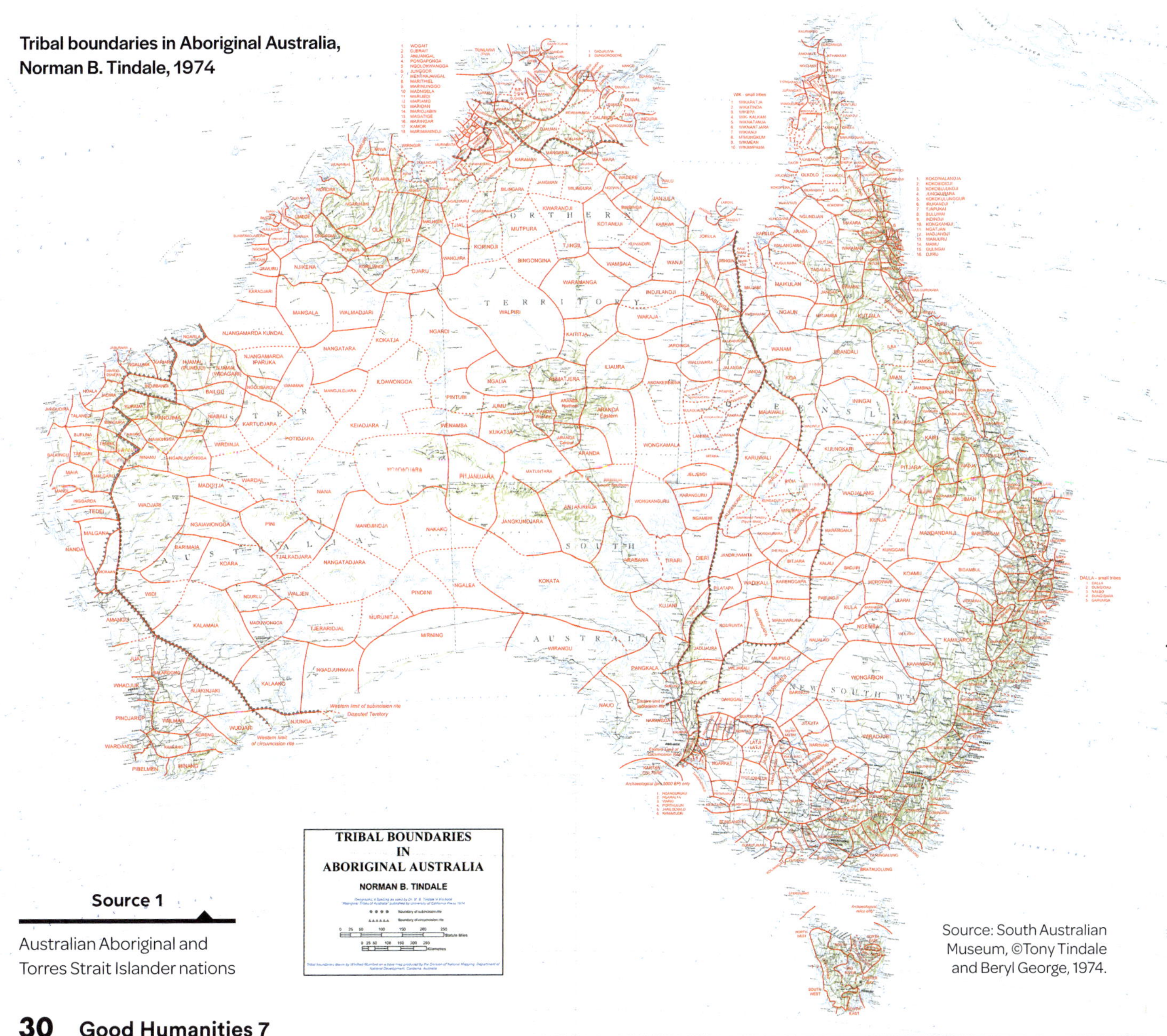

Source 1

Australian Aboriginal and Torres Strait Islander nations

Source: South Australian Museum, ©Tony Tindale and Beryl George, 1974.

Source 2

A warrior from Saibai Island in the Torres Strait off far north Australia. Torres Strait Islanders are of Melanesian origin and have a unique and rich culture.

Different cultural groups

The first Australians were migratory people, moving from place to place with the seasons to care for the land.

Because of the size of the Australian continent, many Indigenous groups never came into contact with one another. As a result, their languages, stories, art and ways of life developed differently. These different cultural groups are referred to as **nations**.

When Europeans arrived in Australia in 1788, it is estimated there were up to 1 000 000 Indigenous people living in 500 different nations around Australia. Aboriginal peoples occupied and used the entire continent. They had adapted successfully to a wide range of environments, from tropical rainforests to dry desert regions.

Population densities ranged from 3 to 21 square kilometres per person in fertile riverine and coastal environments to more than 90 square kilometres per person in deserts.

Torres Strait Islanders

Torres Strait Islander peoples come from the islands of the Torres Strait, which stretches between northern Queensland and Papua New Guinea. Torres Strait Islanders are distinct from Aboriginal people on the Australian mainland. They originated in Melanesia, a region in the Pacific Ocean to the north-east of Australia.

Along with the Aboriginal inhabitants of Australia, the people of Melanesia also emigrated from Africa. Australia and New Guinea were connected by a land bridge because of low sea levels during the Ice Age.

Torres Strait Islander peoples are sea travellers, traders, fishers and gardeners. Their culture is unique and includes elements from Australia, Papua New Guinea and the surrounding region. Torres Strait Islander culture is linked to Tagai, a great sea hero. Tagai is represented in the sky by several constellations. These constellations are used as a guide for the seasonal cultivation of crops and gardens.

Show what you know

1 How many different Indigenous nations were there when European people arrived in 1788?

2 What are two different meanings of the word 'nation'?

3 Source 1: Approximately how many nations would you go through if you walked from Melbourne to Perth?

Historical significance

Step 1: I can recognise historical significance

4 Rank these events in order, from least to most significant:

- Europeans arrived in 1788
- Mungo Man died 43 000 years ago
- the dingo was introduced from Indonesia around 4000 BCE
- first use of ground-edge axes around 45 000 years ago.

Step 2: I can explain historical significance

5 Why was colonisation by Europeans in 1788 a significant moment in Australian history?

Step 3: I can apply a theory of significance

6 Apply Partington's theory of significance (page 8) to Australia's colonisation by Europeans. Show your response in table format, including these sections: Importance, Depth, Number, Time and Relevance.

Step 4: I can analyse historical significance

7 How many Indigenous people lived in Australia in 1788? Undertake research to find out how many Indigenous people live here now. Has this number increased as much as the number of non-Indigenous people? Why or why not?

Historical significance, page 219

How are Indigenous Australians connected to Country?

The deep relationship between Indigenous Australians and the land is often described as connection to Country. Dhanggal Gurruwiwi, a Galpu elder, explains: 'The land and the people are one, 'cause the land is also related. In our kinship system, as a custodian I'm the child of that land.'

Deep links to Country

Aboriginal people have a deep spiritual, physical, social and cultural connection to the land that is often described as **connection to Country**.

Palyku woman Ambelin Kwaymullina explains: 'For Aboriginal peoples, Country is much more than a place. Rock, tree, river, hill, animal, human – all were formed of the same substance by the ancestors who continue to live in land, water, sky. Country is filled with relations speaking language and following Law, no matter whether the shape of that relation is human, rock, crow, wattle. Country is loved, needed and cared for, and Country loves, needs and cares for her peoples in turn. Country is family, culture, identity. Country is self.'

Uluru

One of Australia's most famous tourist attractions is the rock Uluru. However, it is more than just a rock to the Yankunytjatjara and Pitjantjatjara people, who are the **traditional owners** and custodians of the Uluru-Kata Tjuta National Park. The spirits of their ancestors still live in sacred places at Uluru and Kata-Tjuta, making the land deeply important to them.

The Yankunytjatjara and Pitjantjatjara people have a duty to respect and look after their Country.

Source 1

Local artists at Uluru. Uluru used to be called Ayer's Rock: why do you think the name was changed?

Bob Randall, a Yankunytjatjara elder explains: 'Part of land which has been handed down to you by your ancestors, we say the granny law, has given me my responsibility now that I'm grown up to care for my Country, you know, care for my mother. Care for everything that is around me.

'You can never feel lonely in that situation, you know you just can't. How can you when all around you is family members, from this ground up, to all the trees around you to the clouds hanging up around you, the birds flying by, the animals and reptiles that are just hidden in the shrub there for now, but can come out if they want to, hunt around for their little food. And then they can become food for us as well.'

Source 2

Manja Shelter at Gariwerd in Victoria contains rock art of hand stencils. It is believed that the stencils record a visit to the site and renew a person's ties to it.

Learning Ladder H1.6

Show what you know

1 Describe in your own words how Indigenous people feel about the land.

2 Why do Indigenous people stencil their handprints on rocks?

3 What are two different meanings of the word *Country*?

Cause and effect

Step 1: I can recognise a cause and an effect

4 The Yankunytjatjara and Pitjantjatjara people have a duty to respect and look after their Country. What is the cause of this effect?

Step 2: I can determine causes and effects

5 In 2019, the climbing of Uluru became prohibited after traditional landowners voted to ban the activity. What was the cause of Yankunytjatjara and Pitjantjatjara people wanting tourists not to climb Uluru?

Step 3: I can explain why something is a cause or effect

6 For each cause in this table, write a sentence explaining how the cause led to the effect.

Cause	Effect
European colonists and Indigenous people had different views about the land	Europeans didn't respect Indigenous beliefs and colonised the land without permission
People wanted to record their visit to a sacred site and renew their ties to it	Some would stencil their handprint on a rock wall

Step 4: I can analyse cause and effect

7 List three effects of the difference in views between Indigenous people and European colonists. Rank the effects in order, from most important to least important. Now write a short paragraph linking the effects to the cause, using this format:

'Indigenous and European colonists' views on land were very different. This caused a number of things. First, ... This was the most important outcome. The second most crucial was ... A third, less significant effect was ...'

Cause and effect, page 215

economics + business

Who owns the land?

The rights of Aboriginal Australians need to be respected when making financial and economic decisions. Many Indigenous Australians have a different view than non-Indigenous Australians when it comes to the concept of 'ownership'.

Owning the land

The idea of someone owning land and claiming it as their property is not a traditional Indigenous view. For Indigenous Australians, it is the land that owns them – and every part of their lives is connected to the land.

When the British explorer James Cook set foot on the Australian continent in 1770, he declared the land to be ***terra nullius*** ('no one's land'). He was under instruction to take possession of the land if it was unoccupied or with the consent of the native population. Cook ignored the fact that there were up to 1 000 000 people living on the land, and declared it to be the property of King George III of Great Britain.

The late Beryl Beller, an Elder of the Dharawal people, commented:

> 'They were so ignorant, they thought there was only one race on the earth and that was the white race. So when Captain Cook first came, when Lieutenant James Cook first set foot on Wangal land over at Kundul, which is now called Kurnell, he said, "Oh let's put a flag up somewhere, because these people are illiterate, they've got no fences". They didn't understand that we didn't need fences ... that we stayed here for six to eight weeks, then moved somewhere else where there was plenty of tucker and bush medicine and we kept moving and then come back in twelve months' time when the food was all refreshed ...'

Source 1

Beryl Timbery Beller, 2007

Native title

The concept of owning property is an economic or financial relationship. It does not recognise the close relationship between Aboriginal and Torres Strait Islander peoples and their land – or Country. When Europeans invaded Australia, they didn't see the Indigenous population's connection to the land because they defined ownership of property as farmed land with fences and houses.

In 1982, Eddie Mabo and four other Torres Strait Islanders took their claim for traditional land rights to their islands in the Torres Strait to the Queensland Supreme Court, and eventually to the High Court of Australia. They claimed that their islands had been continuously inhabited and possessed by their people. In 1992, the High Court of Australia decided that *terra nullius* should not have been applied to Australia, and that Aboriginal and Torres Strait Islander peoples have rights to the land. The Australian Parliament passed the *Native Title Act 1993*, which recognises the traditional rights to land and waters. Indigenous Australians can apply to the Federal Court to have their **native title** recognised under Australian law.

Source 2

Samuel Calvert and J.A. Gilfillan, *Captain Cook taking possession of the Australian continent on behalf of the British crown, AD 1770, under the name of New South Wales* (1865 CE), illustrated, Royal Society of Victoria, Melbourne, Australia.

Source 3

Jeffrey Lee is the sole surviving member of the Djok clan, and works as a ranger in Kakadu National Park in the Northern Territory. He is the senior custodian of Koongarra land, on the border of Kakadu National Park. The land contains a major deposit of uranium, and the French energy company Areva was pressuring Lee to mine tonnes of uranium from the site.

Instead of accepting millions of dollars from the mining company, Lee offered the land to the federal government to be incorporated into Kakadu National Park. For him, it was more important to protect the important sites in the country for which he was senior custodian. Lee's proposal was accepted by parliament in 2013, protecting the Koongarra land from mining forever.

'When you dig 'em hole in that country, you're killing me,' said Lee. 'Money don't mean nothing to me. Country is very important to me.' Lee takes his role as senior custodian very seriously. 'There are sacred sites, there are burial sites and there are other special places out there which are my responsibility to look after.'

Learning Ladder H1.7

Economics and business

Step 1: I can can recognise economic information

1 What is *terra nullius*, and why did Lieutenant Cook think it applied to Australia?

Step 2: I can describe economic issues

2 What are the differences in the concept of land ownership for Indigenous and non-Indigenous people?

Step 3: I can explain issues in economics

3 What is native title, and how did it become law in Australia?

Step 4: I can integrate different economic topics

4 Find an example of a native title claim and explain what rights under Australian law the Indigenous people have.

Step 5: I can evaluate alternatives

5 Research the action taken by Jeffrey Lee to protect his Country.

a Why was his land of great economic value?

b How could he have become one of Australia's richest people?

c Why did he choose not to become wealthy?

What do Indigenous Australians hold sacred?

Indigenous Australians connect to their ancestor spirits through stories, ceremonies, totems, dance and art, and through all the features and living things in their Country.

Indigenous beliefs

Many Indigenous peoples refer to spirituality rather than religion to help describe their beliefs. Spirituality gives meaning to all aspects of their life and their close relationship with the land. Everything in the environment is living and shares the same spirit as Aboriginal and Torres Strait Islander peoples.

It is not possible to speak of one Aboriginal and Torres Strait Islander belief system, as there are many. A common theme running through the spiritual beliefs of Indigenous Australians is a sense of belonging to the land and the influence of ancestral spiritual beings and their relationships to each other.

Aboriginal use the of the word Country refers to the physical and spiritual features of the land.

The Dreaming

Aboriginal spirituality draws on the stories of the Dreaming, while Torres Strait Islander spirituality is based upon the stories of the Tagai.

The Dreaming gives meaning to everything and establishes the rules for relationships between people, the land and all living things for Aboriginal people. The Dreaming tells about the creation of the world by ancestor spirits who came from the earth and sky to create and shape landforms and all life.

Source 1

Instead of writing, the spiritual knowledge of Indigenous Australians is expressed and passed on through ceremonies, stories, art, music and dance.

Source 2

Images of the cloud and rain spirits known as the Wandjina are shown here in human form in rock paintings in the Kimberley region of north-west Australia. Some artwork in the region dates back 4000 years, and each year religious ceremonies are held to repaint the figures to bring fertility to the land.

Customary law

The Dreaming also gave Aboriginal peoples 'the Law', to ensure each person knows their responsibilities for other people, the land and ancestor spirits. Aboriginal laws are referred to as **customary law**. Like other Australian laws, if you break customary laws, you will be punished.

Oral tradition

Instead of writing, the spiritual knowledge of Indigenous Australians is passed on and expressed through what is called 'oral tradition': ceremonies, stories, art, music and dance. Aboriginal and Torres Strait Islander Elders are respected and trusted to pass important information on to young men and women. Despite the disruption to culture caused by the invasion, many groups have preserved their cultures thanks to their oral tradition.

Learning Ladder H1.8

Show what you know

1 What is the difference between an oral tradition and a written tradition?

2 What kinds of things make up customary law?

3 What is the difference between spirituality and religion?

Historical significance

Step 1: I can recognise historical significance

4 How do Indigenous people connect to their ancestor spirits?

Step 2: I can explain historical significance

5 Source 2: How significant are the Wandjina images to Aboriginal people in the Kimberley region today?

Step 3: I can apply a theory of significance

6 Read the section on Historical significance in the History How-To section on page 219. Then copy and complete this table about Aboriginal customary law.

Question about significance	Response (and reason why)
How important was it to people at the time?	
How deeply were their lives affected?	
How many lives were affected?	
How long were their lives affected for?	
How important is customary law today?	

Step 4: I can analyse historical significance

7 List three reasons Aboriginal customary law is important today. How do each of these reasons make the other two reasons stronger?

Historical significance, page 219

How are Indigenous families connected?

'Kinship' is another word for family relationships, and it is at the centre of Indigenous society in Australia. The kinship system ensures that everyone knows their relationships and shared responsibilities to care for their community and their Country.

The kinship system

Traditional Aboriginal and Torres Strait Islander people have extended family relationships that are important to sharing culture and organising society. This is referred to as the **kinship** system. Wadjularbinna Doomadgee, a Gungalidda elder, explains:

> 'All people with the same skin grouping as my mother are my mothers ... They have the right, the same as my mother, to watch over me, to control what I'm doing, to make sure that I do the right thing. It's an extended family thing ... It's a wonderful secure system.'

Source 1

Wadjularbinna Doomadgee, a Gungalidda elder, quoted in exhibition, Australian Museum 2018

The kinship system organises communities to make sure everyone knows their relationships and shared responsibilities. While kinship rules vary across the different Aboriginal and Torres Strait Islander groups, caring for their community as a family is always very important. Kinship remains important for many Indigenous groups today.

Children could have a close relationship with relatives other than their mother and father. For example, your mother's sister was also called *mother*, and your father's brothers were also called *father*.

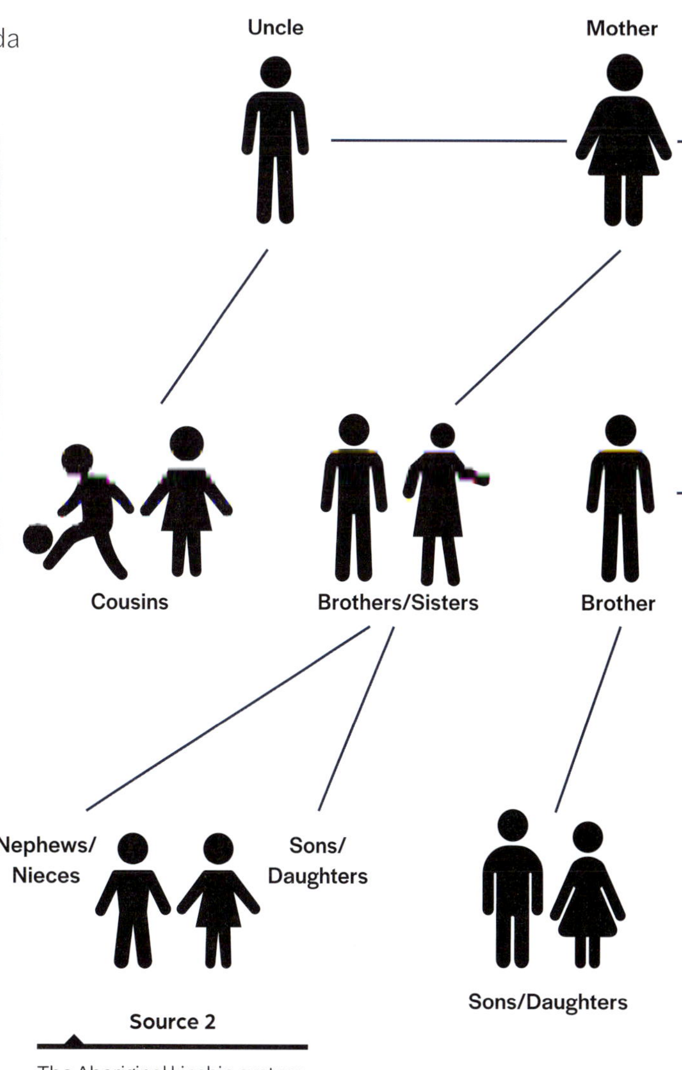

Source 2

The Aboriginal kinship system

Aboriginal kinship system

While kinship systems vary across cultures, Source 2 shows terms from one kinship system.

Kinship terms

Personal names are rarely used in Aboriginal societies. A person is addressed by the appropriate relationship term, such as father, older sister, uncle and so on.

Term	Meaning
Mother	Your birth mother and your mother's sisters.
Father	Your birth father and your father's brothers.
Brother and sister	Term for siblings and the children of your mother's sister and father's brother.
Aunty and uncle	Your father's sisters and mother's brothers and also used to address older people who you might not be related to.
Cousin	Refers to children of your father's sister and your mother's brother. The term 'cousin' can be extended to any relative of your generation who might share the same great grandparent.

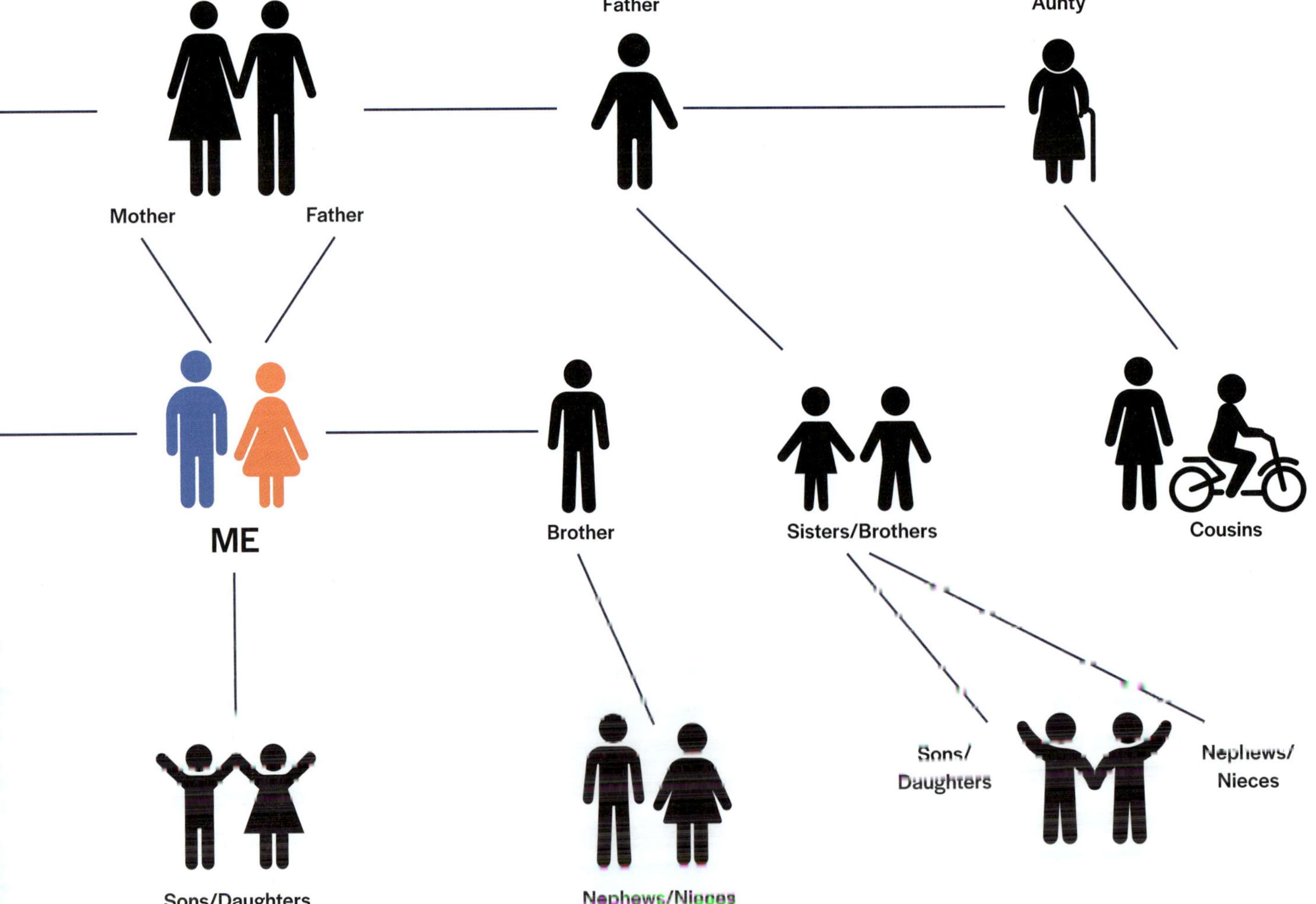

Moiety and totems

For some Aboriginal people, everyone and everything in the environment is split into two halves. This is called **moiety**. In order to understand the whole universe, the two halves must come together. A child's moiety is determined by their mother's or their father's moiety. People who share the same moiety are considered siblings and are forbidden to marry. (In some Aboriginal cultures, kinship terms, or skin names, can also determine who you are able to marry.)

Each person has emblems or **totems** that represent their nation, clan and family group, as well as a personal totem. Totems link a person to their Country – its landforms, air, water and living creatures. Every Indigenous person has a responsibility to protect their totems and pass them on to the next generation.

Totems were split between moieties. For example, the two moiety totems of the Wurundjeri people (pages 44–47) are Bunjil the wedge-tailed eagle and Waa the crow. Wurundjeri clan Law said that there could only be marriage between different moieties: a Bunjil man could only marry a Waa woman and a Waa man could only marry a Bunjil woman.

Elders

Another important connection point within an Indigenous community is an Elder. A person becomes an **Elder** by earning respect and authority in a community. It is an Elder's duty to instil a respect for the land and culture in community members by teaching young people about their natural environment, and sharing knowledge about the Dreaming.

Elders take a lead role in **initiation ceremonies**. With some variation across cultures, initiation ceremonies take place when girls and boys who have proved they are ready to be adults are given the right to pass from childhood to adulthood. In traditional Aboriginal ceremonies, the initiated member would sometimes have a tooth removed, a nose or ear pierced, or their flesh cut with a sacred marking.

Source 3

John Watson, an Aboriginal traditional owner and Nyikina Elder, tells his grandchildren stories about his Country in the Kimberley area. Elders have the authority to share knowledge of the Dreaming.

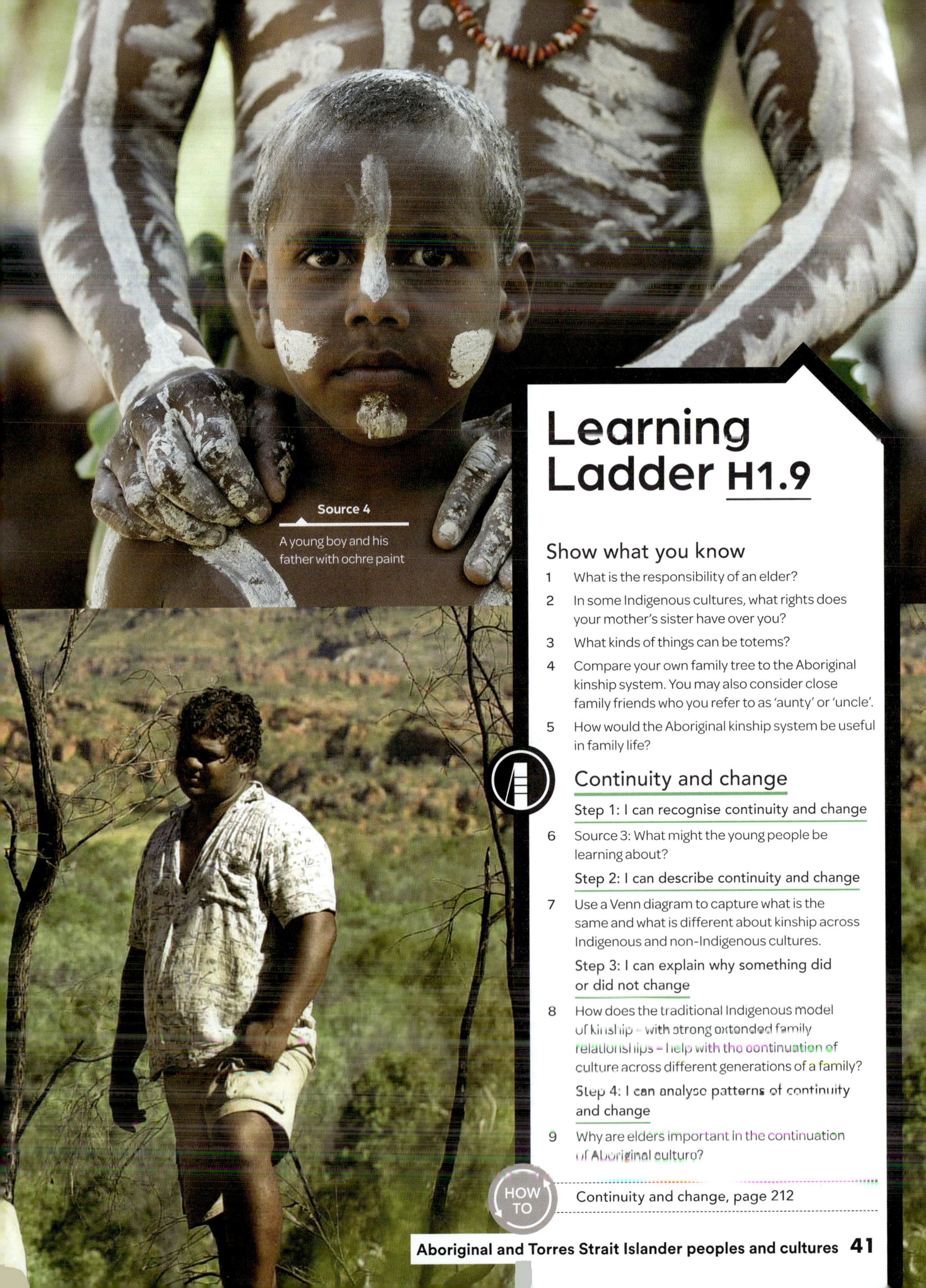

Source 4

A young boy and his father with ochre paint

Learning Ladder H1.9

Show what you know

1 What is the responsibility of an elder?

2 In some Indigenous cultures, what rights does your mother's sister have over you?

3 What kinds of things can be totems?

4 Compare your own family tree to the Aboriginal kinship system. You may also consider close family friends who you refer to as 'aunty' or 'uncle'.

5 How would the Aboriginal kinship system be useful in family life?

Continuity and change

Step 1: I can recognise continuity and change

6 Source 3: What might the young people be learning about?

Step 2: I can describe continuity and change

7 Use a Venn diagram to capture what is the same and what is different about kinship across Indigenous and non-Indigenous cultures.

Step 3: I can explain why something did or did not change

8 How does the traditional Indigenous model of kinship – with strong extended family relationships – help with the continuation of culture across different generations of a family?

Step 4: I can analyse patterns of continuity and change

9 Why are elders important in the continuation of Aboriginal culture?

HOW TO Continuity and change, page 212

H1.10

civics + citizenship

Source 1

Australia is the only country that has three official flags: the Australian National Flag, the Australian Aboriginal Flag and the Torres Strait Islander Flag. Any of the flags can be used interchangeably to officially represent Australia.

How is identity formed?

Identity refers to an individual's sense of belonging to a group or culture, or to a country or region. What it means to be Australian is very difficult to define, as our nation is full of diverse cultures.

Expressing identity

Australia's key values of freedom and inclusion make clear that everyone should feel free to respectfully show their cultural identity without being bullied or victimised. Unfortunately, there are still occasions when Australians are treated unfairly or abused because of their appearance. Understanding and valuing different cultural identities is important in countering any racism.

Australia is a multicultural country with a large range of cultural identities represented. The identities of different cultural groups can be shown in the way people dress, the languages they speak and the music, dance and foods they enjoy.

Indigenous identity

Before Australia was forcibly turned into a British colony in 1788, Indigenous peoples identified themselves by their nation. They would say things like 'I'm a Wurundjeri man', or 'I'm a Waveroo woman'.

Aboriginal people define their Aboriginal identity not by skin colour but by relationships. Kinship, relationships and connections to family are the basis of Aboriginal identity and social organisation (pages 38–41).

Australian flag

The Australian flag features the Union Jack to show our close ties to Great Britain. Indigenous leader Lowitja O'Donoghue said the current flag showed a 'narrow slice of our history' where the rights of Indigenous people were ignored.

'For this reason, most of Australia's Indigenous people cannot relate to the existing flag,' she said. 'For us, it symbolises dispossession [being excluded from our land] and oppression.'

Torres Strait Islander Flag. The green symbolises land, the blue represents sea and the black stripes show the people. The white symbol is a headdress known as a Dhari and a star that represents the Torres Strait's five island groups.

Australian Aboriginal Flag. The red stripe symbolises the land and the black stripe represents the Aboriginal people. The yellow circle represents the life-giving Sun.

Australian National Flag. The British Union Jack represents Australia's history as a British colony (1788–1901). The Southern Cross symbolises Australia's geographic position. A seven-pointed star represents the Federation of six states, with one point representing the territories.

Australia Day

Australia Day is held on 26 January each year. On 26 January 1788, Captain Arthur Philip established the first British colony at Sydney Cove. On Australia Day we are encouraged to reflect on what it means to be Australian and to celebrate all the things we love about Australia.

For Indigenous Australians the date of 26 January is a controversial selection for Australia Day, as it highlights the day that their land was invaded and their way of life was changed forever.

Source 2

Thousands of people gather across Australia on 26 January to protest against celebrating Australia Day on the day the protesters have labelled 'Invasion Day'.

Learning Ladder H1.10

Civics and citizenship

Step 1: I can identify topics about society

1 What are the core values of Australian society?

Step 2: I can describe societal issues

2 Why do some people not relate to the Australian flag?

3 Source 2: Describe what is happening in this photo and suggest why these protesters feel so strongly about this issue.

Step 3: I can explain issues in society

4 How do Aboriginal people define their identity?

5 Why is it difficult to say what it means to be Australian?

Step 4: I can explain different points of view

6 What view does Lowitja O'Donoghue have about the Australian National Flag? Explain why others would like to retain this flag.

Step 5: I can analyse issues in society

7 List arguments for and against keeping 26 January as Australia Day. Try to come up with as many arguments as you can, but at least two of each.

Who are the Wurundjeri?

The clans of the Wurundjeri people lived around the area we know as Melbourne from at least 31 000 years ago. All of their Country was special to them, as it provided them with all the resources they needed. Some sites were of spiritual significance to the Wurundjeri for ceremonies and gatherings with other groups within the Kulin Nation.

Early evidence of Wurundjeri people

A skull of an early man estimated to be up to 15 000 years old was discovered on Wurundjeri Country at Keilor in 1940, and removed to the Museum of Victoria. The discovery led to further archaeological digs that revealed 31 000-year-old fireplaces and the remains of megafauna (see pages 28–29), making Keilor one of Australia's earliest sites of human habitation.

The Keilor archaeological site was renamed Murrup Tamboore, meaning 'Spirit's waterhole' in Woi wurrung, the Wurundjeri people's traditional language. Years later, the skull was returned to its Country. Wurundjeri Elder Bill Nicholson said, 'I held the Keilor skull in my hand. I put it back into *biik*, to Country. This is my connection to this place.'

Wurundjeri people

The Wurundjeri people from the Woi wurrung language group lived across the Yarra Valley, including the northern half of what we now know as the Melbourne metropolitan area. The Wurundjeri are part of the Kulin Nation, an alliance of five language groups.

In Woi wurrung, *wurun* means the manna gum that is common along the Yarra River, and *djeri* is the witchetty grub that is found in or near the tree.

The Wurundjeri people did not 'own' the land, but belonged to it. They moved around in small groups called **clans**, and camped in different areas within clan territories depending on where food was plentiful. People would not marry within their own clan, but were paired with partners from different Kulin clans.

Wurundjeri clans visited the land of other clans and joined in **corroborees** with them. Dances and storytelling were a focus of corroborees, which could involve hundreds of people from within the Kulin Nation. These gatherings were also an opportunity for a council of leaders known as

Source 1

Early Wurundjeri people shared the environment with megafauna such as the marsupial lion (*Thylacoleo carnifex*).

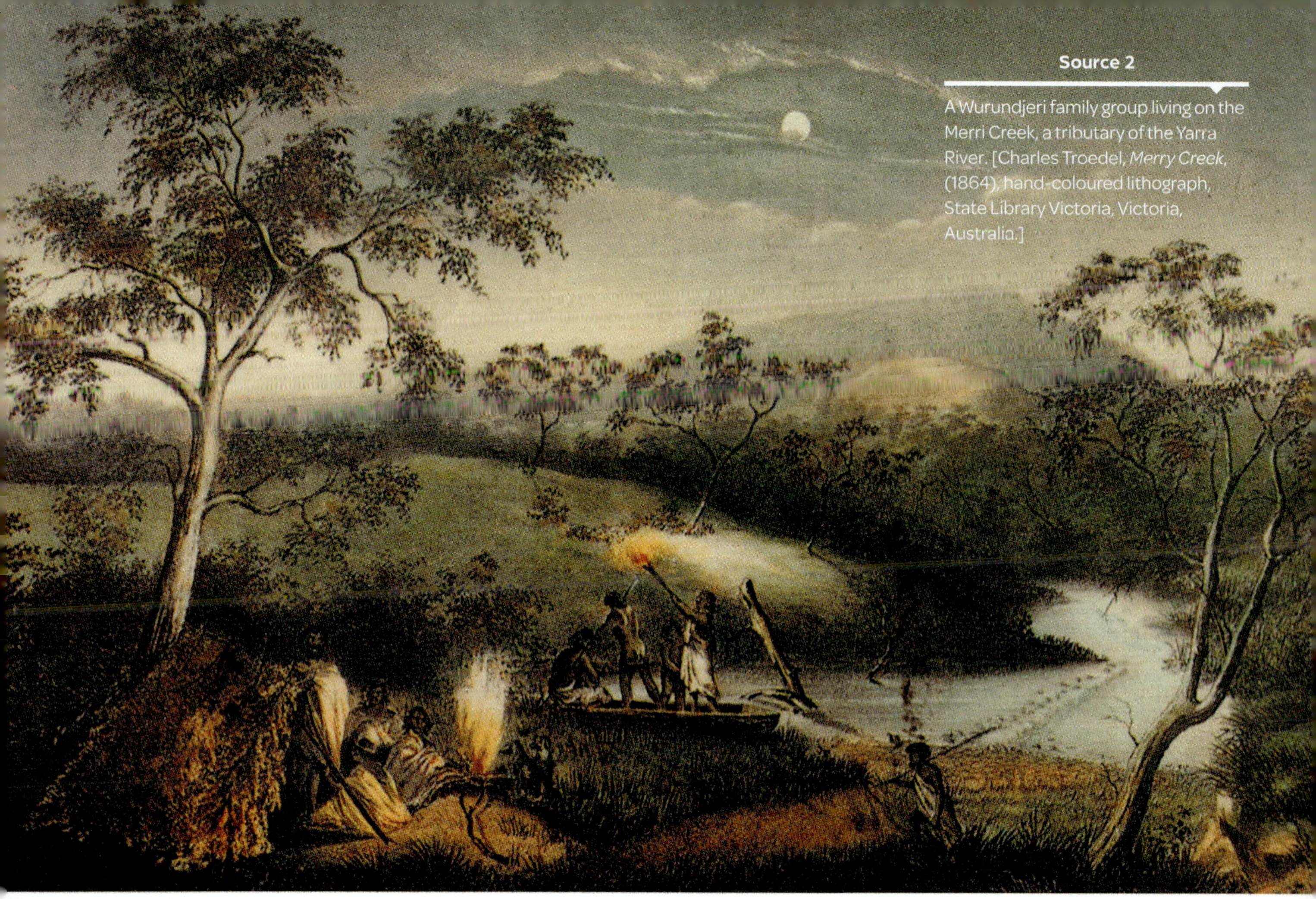

Source 2

A Wurundjeri family group living on the Merri Creek, a tributary of the Yarra River. [Charles Troedel, *Merry Creek*, (1864), hand-coloured lithograph, State Library Victoria, Victoria, Australia.]

Ngurungaeta to be formed. The Ngurungaeta would settle disputes and decide on punishment for those found guilty of crimes against Aboriginal law. These events were also about sharing food, knowledge and celebrating events like marriages.

Official gatherings of the Kulin nation still happen today and are called Tanderrum.

Living on the land

Wurundjeri people adapted to changes in the environment through their long occupation of the Yarra Valley. The Yarra River once flowed across a grassy plain, but that plain flooded at the end of the last Ice Age about 12 000 years ago and became a bay.

The Wurundjeri adjusted their lifestyle to adapt to the weather extremes of their new environment: hot, dry summers and cold winters. Some clans and family groups would shelter in caves on higher land in colder months and keep warm with cloaks and rugs made from possum skins. As the land warmed after the coldest months, the Wurundjeri moved to hunt animals, harvest root vegetables that they had cultivated and collect berries, seeds and birds' eggs.

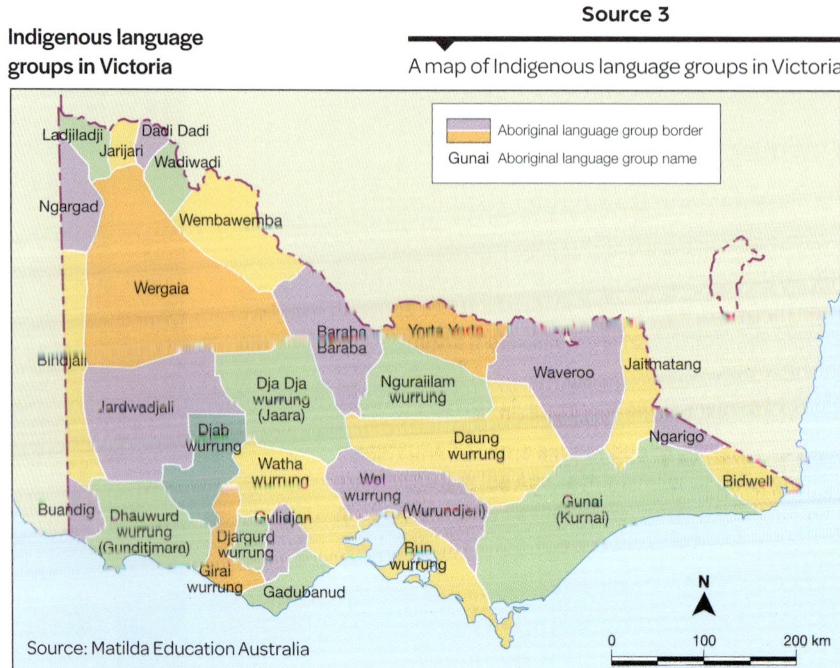

Source 3

A map of Indigenous language groups in Victoria

Indigenous language groups in Victoria

Source: Matilda Education Australia

Source 4

Aboriginal man, woman, child, posed in cloak and with tools, Jackie Logan and Queen Annie. [Source: Museums Victoria]

In the hottest months, they camped for longer periods along rivers, where they hunted fish and eels. Other plants and berries such as banksia flowers and mistletoe berries provided a food source after the first rains following summer.

The Wurundjeri took only what they needed to ensure the ongoing sustainability of the land. Each family group cultivated and sourced food in a particular section of the clan's land. They used firestick farming (see pages 28–29) to promote the growth of new grass to attract game such as kangaroos. Burning grasslands also reduced the risk of bushfires and made travel and access easier. Women used *wulunj* (digging sticks) to cultivate murnong (yams) in the cleared land.

Connection with the land

When travelling throughout their Country, the Wurundjeri would visit sacred sites – places of significance to their culture. Sacred sites were usually marked by natural features such as hills, rivers or rock formations. Some sacred sites were only for men and other sites only for women. Examples of places that are significant to the Wurundjeri include:

- Hanging Rock: a place where male initiation ceremonies were held
- Melbourne Cricket Ground: the site of the MCG was a natural amphitheatre and a place for ceremonies and gatherings
- Dights Falls: a meeting place on the Yarra River at Abbotsford for corroborees, trade and settling disputes.

The Wurundjeri share the same belief system as others in the Kulin Nation. They believe the land and living creatures were created as part of the Dreaming. Stories of the Dreaming have been passed down from generation to generation.

One of the main spiritual ancestors of the Kulin people is Bunjil the wedge-tailed eagle. Bunjil is responsible for the creation of the landscape and all living things. The Wurundjeri believe Bunjil keeps a watchful eye over them and offers guidance to those who choose to listen.

Each clan has a totem that represents its people's spiritual link to the land. Totems for the Wurundjeri are Bunjil the wedge-tailed eagle and Waa the Crow. Within the Wurundjeri clan you are either a crow person or an eagle person, and this totem is inherited from your father.

Source 5

Hanging Rock is a Wurundjeri sacred site.

Creation story: Bunjil the wedge-tailed eagle

At first there was nothing but the stars in the sky at night. A star fell from the sky, and as it fell it formed into the body of a wedge-tailed eagle known as Bunjil.

Bunjil spread his wings as he fell and blew air from his beak to create the earth and all its features. Bunjil took clay from a riverbed and shaped it to form two men. He danced around the figures and blew air into their mouths to fill them with life.

Bunjil's brother, Pallian, controlled rivers, creeks and billabongs. He beat the water with his hands. It became thicker and took on the shape of two women. Bunjil provided each man with a spear and gave each woman a digging stick.

When Bunjil had finished creating the landscape and its living creatures, he insisted that Bellin-Bellin, the musk crow, open a bag filled with wind. One bag was not enough, so Bunjil demanded that Bellin-Bellin open all his bags. A huge rush of wind flowed from the bags, blowing Bunjil and his family high into the sky, where they became stars.

A retelling of the Bunjil the wedge-tailed eagle creation story

Source 6

The 25-metre high sculpture of Bunjil the wedge-tailed eagle on Wurundjeri Way in Melbourne

Learning Ladder H1.11

Show what you know

1 How long have the Wurundjeri lived in the area around Melbourne?

2 How many language groups are there in the Kulin Nation? How many language groups are there in the whole of Victoria? (See page 45.)

3 What happens at a corroboree?

Source analysis

Step 1: I can determine the origin of a source

4 Source 2: Where is the scene in this lithograph and when was it painted?

Step 2: I can list specific features of a source

5 Source 4: Describe the clothing of the Wurundjeri people and suggest why they are worn.

Step 3: I can find themes in a source

6 Source 2: What mood is the painter trying to show? How do they achieve this? In your response, describe elements of the painting.

Step 4: I can use my outside knowledge to help explain a source

7 Source 2: Knowing what you know about kinship groups, who do you think might be in the tent?

HOW TO

Source analysis, page 209

How can we protect our heritage?

It is important to conserve the remains of Australia's ancient past, including the heritage, culture and artefacts of Aboriginal and Torres Strait Islanders.

Ancient Australia

The history and culture of ancient Australians has been difficult to piece together. Their history was not written down.

Artefacts and evidence of human occupation, such as middens and rock art, and structures like fish traps help to give us clues about the daily lives of Indigenous Australians. Consultation with the traditional owners adds knowledge and understanding about historical finds.

Rock paintings

Ancient Australians left behind information about their daily lives and their Dreaming in rock paintings. Traditional art was usually painted in a natural pigment known as ochre.

Discovering how old pictures are can be difficult. By finding out how much carbon the paintings contained (a process called **radiocarbon dating**), archaeologists have dated paintings at Narwal Gabarnmang rock shelter in the Northern Territory to be at least 28 000 years old. This makes them Australia's oldest certified artworks – and among the oldest in the world.

Rock art sites are open-air museums and schools. However, rock art suffers from weathering and erosion, as well as from graffiti and vandalism.

Source 1

A profile showing layers (or strata) of rock and soil.

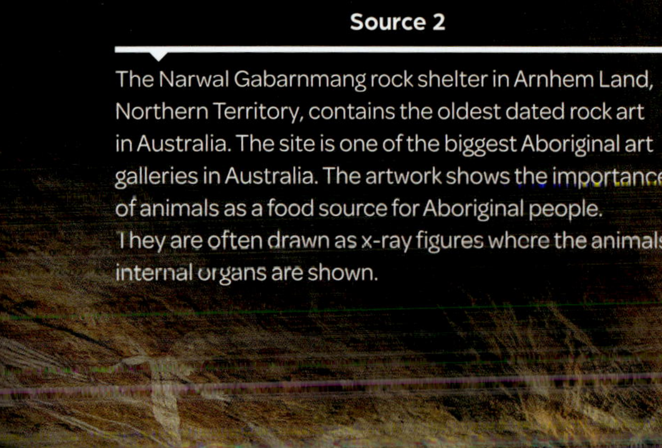

Source 2

The Narwal Gabarnmang rock shelter in Arnhem Land, Northern Territory, contains the oldest dated rock art in Australia. The site is one of the biggest Aboriginal art galleries in Australia. The artwork shows the importance of animals as a food source for Aboriginal people. They are often drawn as x-ray figures where the animals' internal organs are shown.

Archaeological findings

Archaeologists gather evidence about how the first Australians lived from ancient occupation sites such as camps, middens, quarries and burial sites.

Archaeologists use a technique known as **stratigraphy** to learn about changes over time. Different artefacts are found in different strata (or layers) of rock and soil. By dating the strata, archaeologists can assume that the artefacts found there are the same age as the strata. In 2017, archaeologists uncovered 10 000 artefacts in a layer of soil that has been dated to at least 65 000 years ago at the Madjedbebe Rock Shelter in the Northern Territory.

Conserving our past

Many sites important to Indigenous Australians have been destroyed since British invasion in 1788. The sites that remain are rare and fragile, so it is important that they are preserved for future generations. It is also important that descendants of the ancient people are given control of artefacts and sacred sites that are part of their culture.

Institutions like museums play an important role in protecting the artefacts of ancient Australia. The fragile nature and historical significance of many items means they must be handled with great care.

Museums are also places where people can learn about and see artefacts that have survived from ancient Australia. Artefacts are researched by specialists and often interpreted by Indigenous people with traditional knowledge and understanding.

Learning Ladder H1.12

Show what you know

1 What clues did Indigenous people leave that historians can use to figure out how they lived?

2 How do we know how long ago things from Ancient Australia took place?

3 What is the difference between carbon dating and stratigraphy?

4 Describe two benefits and two disadvantages of taking Indigenous artefacts and putting them in a museum.

5 If our society ended right now, what might still be around in 40 000 years for future historians to learn about?

Continuity and change

Step 1: I can recognise continuity and change

6 Source 2: Describe the images you can see. What can we learn about Indigenous culture from what people chose to draw?

Step 2: I can describe continuity and change

7 Source 1: What kind of technology did the society whose objects are in Layer 2 have? What about Layer 3?

Step 3: I can explain why something did or did not change

8 Why are Indigenous artefacts and sites preserved better now than they used to be?

Step 4: I can analyse patterns of continuity and change

9 Why has the amount of change to the Australian landscape increased since European settlement?

Continuity and change, page 212

Masterclass

Work at the level that is right for you or level-up for a learning challenge!

Step 1

a **Chronology: I can read a timeline**

Use the timeline on page 22 to answer these questions:

i Over what time period did Indigenous people migrate to the Australian continent?

ii Over what time period did Indigenous people disperse across the whole continent?

iii How many years were there between the flooding of the Tasmanian and the Papua New Guinean land bridges?

b **Source analysis: I can determine the origin of a source**

Look at Source 1. Is this a primary or secondary source? How do you know? Where is it from? Where are the people in the photo from?

c **Continuity and change: I can recognise continuity and change**

Rank these items in order from 'most changed' to 'least changed' from ancient to modern Australia:

- plants
- animals
- sea level
- weather
- human population.

d **Cause and effect: I can recognise a cause and an effect**

Match the cause with the effect.

Cause	Effect
Indigenous firestick farmers burnt forest	Sea levels were much lower
Indigenous people separated into clans and nations	People are referred to by their relationship term; e.g. mother, aunty
Personal names aren't often used in Indigenous culture	There were 500 nations in Australia when Europeans invaded
Scientists found DNA evidence	There are lots of grasslands in Australia
There was an Ice Age 110 000–12 000 years BP	Australian Aborigines are linked to the first people who left Africa

e **Historical significance: I can recognise historical significance**

Rank these things from least to most historically significant:

- cultivation of crops, like murnong (yam daisy)
- first use of stone axes in Australia
- extinction of megafauna
- trade between Indigenous Australians.

Step 2

a **Chronology: I can place events on a timeline**

Rank these events from earliest to most recent:

Indigenous people living around Sydney since at least 30 000 years BP

16 000–12 000 years BP: rapid sea level rise

18 000–15 000 years BP: dry, cold period

People migrate to Australia from the north from about 65 000–50 000 years BP.

b **Source analysis: I can list specific features of a source**

What are the four individuals in Source 1 doing? Write one sentence for each person.

c **Continuity and change: I can describe continuity and change**

List three things that have changed in Australia since 10 000 years BP and three things that have stayed the same.

d **Cause and effect: I can determine causes and effects**

What conditions allowed people to move freely around the world about 12 000 years ago?

e **Historical significance: I can explain historical significance**

Why is the use of stone axes in Australia considered so important?

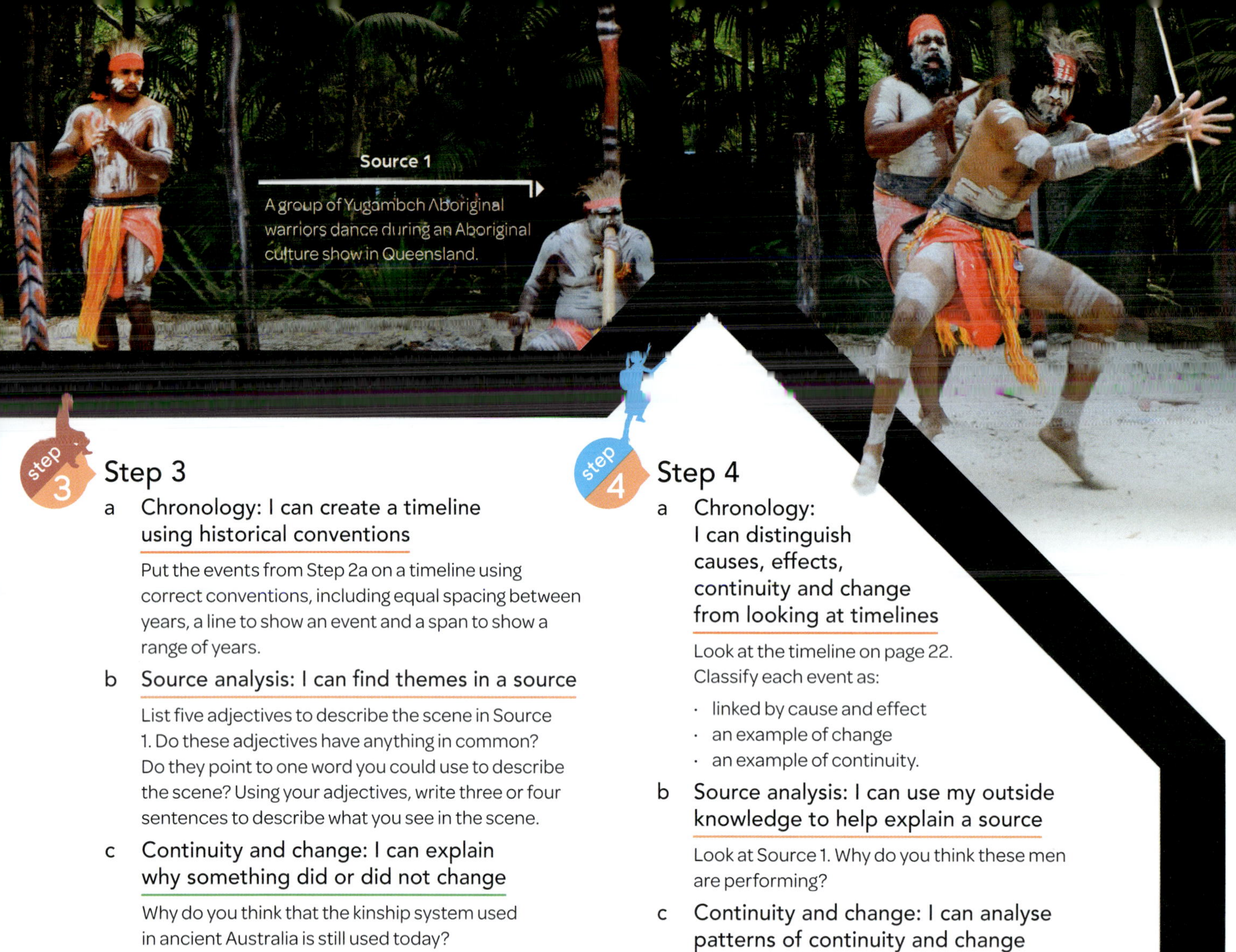

Source 1

A group of Yugambeh Aboriginal warriors dance during an Aboriginal culture show in Queensland.

Step 3

a Chronology: I can create a timeline using historical conventions

Put the events from Step 2a on a timeline using correct conventions, including equal spacing between years, a line to show an event and a span to show a range of years.

b Source analysis: I can find themes in a source

List five adjectives to describe the scene in Source 1. Do these adjectives have anything in common? Do they point to one word you could use to describe the scene? Using your adjectives, write three or four sentences to describe what you see in the scene.

c Continuity and change: I can explain why something did or did not change

Why do you think that the kinship system used in ancient Australia is still used today?

d Cause and effect: I can explain why something is a cause or effect

Why don't traditional owners want people climbing Uluru?

e Historical significance: I can apply a theory of significance

Read the section on Historical significance in the History How-To section on page 219. How significant was the dingo to early Indigenous Australians?

Step 4

a Chronology: I can distinguish causes, effects, continuity and change from looking at timelines

Look at the timeline on page 22. Classify each event as:

- linked by cause and effect
- an example of change
- an example of continuity.

b Source analysis: I can use my outside knowledge to help explain a source

Look at Source 1. Why do you think these men are performing?

c Continuity and change: I can analyse patterns of continuity and change

Look at the timeline on page 22 and the suggestions on page 212 from the skills section. What pattern of continuity or change can you see in ancient Australia?

d Cause and effect: I can analyse cause and effect

List four effects of the rise and fall of the sea levels around Australia. Rank these effects in order, from most significant to least significant. Then write a paragraph detailing the four effects, discussing them in order of importance. Start your paragraph with this topic sentence: 'Sea level changes had a number of effects on ancient Australia.'

e Historical significance: I can analyse historical significance

List reasons why 26 January 1788 is important in Australian history, coming up with at least one for each of these categories: social reasons, political reasons, historical reasons, economic reasons. Which type of reason is the most important, and why?

Masterclass

Step 5

a Chronology: I can describe patterns of change

Look at the timeline on page 22. If you had to summarise Australian Indigenous history in one sentence, what would it be? Write that sentence, then write another four or five sentences explaining it and backing it up using evidence from the timeline.

b Source analysis: I can use the origin of a source to explain its creator's purpose

Look at Source 1 on page 51. Why do you think this photo was taken?

c Continuity and change: I can evaluate patterns of continuity and change

How does an Indigenous person's connection to Country provide continuity?

d Cause and effect: I can evaluate cause and effect

How important were the rises in sea level to world history? Justify your answer with historical examples.

e Historical significance: I can evaluate historical significance

Do you think the date of Australia Day should be changed? Justify your response.

Historical writing

1 Structure

Imagine you are writing an essay with this title: 'Contrast traditional Indigenous and modern food-collecting methods'.

Write an essay plan based on the format on page 225. Include at least two main paragraphs.

2 Draft

Using your essay plan and the drafting suggestions on page 226, write a 300–400 word essay response.

3 Edit and proofread

Using the editing and proofreading guidelines on page 227, edit your draft, then proofread it.

Historical research

4 Organise and present information

Imagine you were asked to complete a research task on this topic: 'Environmental changes from ancient to modern Australia'.

Develop a graphic organiser that would help you take notes for this topic.

What *kinds* of information would go in each box? You don't need to write full notes, just list the kinds of information, e.g. facts, oral histories.

Capstone

How can I understand Aboriginal and Torres Strait Islander peoples and cultures?

In this chapter, you have learnt a lot about Aboriginal and Torres Strait Islander peoples and cultures. Now you can put your new knowledge and understanding together for the capstone project to show what you know and what you think.

In the world of building, a capstone is an element that finishes off an arch, or tops off a building or wall. That is what the capstone project will offer you, too: a chance to top off and bring together your learning in interesting, critical and creative ways. You can complete this project yourself, or your teacher can make it a class task or a homework task.

Scan this QR code to find the capstone project online.

mea.digital/GHV7_H1

Ancient Egypt

WHAT WAS THE SOCIAL ORDER OF ANCIENT EGYPT?

page 62

civics+citizenship

page 64

HOW ARE LAWS MADE?

key individual

page 66

WHO WAS RAMSES II?

key individual

page 82

HOW DID CLEOPATRA RULE EGYPT?

How can I understand ancient Egypt?

Ancient Egypt is one of the most fascinating civilisations that ever existed. No similar civilisation gave rise to it – and no similar civilisation has existed since. Yet Egyptian ideas are all around us, from mathematics to astronomy, science to medicine, from glorious monuments to influences found in modern religions.

learning ladder

Step	Chronology	Source analysis	Continuity and change
step 5	**I can describe patterns of change** I read timelines and see the 'big picture'. I group timeline events and see if they show patterns of change. I know typical historical patterns to look for.	**I can use the origin of a source to explain its creator's purpose** I combine knowledge of when and where a source was created to answer the question, 'Why was it created?'	**I can evaluate patterns of continuity and change** I answer the question, 'So what?' about patterns of continuity and change. I weigh up different things and debate the importance of a continuity or a change.
step 4	**I can distinguish causes, effects, continuity and change from looking at timelines** I read timelines and find events that are linked by cause and effect. I find things that are the same or different from then until later times.	**I can use my outside knowledge to help explain a source** I have enough outside knowledge about ancient Egypt to help me explain a source.	**I can analyse patterns of continuity and change** I see beyond individual examples of continuity and change in ancient Egypt and identify broader patterns, and I explain why they exist.
step 3	**I can create a timeline using historical conventions** When given a set of events, I construct a historical timeline, making sure I use correct terminology, spacing and layout.	**I can find themes in a source** I look a bit closer into a source and find more than just features. I find themes or patterns in the source.	**I can explain why something did or did not change** I answer the question 'why?' something changed or stayed the same between historical periods.
step 2	**I can place events on a timeline** When given a set of ancient Egyptian events, I can put them in order from earliest to latest – the simplest kind of timeline.	**I can list specific features of a source** I look at an ancient Egyptian source and list detailed things I can see in it.	**I can describe continuity and change** I have enough content knowledge about two different historical periods to recognise what is similar or different about them, and can describe it.
step 1	**I can read a timeline** I read timelines with ancient Egyptian events on them and answer questions about them.	**I can determine the origin of a source** I can work out when and where an ancient Egyptian source was made by looking for clues.	**I can recognise continuity and change** I recognise things that have stayed the same and things that have changed from ancient Egypt until now.

Source 1

Fragment of wall painting showing Queen Ahmose-Nefertari from Kynebu's tomb in Thebes, c. 1150 BCE, British Museum, London, UK.

Warm up

Cause and effect	Historical significance
I can evaluate cause and effect I answer the question 'So what?' about cause and effect. I weigh up different things and debate the importance of a cause or an effect.	**I can evaluate historical significance** I answer the question 'So what?' about things that are supposedly historically important. I weigh up events against each other and cast doubt on how important things are.
I can analyse cause and effect I don't just see a cause or effect as one thing. I determine the factors that make up causes and effects.	**I can analyse historical significance** I separate out the various factors that make something historically important in ancient Egypt.
I can explain why something is a cause or an effect I can answer 'how?' or 'why?' a cause led to an effect in ancient Egypt.	**I can apply a theory of significance** I know a theory of significance. I use it to rank the importance of ancient Egyptian events.
I can determine causes and effects Applying what I have learnt about ancient Egypt, I can decide what the cause or effect of something was.	**I can explain historical significance** I answer the question 'why?' about things that were important in ancient Egypt.
I can recognise a cause and an effect From a supplied list, I recognise things that were causes or effects of each other in ancient Egypt.	**I can recognise historical significance** When shown a list of things from ancient Egyptian history, I can work out which are important.

Chronology

1 Source 1 is dated c. 1150 BCE. Explain what these numbers and abbreviations mean.

Source analysis

2 Look at Source 1. Describe the facial expression of Egyptian queen Ahmose-Nefertari in the painting fragment.

Continuity and change

3 What things did the ancient Egyptians develop that still exist today?

4 Suggest why only a fragment of the painting in Source 1 is in the British Museum in London and not the whole painting.

Cause and effect

5 The pyramids were built to house dead bodies. What effect have the pyramids had on modern Egypt? What purpose do they serve now?

Historical significance

6 Is it important to study ancient Egypt if no other civilisations similar to it have existed? Why or why not?

What can we learn from studying ancient Egypt?

Ancient Egyptians were able to build some of the largest monuments ever created without modern technology. Their artistic style is both unique and influential. Their culture is mysterious, their religion and obsession with death and the afterlife intriguing. Egyptians were among the first people to ask the big questions: What happens after we die? Where does the sun go? How was the world made? Many of the answers to these questions filtered into their religion and were passed on to later civilisations.

Learning about a culture so different from our own gives us insights into human nature, and removes the limitations of our modern ways of thinking. If we want to learn lessons from history, studying a civilisation that lasted for thousands of years is essential.

key ideas timeline.

c. 5500 BCE
Early settlers in the Nile Valley

3100 BCE
Hieroglyphic script developed

2560 BCE
Great Pyramid built

2550 BCE
Papyrus used for writing

Early Dynastic Period 2950–2575 BCE

Old Kingdom 2575–2150 BCE

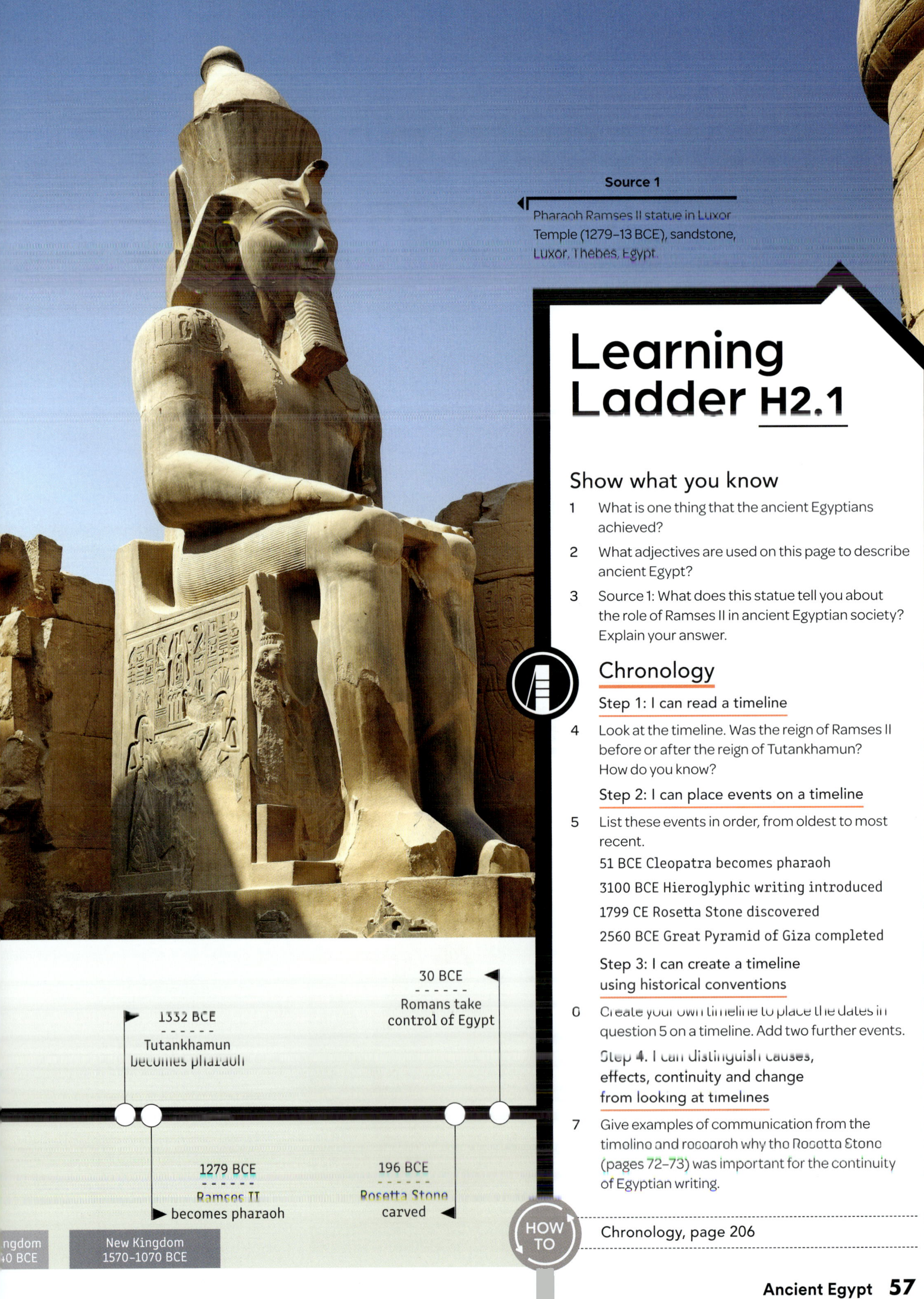

Source 1

Pharaoh Ramses II statue in Luxor Temple (1279–13 BCE), sandstone, Luxor, Thebes, Egypt.

Learning Ladder H2.1

Show what you know

1 What is one thing that the ancient Egyptians achieved?

2 What adjectives are used on this page to describe ancient Egypt?

3 Source 1: What does this statue tell you about the role of Ramses II in ancient Egyptian society? Explain your answer.

Chronology

Step 1: I can read a timeline

4 Look at the timeline. Was the reign of Ramses II before or after the reign of Tutankhamun? How do you know?

Step 2: I can place events on a timeline

5 List these events in order, from oldest to most recent.

51 BCE Cleopatra becomes pharaoh
3100 BCE Hieroglyphic writing introduced
1799 CE Rosetta Stone discovered
2560 BCE Great Pyramid of Giza completed

Step 3: I can create a timeline using historical conventions

6 Create your own timeline to place the dates in question 5 on a timeline. Add two further events.

Step 4: I can distinguish causes, effects, continuity and change from looking at timelines

7 Give examples of communication from the timeline and research why the Rosetta Stone (pages 72–73) was important for the continuity of Egyptian writing.

HOW TO

Chronology, page 206

How could people live in a desert?

The land of the ancient Egyptians was part of the largest desert on Earth. Their location in such a hot and dry environment influenced how people dressed, the houses they lived in and what they did.

Desert life

Ancient Egyptians lived on the edge of the world's largest desert – the Sahara. Living in such a dry environment, the ancient Egyptians needed access to a permanent source of water. They based their civilisation along the banks of the Nile River, the longest river in the world.

Each summer, the Nile River would flood and deposit rich silt that fertilised the land. Because of the flooding, the Egyptians described their region as either black land or red land. The black land was the rich black soil deposited along the banks of the Nile River during flooding (see page 60). This is the land the ancient Egyptians used for farming. The red land was the dry unproductive desert land that surrounded Egypt to the west and east.

However, the desert did have one advantage: it provided a natural barrier against invading armies.

Living in a hot climate

The climate of ancient Egypt was very hot and dry, with little rainfall. From May to September, the temperatures would reach up to 43°C during the day and drop to just 7°C at night.

Egypt's hot weather influenced the design of their houses. Houses were built out of bricks that

Source 1

This wall painting is from a tomb in Thebes. It shows the simple white linen garments worn by men and women. The linen was made from flax plants, which were grown as crops along the Nile River. The flax was spun into thread and then woven into cloth. Poorer men and manual workers worked naked. [Egyptian tomb wall painting from Thebes, Luxor (11th century BCE), paint on plaster, Egypt.]

© Altair4 Multimedia – Altair4.com

Source 2

Archeolgists uncovered 68 houses at Deir el-Medina in 1922, as shown in the model above. Deir el-Medina was a worker's village built around 1500 BCE. These were the homes of the craftsmen who built and decorated the royal tombs in the Valley of the Kings and Valley of the Queens at Thebes (modern-day Luxor). Most of ancient Egypt's mudbrick homes have disintegrated over time. Deir el-Medina is one of the few remaining examples of ancient Egyptian housing and it provides us with insights into how people withstood living in the desert.

were made from mud and dried in the sun. The thick bricks offered insulation against the hot days and cold nights. Poor people used one layer of bricks to build their homes, while wealthy people used two or three layers of bricks. Homes were mostly painted white to help reflect the sun's rays.

Most of the houses were constructed with flat roofs. Stairs or a ramp gave access to the roof via a roof vent, which also let heat escape from the house. Some roofs had canopies made from reeds to provide shade. Reed mats also covered all the windows and doors to keep out heat, dust and flies.

Dressing for the heat

Ancient Egyptians dressed in lightweight clothes that kept them cool in the hot desert conditions. Young children and slaves were generally naked. Men wore linen loincloths around their waist and were generally bare chested. Women wore long linen dresses. The linen was woven from flax plants that were grown along the Nile River.

Both men and women would shave their heads and wear wigs when in public. When at home, the shaved scalp would help keep them cool. On some occasions perfumed fat was placed on the top of the wig and as it melted it dripped over the body. Historians disagree about the purpose of the fat – some say it was used to cool the skin. Men and women also wore heavy eye make-up that helped reduce the glare from the strong sunlight.

Learning Ladder H2.2

Show what you know

1 What problems might there be for people living in the desert?

2 Why was the Nile River so important?

3 What did Egyptians do to stay cool?

4 Describe the differences between the black land and the red land.

5 Source 1: What do you think the women are carrying? What might these goods have been used for?

Cause and effect

Step 1: I can recognise a cause and an effect

6 Match the cause with its effect.

Cause	Effect
Egypt could get extremely hot	People wore heavy eye make-up
Egyptians lived in a dry environment	Mudbrick houses insulated against the heat
Strong sunlight created glare	Rich silt fertilised the land
The Nile flooded every year	They needed a permanent source of water

Step 2: I can determine causes and effects

7 Source 1: Why did some Egyptians work naked?

Step 3: I can explain why something is a cause or an effect

8 Source 2: Why was the discovery at Deir-el-Medina such an important and unique one?

Step 4: I can analyse cause and effect

9 Source 1: What impact did Egypt's climate and environment have on the clothes they chose to wear and make?

HOW TO

Cause and effect, page 215

How did the Nile sustain life?

The Nile was central to life in ancient Egypt. It provided fresh water for drinking and growing crops, wildlife to hunt, fish and water birds to eat, building materials such as mud (see pages 58–59) and papyrus reeds for making paper. It also provided a means of transporting goods.

The life-giving Nile River

Life in the desert region of ancient Egypt relied on the Nile River. Irrigation channels were built to feed water away from the Nile and allow crops to be grown further from the river.

The most important thing the Nile provided was a strip of fertile land where Egyptians could grow crops such as wheat for food and flax for clothing. The papyrus reeds along the riverbanks were used to make baskets, sandals, boats – and even paper (see page 73).

The inundation

Each summer, the Nile rose by up to eight metres as it carried rain that had fallen at its source, almost 7000 kilometres away. The flooding was known as 'the **inundation**', and ancient Egyptians thought it was the work of Hapi, the god of the annual flooding of the Nile River.

The inundation each year carried rich soil that was deposited along the flooded banks of the Nile. Farmers depended on these black lands of rich soil to grow their crops, as much of the dry red lands could not be used for farming (see page 58).

While the inundation brought the life-giving soil, the flooding also disrupted life in Egypt between June and September. Towns were built on higher land above the floodline, but the farmland was under water. During the inundation, farmers moved away from their farms to labour on large building projects such as the pyramids.

Source 1

The Nile River was an important resource for ancient Egyptians.

Source 2

Ancient Egyptians made small fishing boats by strapping together bundles of papyrus reeds. This painting from the tomb of Nebamun (c. 1350 BCE) shows him hunting birds in the marshes by the Nile River. Nebamun is joined by his wife and daughter. [Painting from the tomb of Nebamun (c. 1390 BCE), paint on plaster, The British Museum, London UK.]

Source 3

A farming calendar

Goods barge

Hunting birds

Irrigation channel

Learning Ladder H2.3

Show what you know

1 Why is the Nile River particularly important for the people of Egypt?

2 Source 2: What activity is shown in in this painting?

3 Source 1: Describe three different activities shown in the diagram and suggest why the Nile River was important for each of them.

4 Why do you think Egyptians thought the Nile flooded because of the god Hapi?

Continuity and change

Step 1: I can recognise continuity and change

5 The Nile was the lifeblood for many Egyptians. What performs a similar role for Australians?

Step 2: I can describe continuity and change

6 What changes did the inundation bring each year? Create a table listing positive and negative aspects of the annual inundation.

Step 3: I can explain why something did or did not change

7 Approximately 95 per cent of the Egyptian population still live within a few kilometres of the Nile River. Explain why the river remains the lifeblood of Egypt.

Step 4: I can analyse patterns of continuity and change

8 Do some research about modern Egyptian farming practices. Describe similarities and differences between crops grown along the Nile in ancient and modern times.

HOW TO

Continuity and change, page 212

What was the social order of ancient Egypt?

The society of ancient Egypt was organised by class, in a system called a hierarchy. At the top was the royal family: the pharaoh and his family. At the bottom were the slaves and the farmers, who made up the bulk of the population.

Ruled by a god-king

The **pharaoh** was the leader of ancient Egyptian society, and nearly all pharaohs were men. People believed the pharaoh to be both a king and the human form of the god Horus. In Egyptian society, religion played a key role in people's lives – as well as in the **afterlife** (see page 75). The pharaoh was the link between the gods and the people, and therefore he had both earthly and divine responsibilities.

- His earthly responsibilities included protecting the people and maintaining order, defending Egypt from invaders, and managing the requirements of building, farming, mining and trading.
- His divine responsibilities included running religious ceremonies, selecting the sites of new temples and, most importantly, maintaining ***ma'at*** – the order and balance in the world and the heavens.

Source 1

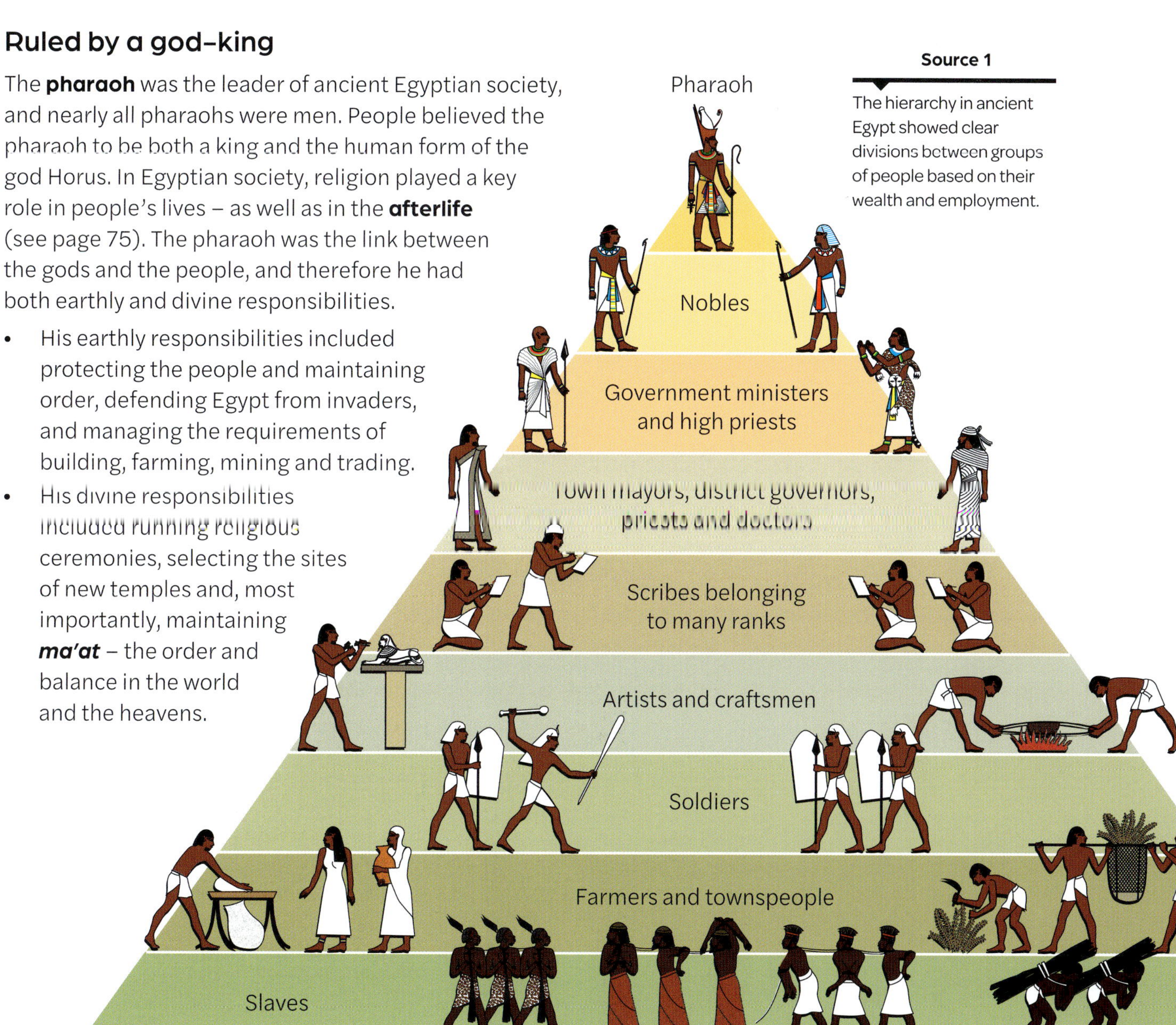

The hierarchy in ancient Egypt showed clear divisions between groups of people based on their wealth and employment.

Power of the pharaoh

The pharaoh was an all-powerful being. People knelt down before the pharaoh and kissed the ground. Ordinary people could be killed for accidentally touching a pharaoh. Huge monuments were built for pharaohs in the form of temples, statues and tombs. Pharaohs wore symbols of their power, including:

- a cobra headpiece that showed a readiness to strike at enemies
- an animal tail that showed strength and fertility
- a false beard that symbolised his status as a god
- a flail (or whip) that demonstrated authority
- a shepherd's crook that showed that the pharaoh looked after his people.

The pharaoh was the chief priest and the commander of the army. He made all the laws and made the most important decisions in courts. The power of the pharaoh was enhanced by his wealth. The pharaoh owned all of the land and its resources. He could also demand heavy taxes from the people.

Source2

Pharaoh Tutankhamun's coffin was decorated with gold and precious stones and placed in an elaborate tomb (see page 74). [Inner sarcophagus of Tutankhamun (18th Dynasty), gold and precious stones, Egypt.]

The rule of the pharaohs lasted 3000 years over 31 dynasties. A **dynasty** was when one family maintained power, and then handed down the throne to an heir. The heir was the oldest son of the pharaoh's main wife. If she were unable to supply a male heir, the son of a lesser wife would become heir to the throne. This provided great stability and continuity for the Egyptian civilisation.

Learning Ladder H2.4

Show what you know

1 How did a pharaoh show he was powerful by what he wore?

2 How were a pharaoh's earthly and divine responsibilities different?

3 Source 2: What features on the coffin tell you that pharaohs were wealthy?

4 Source 1: Describe the difference between lower-class and upper-class Egyptians in terms of the work they did.

5 Source 1: Who were the scribes and why were they well regarded in ancient Egyptian society (page 72)?

Continuity and change

Step 1: I can recognise continuity and change

6 What did the pharaohs have built to ensure they were remembered well after they died?

Step 2: I can describe continuity and change

7 How did dynasties ensure there would be continuous rule? How long did dynastic rule last in Egypt?

Step 3: I can explain why something did or did not change

8 Source 1: What social groups do we have today that are most like those shown in the diagram?

Step 4. I can analyse patterns of continuity and change

9 Imagine you have been given this essay question: 'Why does a strict social hierarchy promote continuity rather than change?'
   a What is this question asking you?
   b Take six dot-point notes that would help answer the essay question above.

Continuity and change, page 212
Historical writing, page 223

How are laws made?

Without laws, people could do whatever they liked. Laws determine the rights and responsibilities of people based on a code of accepted behaviour.

Law in ancient Egypt

The goddess Ma'at regulated the actions of people. Over time, Ma'at's principles formed the Egyptian concept of truth, order, balance and justice – also called *ma'at*.

On earth, the pharaoh was the supreme judge and lawmaker. Pharaohs followed the principles of *ma'at* in judgements. The pharaoh decided the most important criminal cases (acts that intentionally cause harm to others) and delegated legal power to his chief advisors, **viziers**, who often acted as judges. Minor **criminal** and **civil cases** (disputes between people and acts that cause loss to others) were tried by elders in each town.

Everyone was viewed as equal under the law, regardless of their social position – apart from slaves. Legal judgements were based on common sense. Punishment for theft might be the return of the goods and a fine of twice their value. Harsher punishments involved being beaten with a cane, or the loss of a hand or ear – or even the nose or tongue. In extreme cases, the guilty person was drowned, beheaded or burned alive.

Source 1

This image shows the judgement of the dead in the underworld. The heart of a dead person (left) is being weighed against an image of Ma'at (right). Scales are still used today as a symbol of justice. The feather on Ma'at's head represents truth. The Egyptians believed that a dead person's heart was weighed against a feather from Ma'at's headdress. If the heart was lighter than the feather, the dead person would be admitted to the afterlife, while if it was heavier, the dead person's heart was eaten by Ammut – the devourer of the dead who had the head of a crocodile, body of a lion and hindquarters of a hippopotamus – and they were not allowed into the afterlife. This is why hearts were left in Egyptian mummies while all other organs were removed – so Ma'at could make her judgement. [Judgement of the dead in underworld court of Osiris, from *Book of the Dead* (c. 1450 BCE), papyrus.]

Source 2

Prime Minister Scott Morrison listens to a question from an Opposition member (the Opposition is the party with second-largest number of members of parliament). Debate in parliament can often become heated.

Law in modern Australia

Like ancient Egypt, modern Australian law-making is based on key principles. All Australians have the responsibility to obey the law and the right to be treated fairly.

Australia's laws are made in two ways, as **statutory laws** or **common laws**.

Statutory law

Statutory laws are made and passed by **parliament**. State parliaments make laws about things they control, such as education and health. Federal parliament makes laws about things that apply to the whole nation, such as taxation and migration.

Common law

Common law is created when a judge makes a decision in a case that is not clearly covered by statutory law. The judge must make a decision on how to resolve the issue. This decision sets a **precedent**, which means that similar cases in the future should follow this common law judgement.

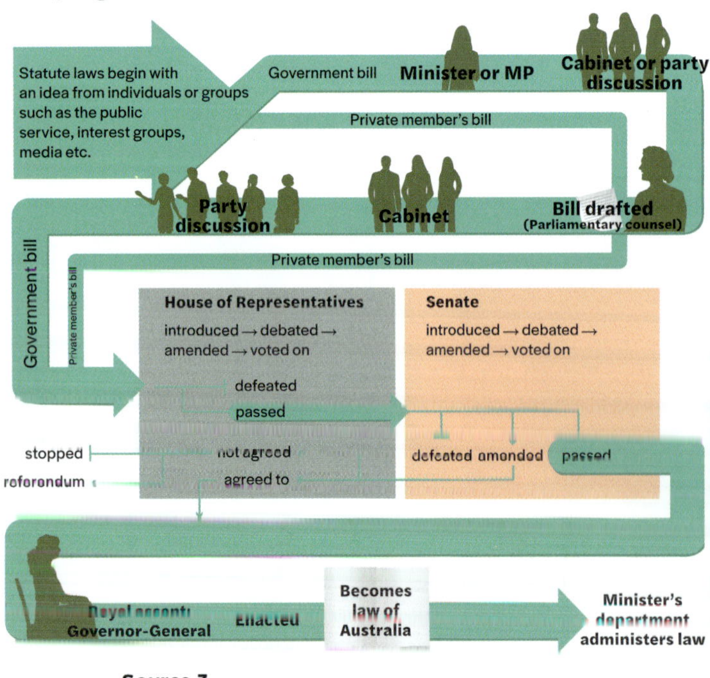

Source 3

Passing statute laws in Australia

Learning Ladder H2.5

Civics and citizenship

Step 1: I can identify topics about society

1 Source 1: Why do you think scales are used as a symbol of justice in ancient Egypt and in modern times?

2 Look at Source 3:
  a What type of laws are being made here?
  b How else are laws made?

Step 2: I can describe societal issues

3 Why is it important that everyone be treated equally under the law? Was everyone treated the same in ancient Egypt?

Step 3: I can explain issues in society

4 What were the important ideals of the *ma'at* system of justice that are still important today?

5 Common sense was applied in ancient Egyptian legal decisions as it is today. Imagine you were a judge in ancient Egypt. Suggest a crime where you would give these punishments:
- return of goods with a fine
- cutting off a hand
- cutting out a tongue.

Step 4: I can explain different points of view

6 Name some of the harsh punishments under ancient Egyptian law.

7 Do we still have these punishments in Australia today? If not, what is a harsh punishment for someone who has broken the law?

8 Why do you think there is a difference between punishments in ancient Egypt and modern Australia?

Step 5: I can analyse issues in society

9 Why do we need laws and how do we ensure people obey them?

key individual

Who was Ramses II?

Ramses II was one of the most powerful pharaohs of ancient Egypt. Ramses II was worshipped by his people and feared by his enemies. He built more monuments and temples – mostly celebrating himself – than any other pharaoh.

The rise of a pharaoh

Ramses II ruled Egypt from 1279 to 1213 BCE. He was groomed from birth to become a pharaoh. His father, Seti I, was the pharaoh, and Ramses II was raised in royal palaces, waited on by handmaids and nurses, and educated in writing, poetry and art. However, Ramses II's destiny was to be a great warrior, and he was also trained in the art of combat. He became commander-in-chief of the army at the age of 10, and at 14 he accompanied his father on military campaigns.

Wives and children

At the age of 15, Ramses II was the heir to the throne of Egypt and he married two of his main wives, Nefertari (also known as Nefertiti) and Isetnofret. Ramses II's mummy shows he stood over six feet in height. He became pharaoh at the age of 25, following the death of his father.

Pharaohs were expected to provide heirs to the throne to continue their own dynasty. Ramses II gave special attention to this responsibility – he fathered 103 children! He had eight main wives, dozens of lesser wives and countless numbers of unmarried partners (called concubines).

Ramses II gave his male heirs important positions and trained his first 12 sons as possible successors. But he lived to the age of 91 and none of his first 12 sons outlived him. His 13th son, Merenptah, became pharaoh after Ramses II died in 1213 BCE. The dynasty ended 150 years later.

Source 1

Huge statues of Ramses II and Queen Nefertari at the entrance to the Abu Simbel temple complex. The temples were [illegible] during Ramses II's reign to commemorate his victory at the Battle of Kadesh. [Abu Simbel temples (13th century BCE), sandstone, Aswan, Egypt.]

Source 2

Mummy of Ramses II who ruled Egypt for 66 years. Examination of the mummy suggests that Ramses II may have died from infection caused by a dental abscess. [Mummy of Ramses II (discovered in 1881), National Museum of Egyptian Civilization, Cairo, Egypt.]

Was Ramses an effective pharaoh?

Historians disagree about the reign of Ramses II. Some claim he was an effective pharaoh, while others suggest he was an egotistic showman. The physical and written evidence suggest that Egypt was a very stable and prosperous place during his reign. Ramses II defended the country's borders and recovered lands lost in previous battles; he also built more temples than any other pharaoh. As a result, he was loved by his subjects and called Ramses the Great by some historians in the 19th century.

Source 3

A representation of Ramses II wearing the special blue crown, known as the *khepresh*. This illustration, by Evelyn Paul, was drawn in the early 1900s.

Source 4

Ramses II sent strong simple messages in the visual images he created in public places and his temples. He promoted himself as superhuman and god-like, as this image suggests. The propaganda states that Ramses II always won his wars – even when the facts of history suggest otherwise. [Limestone relief depicting Ramses II smiting enemies (13th century BCE), limestone, Egyptian National Museum, Cairo, Egypt.]

The Battle of Kadesh

According to the Poem of Pentaur, Ramses II single-handedly defeated an army of 2500 Hittite warriors, cutting down his enemies with spear and sword and running over their bodies with his horse-drawn chariot. The Hittites surrendered following his superhuman display.

However, historians do not believe the account of the Battle of Kadesh told in the Poem of Pentaur. They do agree that Ramses II was a master of **propaganda**. Propaganda is biased or misleading information used to promote a point of view.

Poem of Pentaur

The Poem of Pentaur is the 'official' account of the Battle of Kadesh. Ramses II liked it so much he had the poem inscribed on the walls of some of his temples.

Like Monthu, god of war, was I made,
With my left hand hurled the dart,
With my right I swung the blade,
Fierce as Baal in his time, before their sight.
Two thousand and five hundred pairs of horses were around,
And I flew into the middle of their ring,
By my horse-hoofs they were dashed all in pieces to the ground,
None raised his hand in fight,
For the courage in their breasts had sunken quite;
And their limbs were loosed for fear,
And they could not hurl the dart,
And they had not any heart
To use the spear;
And I cast them to the water,
Just as crocodiles fall in from the bank,
So they sank.
And they tumbled on their faces, one by one.
At my pleasure I made slaughter,
So that none
E'er had time to look behind, or backward fled;
Where he fell, did each one lay
On that day,
From the dust none ever lifted up his head.
Then the wretched king of Khita, he stood still,
With his warriors and his chariots all about him in a ring,
Just to gaze upon the valour of our king.

Source 5

Poem of Pentaur, 1270 BCE, *Notes for the Nile*, GP Putnam's Sons, Hardwicke D Rawnsley, 1892.

Learning Ladder H2.6

Show what you know

1 Why did Ramses II have so many wives and children?

2 What is the difference between a wife and a concubine?

3 Source 5: According to the poem, what was Ramses II like? Give examples from the text.

4 Source 4: Describe the figures portrayed in the large carving. Why is this propaganda?

Historical significance

Step 1: I can recognise historical significance

5 Rank these events in order, from most to least important:
- Ramses II had two main wives
- Ramses II had 103 children
- Ramses' son Merenptah became pharaoh
- the Battle of Kadesh

Step 2: I can explain historical significance

6 Why is Ramses II remembered?

Step 3: I can apply a theory of significance

7 An important aspect of historical significance is the length of time the people were affected. How did Ramses II ensure the rule of his dynasty continued well after his death?

Step 4: I can analyse historical significance

8 Sources 1, 4 and 5: How did Ramses II use propaganda to promote his own significance? Based on the position of the pharaoh in society, how successful do you think he would have been convincing the rest of society?

Historical significance, page 219

How did Egyptians connect with others?

Egypt's geography helped to protect and connect it. The deserts that surrounded Egypt provided a barrier to invaders, while the Nile River opened opportunities to trade with other peoples – or attack them.

Trade

Ancient Egypt did not have all the goods it needed to become a powerful nation. Egyptians traded goods such as grain, papyrus, flax, sandals, beer, wine, cheese, dried fish and linen cloth. Egypt imported a range of luxury goods, including silver, spices, cedar wood, incense, precious stones, gold, copper, jewels, ivory, ostrich feathers, slaves, exotic animals and ebony.

While most trade occurred peacefully, some was established by force. Pharaoh Djer (3050–3000 BCE) attacked Nubia to secure valuable trade centres, principally gold mines. The pharaohs who came after Djer kept a strong Egyptian border and used patrols to protect these resources and trade routes.

Conflict

Ancient Egypt's natural protection was the surrounding desert, and at first they didn't see the need for an organised army. If the pharaoh needed men to defend the country, he would call up the farmers to fight.

Around 1720 BCE, the Hyksos people conquered Lower Egypt using superior weaponry. The Hyksos had bows that could shoot further, bronze axes that were harder and armour that was stronger than that of the Egyptians. They used chariots that carried two soldiers: one to drive and one to shoot his bow or throw his spear.

The introduction of the horse and chariot, along with new bronze weapons, transformed Egypt's army into one of the largest and most powerful armies in the world. After finally defeating the Hyksos, the Egyptians secured their borders and launched attacks on their neighbours using the new weaponry. Ramses II (see page 66) marched his army into Syria and to the Battle of Kadesh, which involved about 6000 chariots altogether – the largest chariot battle of all time.

Trade routes of ancient Egypt

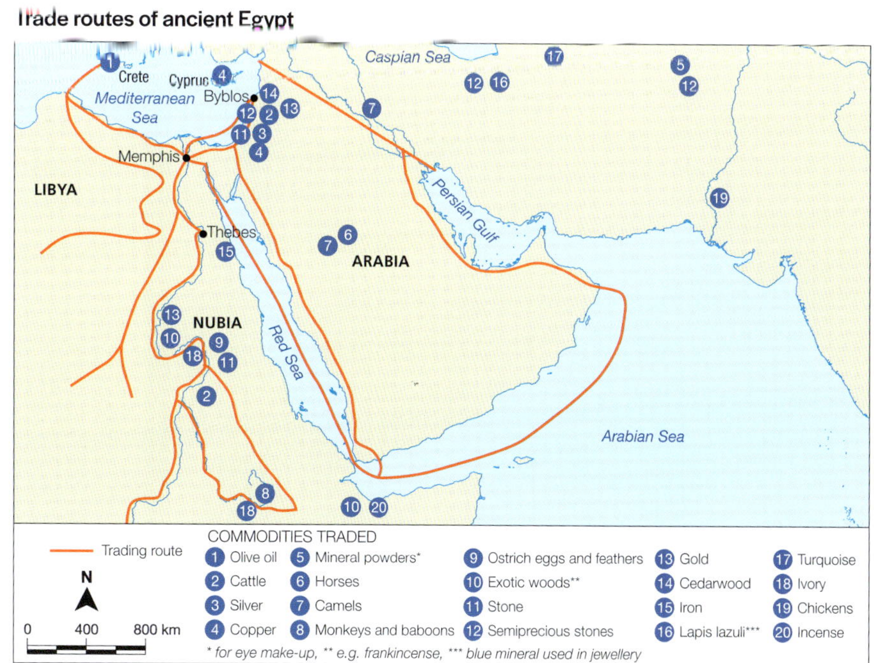

Source: Matilda Education Australia

Source 1

Ancient Egyptians traded goods such as grain and papyrus as well as luxury goods such as silver and spices.

Source 2

A wall painting from the tomb of Rekhmire in Luxor. [Detail from the Tomb of Rekhmire, Sheikh Abd el Qurnah Necropolis (18th Dynasty), Luxor, Thebes, Egypt.]

Source 3

The Egyptians adopted a new way of fighting, using horses, chariots and bronze weapons, after being invaded by the Hyksos. Pharaohs wore a special blue crown, known as the *khepresh*, during battle. [Herbert M. Herget, *King Thutmose III drives his chariot into battle near Har Megiddo* (1941), colour lithograph.]

Learning Ladder H2.7

Show what you know

1 How did the geography of Egypt (the desert and the Nile) affect trade and conflict?

2 Source 1: If you were based in Memphis, from which directions did most of the goods you traded come from?

3 Source 2: What is shown in the wall painting? When was it painted? Why do you think it was painted in the tomb of a pharaoh?

4 Source 3: How did Egyptian warfare change after their defeat by the Hyksos people?

Cause and effect

Step 1: I can recognise a cause and an effect

5 What was the cause of Djer's attack on Nubia?

Step 2: I can determine causes and effects

6 Source 3: What were the effects of the new weapons introduced by the Hyksos people?

Step 3: I can explain why something is a cause or an effect

7 For each cause in this table, write a sentence explaining how the cause led to the effect.

Cause	Effect
Egypt did not have all the resources it needed to become a strong nation	It sold goods to other nations and bought silver, cedar wood, ivory and gold
Egypt adopted the horse, chariot and bronze weapons	Egypt became one of the most powerful nations in the world

Step 4: I can analyse cause and effect

8 What were the many effects of Egypt's location on the Nile? Write three to four sentences explaining how these effects together were beneficial for Egypt. Use phrases such as 'in combination…', 'put together…', and 'when looked at simultaneously …'.

HOW TO

Cause and effect, page 215

How were pictures used to communicate?

The picture writing found on many statues, tombs and temples is known as hieroglyphics. Once the code to hieroglyphics was deciphered following the translation of the Rosetta Stone, historians learnt much about ancient Egyptian society.

Reading and writing

Ancient Egyptians used symbols and pictures known as **hieroglyphics** to read and write. The complicated system involved more than 750 symbols – some represented sounds and others represented entire words. The hieroglyphics writing system was introduced around 3100 BCE.

With such a complicated writing system it is not surprising that very few people in ancient Egypt could read and write. The people who trained over many years to write using hieroglyphics were called **scribes**, and they were mostly men. Men trained to be scribes from the age of six or seven. They were well regarded in ancient Egyptian society because of their rare and important skill (see page 62).

Source 2

The Rosetta Stone was created in 196 BCE and rediscovered in 1799. It was decoded 20 years later, unlocking the ancient hieroglyphics used in Egypt.

Source 1

Hieroglyphics are shown here on the first form of paper, made from papyrus reeds.

Source 3

Hieroglyphics used a combination of symbols and pictures.

Hieroglyphics took a long time to write, so they were mainly used in special circumstances such as on the walls of tombs and temples. Simpler scripts were used in daily writing.

Writing was an important part of the curriculum in ancient Egyptian schools, but formal education was only for males who were training to be scribes or officials. Writing was practised on cheap materials like flakes of limestone until the young scribe graduated to **papyrus**. Papyrus was the first paper-like material used by a civilisation. It was made from the river plant of the same name. Our word 'paper' comes from the word *papyrus*.

Unlocking the code

After Egypt became part of the Roman Empire, people gradually forgot how to use hieroglyphics.

In 1799, a French soldier found a piece of black granite with symbols on it near the ancient Egyptian town of Rosetta. Scholars tried for many years to decipher the text. Twenty years after its discovery, Jean François Champollion worked out that the Rosetta Stone contained the same message written three times: in hieroglyphics, in ancient Egyptian handwriting script and in ancient Greek. This became the key to unlocking the ancient language.

Learning Ladder H2.8

Show what you know

1 How long ago were hieroglyphics introduced?
2 Why could only a few people read and write?
3 What different things did Egyptians write on?
4 Where do we get the word 'paper' from?
5 What modern invention do we have that is like the Rosetta Stone?
6 Use signs and symbols to create your own 'hieroglyphics'. Write a message, then exchange messages with a friend. See if you can interpret each other's hieroglyphics.

Chronology

Step 1: I can read a timeline

7 Look at the timeline on page 56. When were hieroglyphics first developed?

Step 2: I can place events on a timeline

8 Put these events in order, from earliest to most recent.

Frenchman Champollion translates the Rosetta Stone in 1822 CE
Hieroglyphics developed in Egypt 3100 BCE
Rosetta Stone was made in 196 BCE.
Minoans (early Greeks) started using hieroglyphics 1200 years after the Egyptians.

Step 3: I can create a timeline using historical conventions

9 Use the list of events in question 8 to create your own timeline.

Step 4: I can distinguish causes, effects, continuity and change from looking at timelines

10 How did the complex nature of hieroglyphics affect the role of scribes in Egyptian society and the importance of the Rosetta Stone?

Chronology, page 206

What did the ancient Egyptians believe?

Religion played a central role in people's lives and their afterlives. Just about everything was seen to be controlled by the gods or goddesses, who were closely linked to the natural world.

The importance of religion

Ancient Egyptians believed in many gods and goddesses, who usually took the form of animals. Belief in these gods helped explain the things that happened around them. The gods of ancient Egypt controlled the environment and all aspects of people's lives, including knowledge, love, health and death.

For the ancient Egyptians, the sun god Ra was the most important of the gods. Ra was represented as a man with a falcon's head and a headdress in the shape of the sun. As god of the sun, Ra would ride across the sky every day in a boat. Ra was the ruler of the gods. He created all forms of life on Earth and also created the other gods and the seasons of the year.

Source 2

The god Ra-Horakhty was a combination of Horus and Ra. Here, Ra-Horakhty is shown with the sun headdress of Ra and the ankh and sceptre of Horus. [Detail of painting from the tomb of Nefertari (13th century BCE), Valley of the Queens, Egypt.]

Source 1

Pharaoh Tutankhamun's burial chamber

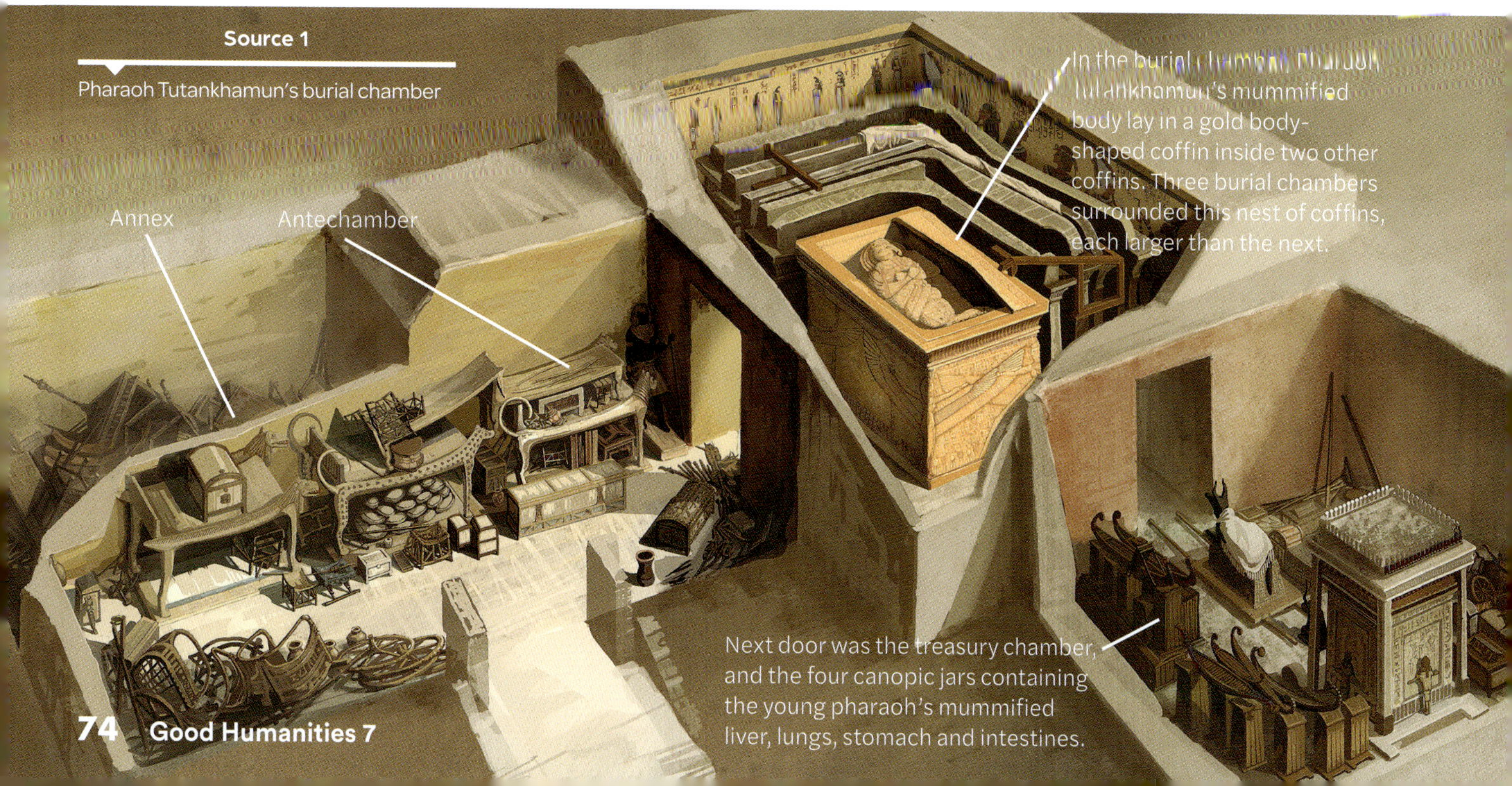

Annex

Antechamber

In the burial chamber, Pharaoh Tutankhamun's mummified body lay in a gold body-shaped coffin inside two other coffins. Three burial chambers surrounded this nest of coffins, each larger than the next.

Next door was the treasury chamber, and the four canopic jars containing the young pharaoh's mummified liver, lungs, stomach and intestines.

Animal gods

The form taken by Egyptian gods was usually a human body with the head of an animal. The animal chosen was usually linked to one of the characteristics of that animal. For example, Sobek was the crocodile-headed god of water and Bastet was the cat-headed protector of homes.

Horus, the god of the sky, was shown as a royal man with the head of a falcon. Horus was closely linked to the pharaoh, who was thought to be the living Horus on Earth. Horus often holds a sceptre (a long pole) and an ankh (a life symbol). He also wears the pharaoh's white and red crown – a symbol of the unity between Upper Egypt and Lower Egypt.

As gods on Earth, a pharaoh could be portrayed as a sphinx with a human head and an animal body, such as that of a lion.

The afterlife

The ancient Egyptians believed in life after death. The deceased person needed a body and other items for transportation and comfort in the afterlife. Preparing a body for the afterlife meant preserving the body through a process called **mummification** (see pages 70–77).

A special book, called the Book of the Dead, was placed in the coffin. It contained spells to protect the person or to guide them through the dangerous underworld on their way to the afterlife.

An Egyptian creation legend

According to ancient Egyptian myth, in the beginning there was a huge ocean called Nun. Ra, the creator god, let out a deep breath and created his son Shu, the god of the air. From his saliva, he created his daughter Tefnut, the goddess of moisture. He then moved across the water and land appeared. He stood on the land and ordered animal and plant species to come out of the ocean. Later, he created people from his tears.

Translated from an ancient Egyptian creation legend

Source 3

Horus the falcon-headed god of the sky. [Statue of Horus (c. 730–709 BCE), bronze, Musee du Louvre, Paris, France.]

Learning Ladder H2.9

Show what you know

1 Why did the Egyptians think Ra flew across the sky?

2 What is the link between the god Horus and the pharaoh?

3 Why were people mummified?

Source analysis

Step 1: I can determine the origin of a source

4 Source 3: How and when do you think this statue was made? Is it a primary or secondary source?

Step 2: I can list specific features of a source

5 Source 2: Describe this image in detail. What tells you it is a painting of a god?

Step 3: I can find themes in a source

6 Source 1: Use this secondary source to make a list of the items buried with Tutankhamun to use in the afterlife.

Step 4: I can use my outside knowledge to help explain a source

7 Source 2: List the separate features of Ra and Horus that are combined in this painting.

Source analysis, page 209

How were people mummified after death?

Ancient Egyptians believed that when a person died, their spirit left their body and was reunited with the body after burial. This could only happen if the body was preserved in a lifelike form, so the ancient Egyptians treated bodies to stop them from rotting, in a process called embalming. The bodies were then wrapped in strips of linen, in a process called mummification.

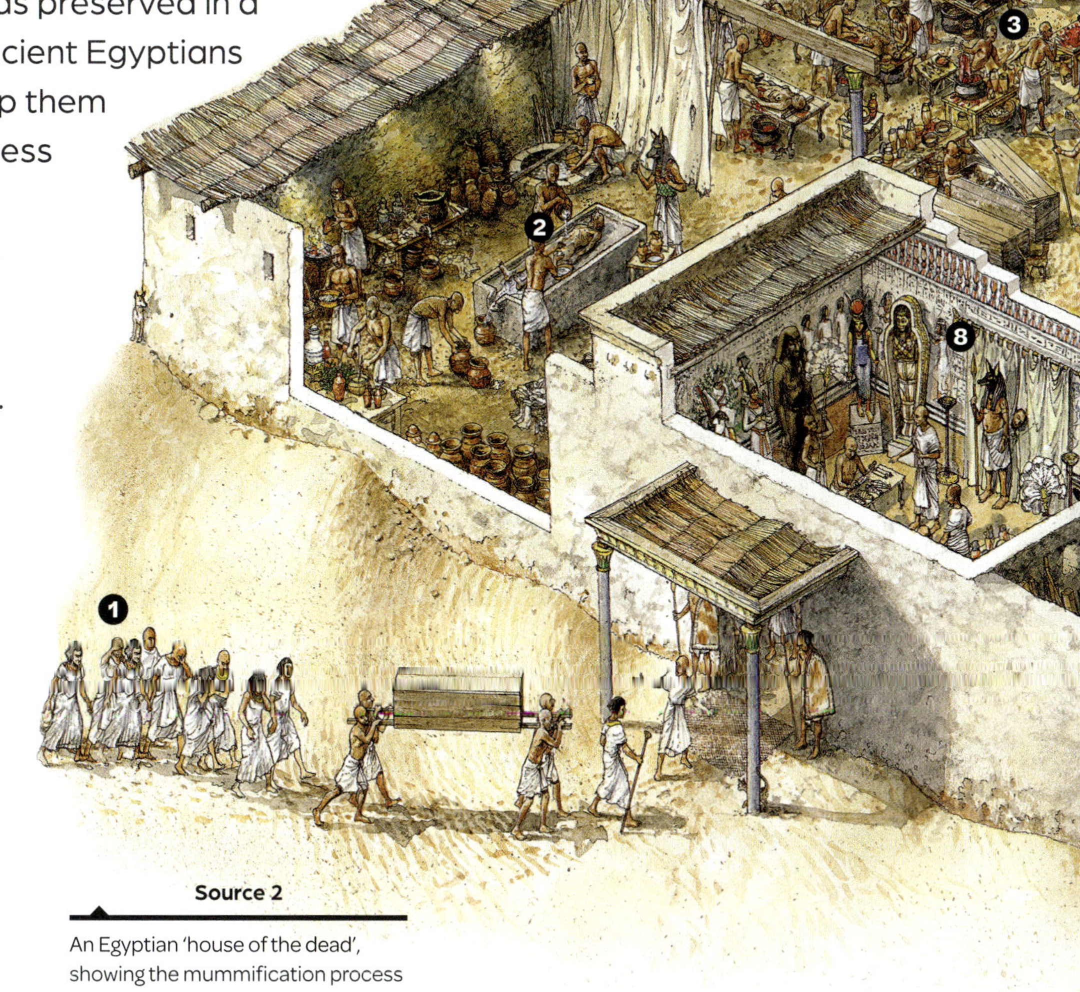

Source 2

An Egyptian 'house of the dead', showing the mummification process

1 Mourners follow the dead body as it is brought to the House of the Dead.

2 The body is washed as a jackal-headed priest reads from the Book of the Dead.

3 Only the heart is left in the body. The lungs, stomach, liver and intestines are removed. The brain is removed by pushing a hook up through the nostrils.

4 The organs are dried, rubbed with scented oils and put into four canopic jars:
- the lungs jar has a baboon's head
- the stomach jar has a jackal's head
- the liver jar has a man's head
- the intestines jar has a falcon's head.

5 The body is covered in salt for 70 days to remove any moisture. It is then washed in water and covered in scented oil.

Source 1

The internal organs removed from the body during mummification were placed in special jars, called canopic jars. [Field Museum, Chicago, US, 2018.]

Source 3

Cat mummies have been found in the tombs of ancient Egyptians. This mummified cat is from 332–330 BCE. Cats were very important, as they killed the rats and mice that got into a family's grain stores. Cats were also linked with the goddess Bastet, the protector of homes (see page 75). Harming or killing a cat could be punished by death.

Learning Ladder H2.10

Show what you know

1 Source 3: Why were cats important to Egyptians?

2 Source 2: Identify three different jobs people did during the mummification process?

3 Source 1: Why do you think the canopic jars are in a US Museum even though they are Egyptian?

4 How would historians know about all the steps in the mummification process?

Source analysis

Step 1: I can determine the origin of a source

5 Source 2: Why was this modern secondary source created?

Step 2: I can list specific features of a source

6 Source 1: Use Source 2 to help you identify the contents of each of these canopic jars.

Step 3: I can find themes in a source

7 List five adjectives that come to mind when you look at Source 2.

Step 4: I can use my outside knowledge to help explain a source

8 Why were pharaohs mummified the most carefully?

9 Which social groups (page 62) do you think did or did not get mummified? Give evidence for your answer.

Source analysis, page 209

6 The embalmed body is wrapped in bandages of linen, and amulets (objects believed to have magical powers) are wound into the bandages. Prayers from the Book of the Dead are said.

7 The mummy is wrapped in up to 20 layers of bandages and painted with sticky resin from pine trees.

8 The completed mummy is put into a coffin. Important people had several coffins, one inside the next, which were placed inside a stone box called a sarcophagus.

What are pyramids for?

How could the ancient Egyptians have constructed the Great Pyramid of Giza without modern tools and machinery? The pyramid is 147 metres tall and made of 2.3 million blocks, each weighing at least 2.5 tonnes. The blocks are carved and placed so precisely that not even a knife blade can fit between them. The base is an almost perfect square, with each side of the pyramid facing one of the four points of the compass.

The amazing pyramids of ancient Egypt

The **pyramids** of ancient Egypt were built as tombs for pharaohs. More than 160 pyramids have been discovered, the largest being the Great Pyramid of Giza. Built around 2560 BCE as the tomb of the pharaoh Khufu, it is the last of the Seven Wonders of the Ancient World that is still standing.

Huge ramps were probably built to help drag stone blocks up to higher levels.

Source 1

Blocks were accurately cut by driving wooden stakes into the stone. The stakes were then made wet to help them expand and split the rock. About 20 000 men worked for 20 years to build the Great Pyramid.

Inside the pyramids

Inside the Great Pyramid of Giza are two main chambers: the King's Chamber and the Queen's Chamber. The Grand Gallery, with walls of polished limestone, is a huge hallway leading to the King's Chamber.

Tomb robbers would often break into pyramids searching for gold, and they would tear at the wrappings of pharaohs and their queens, often ripping off their heads, arms and hands. To stop tomb robbers from stealing treasures buried with the pharaohs, the Egyptians included false doors, dead-end passages, low ceilings and passages blocked with rubble and statues of the most feared gods.

The ancient Egyptians didn't have wheels or work animals. They relied on human power to move the giant blocks, each weighing at least 2.5 tonnes. Until recently, nobody really knew how they moved the blocks. Log rollers, sledges, ramps, oil-slicked slipways (and even alien help!) have all been suggested as methods used to build the pyramids.

In 2014, scientists from the University of Amsterdam built a sled and tested if it could pull heavy stone across sand. They found that pulling a sled across damp sand requires half the force of hauling it in dry sand. Wet sand doesn't pile up in front of the sled as it moves along.

Source 3

Inside the Great Pyramid of Giza are king's and queen's chambers and an unfinished underground chamber. The King's Chamber has two air shafts connected to the outside. Khufu's burial ship is a full-size vessel that was sealed into a pit at the foot of the pyramid. It was intended for use in the afterlife (see page 75).

A capstone called a pyramidion sat on top of the pyramid. It was made from polished granite and covered with glittering metal, probably gold.

The original pyramid was covered with white limestone panels.

Historians think that the blocks were dragged on sleds across the sand or on rolling logs across harder surfaces.

Heavier blocks of granite were used around the burial chamber. The granite was transported on barges from Aswan, about 800 kilometres away.

Source 2

A representation of how pyramids were built

Source 4

A wall painting from the tomb of Djehutihotep (c. 1932–1842 BCE). It shows his huge statue being dragged from the quarries with a person pouring water in front of it. It is likely that water was poured in the path of the sled to reduce friction.

Learning Ladder H2.11

Show what you know

1 Why did ancient Egyptians build pyramids?

2 Approximately how much did the Great Pyramid weigh?

3 Sources 1 and 2: How were the pyramids built?

4 Why do you think it took historians so long to figure out how the the blocks were moved?

5 In your opinion, is it a good idea to use so many resources building tombs? Explain your response.

Historical significance

Step 1: I can recognise historical significance

6 Identify which of the Seven Wonders of the Ancient World are still standing today.

Step 2: I can explain historical significance

7 What impact do you think pyramids would have had on people from other civilisations who saw them?

Step 3: I can apply a theory of significance

8 Think about how the pyramids were built. Write a sentence for each point.

- How important were the pyramids to people at the time?
- How deeply were their lives affected by the pyramids?
- How many people did it affect?
- How long were they affected for?

Are the pyramids relevant today? If so, how?

Step 4: I can analyse historical significance

9 Compare the importance of the pyramids to the importance of the Great Wall of China.

HOW TO

Historical significance, page 219

economics + business

How do businesses act responsibly?

Evidence suggests that ancient Egyptians showed some care for their workers, but today there are strict laws and regulations in place to ensure businesses provide a safe, healthy and fair work environment.

Egyptian workers

More evidence is coming to light to suggest that the tens of thousands of workers who built the pyramids and other royal tombs were free men, not slaves. The workmen were divided into groups, each with an overseer. Evidence of worker villages such as Deir el-Medina (see page 59) have been uncovered. They housed the craftsmen who built and decorated the royal tombs.

In late summer and early autumn, when the Nile inundation (see pages 60–61) flooded farms, a large labour force would move to big building projects such as the pyramids at Giza (see pages 78–79). They are thought to have been willing to work for food and shelter, for the benefit of the pharaoh and to improve their chances of a better afterlife (see page 75).

The skeletons of the Egyptian workers at Giza show some care for injured workers. Those with hand injuries had wooden supports on each side of the hand. Another had evidence of surgery on a damaged leg.

Responsible businesses

The ancient Egyptians showed some responsibilities to keep their workforce healthy and well organised to construct large buildings. Today, laws have been developed to protect workers from being treated unfairly by employers. Businesses have both legal and social responsibilities to their staff and to the environment.

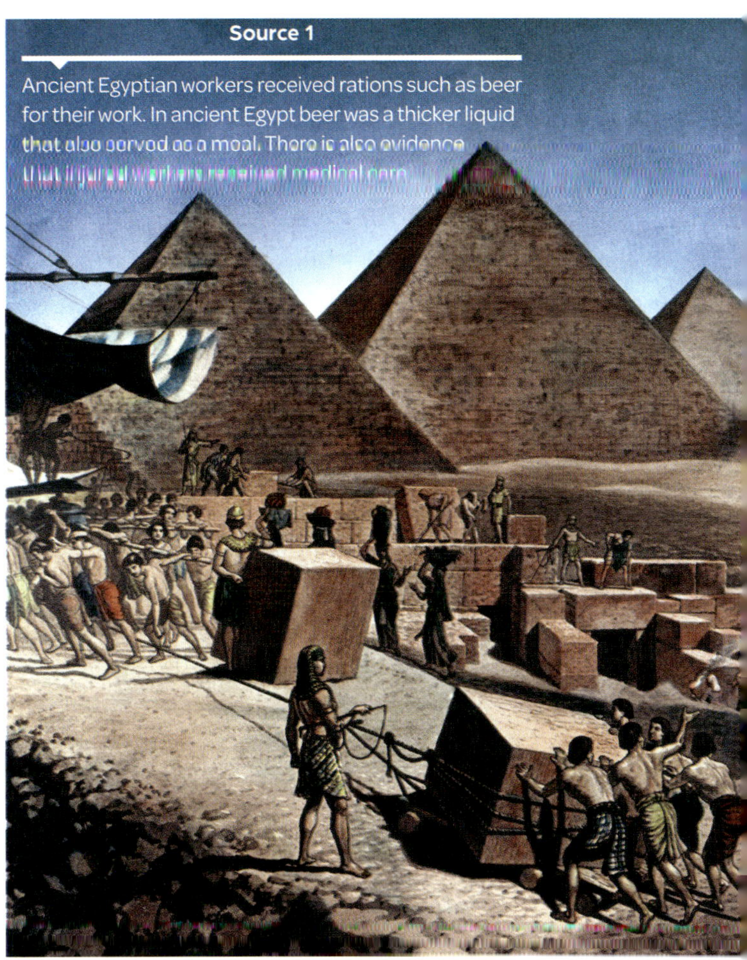

Source 1

Ancient Egyptian workers received rations such as beer for their work. In ancient Egypt beer was a thicker liquid that also served as a meal. There is also evidence that injured workers received medical care.

Safety

It is the responsibility of every employer to provide a safe and healthy workplace for its employees. WorkSafe Victoria is a government agency that regulates health and safety in the workplace. To make sure businesses are following **occupational health and safety (OHS)** laws and regulations, WorkSafe Victoria inspects workplaces to enforce the law and embed a proactive, prevention-led approach to workplace health and safety.

WorkSafe Victoria also provides information to businesses, employees and the general public through its website and television advertisements,

Source 2

WorkSafe Victoria has used graphic advertisements to remind businesses and workers of the importance of safety at work.

and by providing workplaces with posters and other materials that can be displayed in communal areas such as staffrooms and kitchens.

If a workplace incident occurs, WorkSafe Victoria manages Victoria's **workers compensation scheme**. This ensures that employees who are injured while conducting their work can receive medical treatment and other assistance including rehabilitation to help them get back to work quickly.

Discrimination

Employers have a responsibility to ensure all employees are treated fairly and without **discrimination**, **harassment** or **bullying**. These responsibilities are part of Australian federal and state laws such as the *The Australian Human Rights Commission Act 1986* and *The Fair Work Act 2009*.

Discrimination can be described as one person being treated less favourably than another because of their background, appearance or characteristics. Discrimination can happen to a person when they are applying for a job, when asking for a promotion or pay rise, when selecting people for training or special programs and in many other workplace situations.

In Australia it is illegal to discriminate in the workplace due to race, religion, nationality or ethnic origin, sex, marital status, pregnancy, age, disability, sexual orientation or gender identity.

Social responsibility

A business demonstrates **social responsibility** by considering the impact of decisions on people and the environment. Socially responsible businesses implement environmentally friendly practices such as reducing waste and recycling. They can also work with charitable organisations to contribute to communities.

Learning Ladder H2.12

Economics and business

Step 1: I can recognise economic information

1 What evidence is there that ancient Egyptians took responsibility for their workers?

Step 2: I can describe economic issues

2 What is the difference between the legal and social responsibilities of businesses? Give an example of each.

Step 3: I can explain issues in economics

3 What types of discrimination can occur in the workplace?

4 What action to reduce discrimination in the workplace is undertaken by:
   a governments?
   b employees?
   c businesses?

Step 4: I can integrate different economic topics

5 Source 2: Look at the scene from the WorkSafe advertisement.
   a Who do you think the advertisement is targeting?
   b Why do you think there is a need for a regulator like WorkSafe to enforce safety laws?
   c Your school is a work environment. What safety rules are in place?

Step 5: I can evaluate alternatives

6 Source 1: What does this image suggest about conditions for ancient Egyptian workers building the pyramids?

7 Go to the Safe Work Australia 'Construction' web page (http://mea.digital/gh7_h2_1).
   a What is high-risk construction work in Australia today?
   b What are the five most common work-related fatalities?
   c What requirements are there for people to be able to undertake construction work in Australia?

key individual

How did Cleopatra rule Egypt?

Cleopatra was smart, cunning and determined to remain powerful, whatever the cost.

Queen Cleopatra was the last pharaoh of ancient Egypt. She became pharaoh in 51 BCE – at the age of 18 – and ruled until 30 BCE before Egypt fell into the hands of the Romans.

Cleopatra's family, the Ptolemy dynasty, had ruled Egypt for 300 years. Her ancestry was Macedonian, and she grew up speaking, reading and writing Greek. Cleopatra could also speak many other languages including Egyptian and Latin.

Cleopatra was the favourite child of her father, Ptolemy XII. When he died in 51 BCE, his will set out that 18-year-old Cleopatra and her 10-year-old brother, Ptolemy XIII, would be joint rulers. As was customary in Egypt, Cleopatra married her brother when he was 12 years of age.

Cleopatra takes control

Cleopatra quickly took control as the main ruler of Egypt, making it clear she had no intention of sharing power. Ptolemy XIII's name was dropped from official documents and Cleopatra's image was the sole face on coins.

Source 1

Bronze 80 unit coin of Cleopatra VII of Egypt. 51-30 BCE. Cleopatra stamped her authority by taking her brother's name off official documents and appearing without her joint ruler on coins.

Source 3

Bas relief fragment portraying Cleopatra (3rd–1st century BCE)

As Ptolemy XIII grew older he demanded more power. Eventually he forced Cleopatra from the palace and took over as pharaoh. Cleopatra needed support from the Romans to regain power in Egypt. In 48 BCE, the powerful Roman general Julius Caesar arrived in Egypt.

Cleopatra saw an alliance with Caesar as her chance to regain power. After taking power in Egypt, Caesar set himself up in a royal palace at Alexandria. Cleopatra could not walk into the palace unrecognised, so legend says she had herself rolled in a rug as a gift for Caesar, which was carried through the enemy lines to his room. Caesar and Cleopatra became lovers. At the time, she was 21 years old and Caesar was 52.

Cleopatra persuaded Caesar to overthrow her brother's forces. Caesar's Roman forces defeated Ptolemy XIII's army at the Battle of the Nile. Ptolemy XIII drowned in the Nile and Cleopatra became the sole pharaoh of Egypt.

Cleopatra and Caesar

In 47 BCE, after the drowning of Ptolemy XIII, Cleopatra married her brother, Ptolemy XIV, who was next in line to the crown. She continued her union with Caesar, which she saw as important in helping to restore Egypt as a world power. Cleopatra had a son with Caesar, named Caesarion ('little Caesar').

Cleopatra wanted Egypt to remain independent of Rome. She improved the Egyptian economy and established trade with many Arab nations. As sole leader of a prosperous Egypt, Cleopatra was popular.

Cleopatra was in Rome when Caesar was assassinated in March 44 BCE. She returned to Egypt, and a few months later Ptolemy XIV was poisoned. It is thought that Cleopatra was behind the murder. By September 44 BCE, Caesarion was proclaimed co-ruler of Egypt alongside Cleopatra.

Source 2

Frederick Arthur Bridgman, *Cleopatra on the Terraces of Philae* (1896), oil on canvas, 29 7/8 x 46 1/8 inches, Dahesh Museum of Art, New York, USA

Source 4

Cleopatra Testing Poisons on Condemned Prisoners is an 1887 painting by French artist Alexandre Cabanel. It shows Cleopatra testing the effects of poisons on prisoners who had been condemned to death in order to choose the most effective and painless poison to keep with her in case she was forced to take her own life.

Marc Antony

In 41 BCE, Cleopatra met and fell in love with Marc Antony, who was likely to succeed Caesar as ruler of Rome. Cleopatra wanted her son, Caesarion, to eventually rule Rome. Cleopatra and Antony combined their armies to fight Octavian, who was Julius Caesar's legal heir.

In 31 BCE Octavian defeated the forces of Cleopatra and Marc Antony at the Battle of Actium, and the pair retreated to Egypt.

The end of the pharaohs

Cleopatra was determined not to let Octavian execute her. She began to test poisons on prisoners to see which would be best to take her own life. Octavian's forces entered Alexandria, the capital of Egypt, to face Marc Antony. Understanding that he was going to be captured by Octavian, and thinking that Cleopatra had died, Marc Antony stabbed himself with his sword. Before he died, a messenger arrived with information that Cleopatra still lived. Antony was taken to Cleopatra to try and convince her to make peace with Octavian before dying in her arms.

As Octavian's forces entered Alexandria in 30 BCE, Cleopatra and her servants barricaded themselves inside her mausoleum (burial chamber) with stores of gold, pearls, art and other treasures for the afterlife.

Legend tells us that Cleopatra took her own life by allowing a venomous snake to bite her. Octavian took control of Egypt and had Cleopatra's son, Caesarion, strangled. Egypt became part of the Roman Empire. The Ptolemy dynasty and the Egyptian empire came to a close. Cleopatra was the last pharaoh of Egypt.

Did Cleopatra die from a snake bite?

Cleopatra is shown in movies and in paintings taking her own life by allowing a snake to bite her.

It's a fabulous story, but is it true?

It's true that the bite of a snake called an asp can be fatal. But unless the asp's venom hits a vital spot, the victim is likely to recover. So it's a complex means of death – and with no guarantee that it will work.

Even at the time of Cleopatra's death there were conflicting reports about how she died: some said an asp bite; others said she killed herself with poison. But within a few years, Roman poets were writing about Cleopatra allowing an asp to bite her … and some claimed she even allowed two asps to bite her. There was no mention of poison.

Obviously, letting two asps bite you was much more dramatic, and the story stopped being history and entered the world of drama. Other parts of the story that may or may not be true were that Cleopatra carried asps on her ships as weapons, and that she left a suicide note saying she would end her life with an asp bite.

In the end, the asp story became accepted as truth, either because people wanted to believe it or because it was suitably dramatic.

Source 5

Reliefs of Cleopatra VII and Caesarion, her son with Julius Caesar, at the Temple of Hathor at Dendera in Egypt. The main temple was built by Cleopatra's father, Ptolemy XII, and had nearly been completed by Cleopatra when she died in 30 BCE. The construction of the Temple of Hathor was completed around 20 BCE. Hathor is the goddess of the sky, women, fertility and love.

Learning Ladder H2.13

Show what you know

1 Describe Cleopatra's personality in your own words. Give evidence from the section.

2 What actions did Cleopatra take to get power and stay in power?

3 Source 5: Who are the figures shown on the wall of the Temple of Hathor?

4 Source 4: Explain what is happening in this scene and why it is happening.

Chronology

Step 1: I can read a timeline

5 Look at the timeline on pages 56–7. In what year did the Romans take control of Egypt to end the rule of the pharaohs?

Step 2: I can place events on a timeline

6 Find dates for these events in the text and place them in order from the earliest to the latest.

Caesar is assassinated
Temple of Hathor completed
Cleopatra marries Ptolemy XIV
Ptolemy XII dies

Step 3: I can create a timeline using historical conventions

7 Create a timeline of the following events. Be sure to use correct historical conventions such as equal spacing between years, lines for events and spans for things taking place across several years.

Cleopatra ruled Egypt 51 BCE–30 BCE
Cleopatra was 18 when she became pharaoh
Caesar arrived in Egypt in 48 BCE
Caesar was born (he was 31 years older than Cleopatra)
Cleopatra died age 39 in 30 BCE

Step 4: I can distinguish causes, effects, continuity and change from looking at timelines

8 Examine the events listed in this section. Do you think there is a causal link between the love affairs of Cleopatra and the fall of Egypt? Or would the fall of Egypt have happened anyway? Use historical dates in your answer.

HOW TO
Chronology, page 206

Masterclass

Learning Ladder

Source 1

A statue of Ramses II crafted from different types of granite (front view). [*The Younger Memnon* (19th Dynasty), granite, 203.3 x 266.8 cm, British Museum, London, UK.]

Step 1

a **I can read a timeline**

Look at the timeline on pages 56–7.

i How much longer after Cleopatra became pharaoh did Egypt become part of the Roman Empire?

ii Roughly how many years were there between early settlers in the Nile valley and the building of the Great Pyramid?

iii Which of the three kingdoms of ancient Egypt lasted the longest and how long did it last for?

iv How long after the early settlers arrived in the Nile Valley did the first dynasty emerge?

b **I can determine the origin of a source**

Source 1: Roughly when was this statue carved and what is it made from?

c **I can recognise continuity and change**

Rank these things in order from most to least changed since ancient Egyptian times:

- religion
- family
- farming.
- food
- transportation

d **I can recognise a cause and an effect**

Copy this table and match the cause with its effect.

Cause	Effect
Belief in afterlife	Bodies are mummified
Flooding of Nile River	Egypt develops new weapons
Hyksos invades Egypt	Union with Julius Caesar of Rome
Cleopatra replaces Ptolemy XII as pharaoh	Farmland is flooded

e **I can recognise historical significance**

Rank these things from least to most historically important:

- use of hieroglyphics
- use of eye make-up
- use of Nile river for trade
- use of bronze weapons
- use of wigs.

Source 2

A statue of Ramses II crafted from different types of granite (rear view). [*The Younger Memnon* (19th Dynasty), granite, 203.3 x 266.8 cm, British Museum, London, UK.]

Work at the level that is right for you or level-up for a learning challenge!

Step 2

a I can place events on a timeline

Put these events in order, from earliest to most recent.

Ramses II becomes pharaoh in 1279 BCE

525–425 BCE Persians rule Egypt

Alexandria is founded by Alexander the Great after he conquers Egypt in 332 BCE

Cleopatra dies in 30 BCE

Rosetta Stone carved in 196 BCE

Romans take control of Egypt in 30 BCE

b I can list specific features of a source

What features of the statue in Source 1 tell you that it is a pharaoh? How is the pharaoh made to look?

c I can describe continuity and change

Create a social structure pyramid for modern Australia like the one in Source 1 on page 62.

d I can determine causes and effects

i What caused people to wear dark eye make-up?
ii What was the effect of regular inundations of the Nile?

e I can explain historical significance

Why do we remember Cleopatra?

Step 3

a I can create a timeline using historical conventions

Put the dates from Step 2a in a timeline using correct historical conventions. Make sure you follow the convention of equal spacing for equal years.

b I can find themes in a source

Source 1: List three adjectives you would use to describe Ramses II. Putting these three ideas together, what do you think a first-time visitor to the original site of this statue might think when looking at it?

c I can explain why something did or did not change

Why is wheat still the main staple food for many people in Australia today, just as it was in ancient Egypt?

d I can explain why something is a cause or an effect

Why did Ramses II have so many children?

e I can apply a theory of significance

How important do you think the pyramids were just after they were built? How important are they now? Why has this changed?

Step 4

a I can distinguish causes, effects, continuity and change from looking at timelines

Look at the timeline on pages 56–7. Which events show that Egypt was ruled continuously, with little change?

b I can use my outside knowledge to help explain a source

Sources 1 and 2: What kind of ruler was Ramses II? What do you know about his character already that you can see evidence of in this statue?

c I can analyse patterns of continuity and change

Egypt was invaded many times. Has this happened in Australia? Do Australians fear attack like Egyptians did? Explain your answer.

d I can analyse cause and effect

Early pharaohs didn't have a permanent army. This meant it was cheaper for the pharaoh. It also left Egypt undefended from a surprise attack. Which effect was more significant? Explain your response.

e I can analyse historical significance

List three reasons why the Rosetta Stone is historically important. Rank your reasons in order of importance. Do any of the reasons support the other reasons? Are the reasons together more convincing than alone? Support your answer with historical evidence.

Masterclass

Step 5

a I can describe patterns of change

Describe how Egypt developed greater monuments and technology over time. Use evidence from the timeline on pages 56–7 in your answer.

b I can use the origin of a source to explain its creator's purpose

Ramses II built many statues to himself. Why do you think he did so?

c I can evaluate patterns of continuity and change

Some historians see a link from ancient Egypt to modern Australia: Egypt influenced Greece → Greece influenced Rome → Rome influenced the development of European countries → European countries influenced Australia. Do you agree or not? Justify your answer.

d I can evaluate cause and effect

Romans conquered Egypt, ending thousands of years of dynastic rule. Is this a good thing or a bad thing? Use historical evidence in your answer.

e I can evaluate historical significance

Is the Egyptian civilisation really that significant? Is it important? Or is it just interesting? Justify your response.

Historical writing

1 Structure

Imagine you are answering an essay question: 'Egyptian civilisation was obsessed with death. Discuss.' Write an essay plan for this topic. Have at least three main paragraphs.

2 Draft

Using the drafting and vocabulary suggestions on page 226, draft a 400–600 word essay responding to the topic.

3 Edit and proofread

Use the editing and proofreading tips on page 227 to help edit and proofread your draft.

Historical research

4 Organise and present information

Imagine you are conducting a research report on the topic: 'Social groups in ancient Egypt'. Take 10 short dot points as notes from this chapter.

Write a contents page for the research topic. You should have at least three sections to your report.

Capstone

How can I understand ancient Egypt?

In this chapter, you have learnt a lot about ancient Egypt. Now you can put your new knowledge and understanding together for the capstone project to show what you know and what you think.

In the world of building, a capstone is an element that finishes off an arch or tops off a building or wall. That is what the capstone project will offer you, too: a chance to top off and bring together your learning in interesting, critical and creative ways. You can complete this project yourself, or your teacher can make it a class task or a homework task.

Scan this QR code to find the capstone project online.

mea.digital/GHV7_H2

H3

Ancient Greece

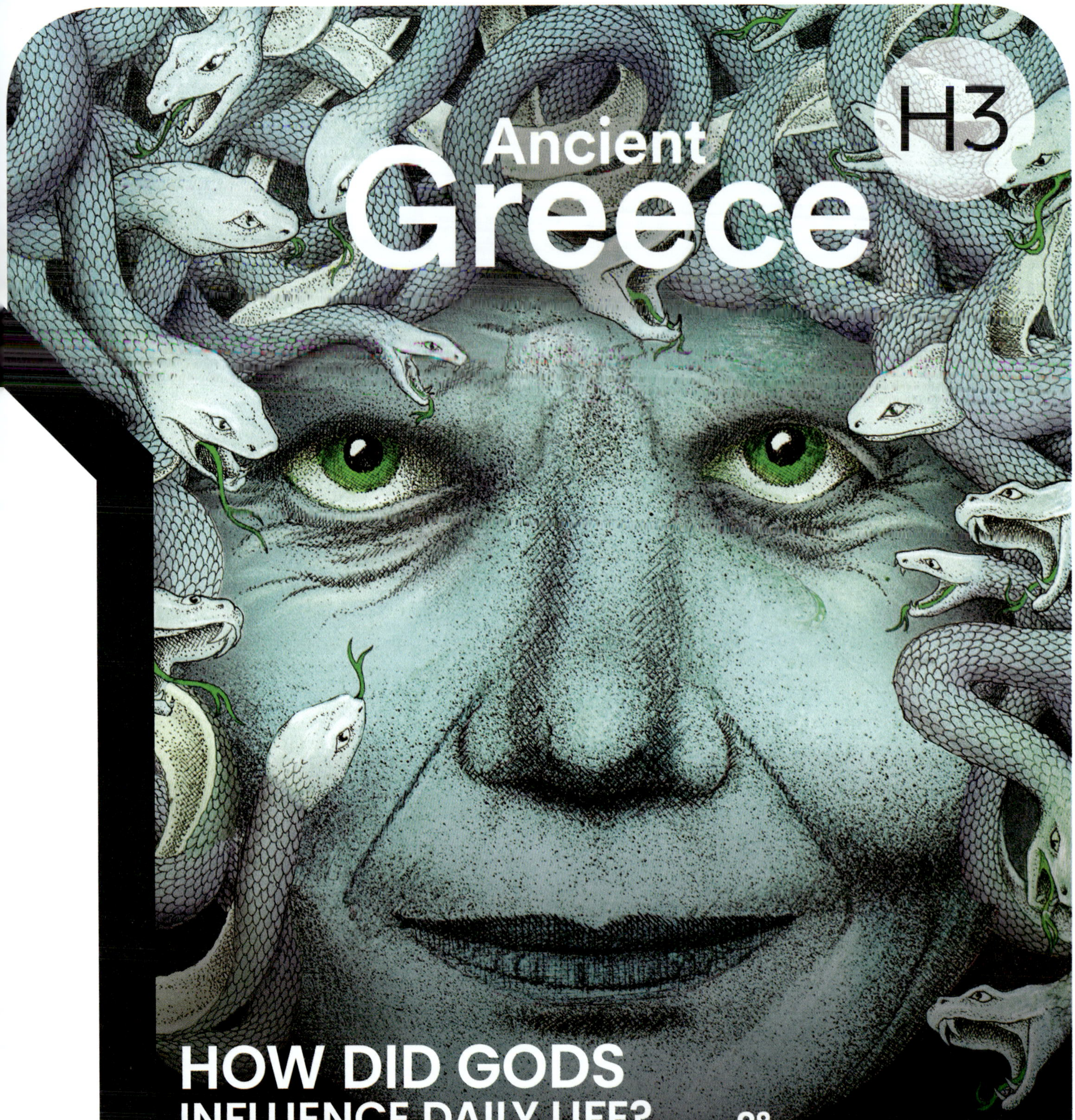

HOW DID GODS INFLUENCE DAILY LIFE?

page 98

cause + effect

page 96

HOW DID THE **ANCIENT GREEKS** BECOME SEAFARERS?

civics + citizenship

page 104

WHAT IS **DEMOCRACY?**

key individual

page 106

HOW DID **PERICLES** FORGE THE GOLDEN AGE OF ATHENS?

How can I understand ancient Greece?

The ancient Greek civilisation dominated the Mediterranean region thousands of years ago – and much of what they did still has an impact on the world today. Ancient Greece influenced everything from government, philosophy and mathematics to architecture, literature and sport. Alexander the Great's conquest of other lands spread Greek culture into Asia. Greek culture also influenced the Roman Empire, which came after it.

learning ladder

Step	Chronology	Source analysis	Continuity and change
step 5	**I can describe patterns of change** I read timelines and see the 'big picture'. I group timeline events and see if they show patterns of change. I know typical historical patterns to look for.	**I can use the origin of a source to explain its creator's purpose** I combine knowledge of when and where a source was created to answer the question, 'Why was it created?'	**I can evaluate patterns of continuity and change** I answer the question, 'So what?' about patterns of continuity and change. I weigh up different things and debate the importance of a continuity or a change.
step 4	**I can distinguish causes, effects, continuity and change from looking at timelines** I read timelines and find events that are linked by cause and effect. I find things that are the same or different from then until later times.	**I can use my outside knowledge to help explain a source** I have enough outside knowledge about ancient Greece to help me explain a source.	**I can analyse patterns of continuity and change** I see beyond individual examples of continuity and change in ancient Greece and identify broader patterns, and I explain why they exist.
step 3	**I can create a timeline using historical conventions** When given a set of ancient Greek events, I construct a historical timeline, making sure I use correct terminology, spacing and layout.	**I can find themes in a source** I look a bit closer into a source and find more than just features. I find themes or patterns in the source.	**I can explain why something did or did not change** I answer the question 'why?' something changed or stayed the same between historical periods.
 step 2	**I can place events on a timeline** When given a list of events in ancient Greece, I put them in order from earliest to latest, the simplest kind of timeline.	**I can list specific features of a source** I look at an ancient Greek source and list detailed things I can see in it.	**I can describe continuity and change** I have enough content knowledge about two different historical periods to recognise what is similar or different about them, and can describe it.
 step 1	**I can read a timeline** I read timelines with ancient Greek events on them and answer questions about them.	**I can determine the origin of a source** I can work out when and where an ancient Greek source was made by looking for clues.	**I can recognise continuity and change** I recognise things that have stayed the same and things that have changed from ancient Greece until now.

Source 1

The ruins of the Parthenon in modern-day Athens. If you look closely you can see the scaffolds used in ongoing restoration work. Its original construction was completed in 432 BCE.

Cause and effect	Historical significance
I can evaluate cause and effect I answer the question 'So what?' about cause and effect. I weigh up different things and debate the importance of a cause or an effect.	**I can evaluate historical significance** I answer the question 'So what?' about things that are supposedly historically important. I weigh up events against each other and cast doubt on how important things are.
I can analyse cause and effect I don't just see a cause or effect as one thing. I determine the factors that make up causes and effects.	**I can analyse historical significance** I separate out the various factors that make something historically important in ancient Greek history.
I can explain why something is a cause or an effect I can answer 'how?' or 'why?' a cause led to an effect in ancient Greece.	**I can apply a theory of significance** I know a theory of significance. I use it to rank the importance of ancient Greek events.
I can determine causes and effects Applying what I have learnt about ancient Greece, I can decide what the cause or effect of something was.	**I can explain historical significance** I answer the question 'why?' about things that were important in ancient Greece.
I can recognise a cause and an effect From a supplied list, I recognise things that were causes or effects of each other in ancient Greece.	**I can recognise historical significance** When shown a list of things from ancient Greek history, I can work out which are important.

Warm up

Chronology

1 Temples were important buildings in ancient Greece. The Parthenon is the most famous temple. Its construction was completed in 432 BCE. Next to it is the Temple of Athena Nike, which was constructed around 426 BCE. Which temple is older?

Source analysis

2 Examine Source 1 and its caption. Approximately when and where was this photo taken? How do you know?

Continuity and change

3 Looking at Source 1, how has the Parthenon changed over time? What clues in the photograph led you to your answer?

Cause and effect

4 Which of these things is an effect of Greece having lots of coastline?
   a Greece has more than 13 000 km of coastline.
   b Ancient Greeks became skilled sailors.
   c Sailors prayed to Poseidon, the god of the sea.

Historical significance

5 Using the introductory paragraph at the top of page 90, copy and complete this sentence: 'Ancient Greece has influenced modern-day ...'

How did Greek civilisation develop?

Ancient Greece was surrounded by the Aegean Sea. Its area covered modern-day Greece, along with parts of modern Turkey and settlements around the Mediterranean Sea and the Black Sea. The mountainous landscape meant that settlements were isolated and they grew into hundreds of separate city-states rather than growing as one single Greek nation. Before city-states, the Minoans on the island of Crete and then the Myceneans on the mainland were two of the earliest Greek civilisations to develop.

Minoans

The Minoan civilisation emerged in Crete in the early **Bronze Age**. Around 3000 BCE, the Minoans were growing crops and herding cattle. By 2500 BCE they were building towns, and making tools, weapons and ornaments from bronze. They also created detailed artworks on pottery that have given modern historians a good idea of what life was like in this ancient civilisation.

Source 1

This octopus flask was made by Minoans around 1500 BCE. [Flask depicting octopus (c.1500 BCE), clay, Heraklion Archaeological Museum, Crete, Greece.]

Source 2

The palace at Knossos was home to the Minoan king on the island of Crete. It had 1400 rooms. The beautiful frescoes on its walls have helped historians learn about the Minoans.

Source: Matilda Education Australia

Source 3

Early settlements of ancient Greece

In 1900 CE, archaeologist Arthur Evans excavated the remains of an ancient civilisation on the island of Crete. He called these people Minoans after their legendary King Minos. Much of Evans' work involved restoring the huge Knossos palace that had been rebuilt about 1700 BCE, possibly after it was destroyed by a tsunami triggered by a volcanic eruption on Thera (modern-day Santorini). The palace survived until about 1350 BCE when it was destroyed by fire, perhaps as part of a Mycenaean invasion.

Historians have learnt a great deal about the Minoans from the excavation of the remains of the palace at Knossos. Large frescoes (paintings) on the palace walls provide evidence that the Minoans used ships to trade by sea. Further evidence shows that Minoans set up trading bases on many Greek islands and traded with Egypt, Syria and other city-states.

Minoans used their own written language, which historians have labelled 'Linear A'. It disappeared suddenly about 1450 BCE and has still not been decoded.

Source 4

The Bull-Leaping Fresco from the palace at Knossos shows a gymnast grabbing a charging bull by the horns and tumbling onto the bull's back before landing at its rear. Modern attempts to try this have resulted in a number of deaths. Experiments have shown that a bull is too fast and powerful to allow its horns to be grabbed even before attempting to flip over its neck. [Bull-Leaping Fresco (c.1400 BCE), stucco, 104.5 × 78.2 cm, Heraklion Archaeological Museum, Crete, Greece.]

Source 5

Mycenaean pottery from 1300 BCE, with images of chariot warriors. [Crater portraying chariot warriors (c.1300 BCE), clay.]

Mycenaeans

By about 1400 BCE, Mycenaean invaders from the Greek mainland had invaded Crete and ended the Minoan civilisation. Unlike Minoan cities, the Mycenaean settlements were surrounded with high perimeter walls. There are still remains of the ancient walls and gates such as the Lion Gate, which was the main entrance to Mycenae.

Scenes of hunting and warfare were depicted on Mycenaean artworks, in contrast to the peaceful themes in Minoan art. As a strong military society, the Mycenaean civilisation became powerful and grew between about 1400 BCE and 1200 BCE.

Like the Minoans, the Mycenaeans took advantage of their closeness to the Mediterranean Sea by becoming great seafarers and sea traders. They took over Minoan trading routes and expanded them. The Mycenaean palaces imported metals, ivory and glass, and exported objects made from these materials. They also exported locally produced products such as olive oil, perfume, wool and wine.

The Mycenaean language is believed to be the most ancient form of Greek. Their written language, known as 'Linear B', has been translated, allowing historians to learn more about Mycenaean culture.

The Mycenaean civilisation was at its peak around 1200 BCE – then it suddenly collapsed. Historians don't know why, but it may have been due to an invasion, an uprising by the people, or a natural disaster such as a volcano, earthquake or drought.

Source 6

Lion Gate was the main entrance to Mycenae. It was built around 1250 BCE and named after the sculpture of two lions above the entrance. It is the only surviving piece of Mycenaean sculpture. [*Lion Gate* (c.1250 BCE), stone, Mycenae, Greece.]

Troy

Homer was the Greek author of the poem the *Iliad*, which tells the story of how a Greek force led by Mycenaeans captured and destroyed the city of Troy.

The war began because Prince Paris of Troy abducted Helen of Sparta, who was the wife of Menelaus, the king of Sparta. Menelaus asked his brother, Mycenaean king Agamemnon, to get Helen back. Agamemnon put together a great Greek navy and led a 10-year siege of Troy, which included great battles outside Troy's city walls.

key ideas timeline.

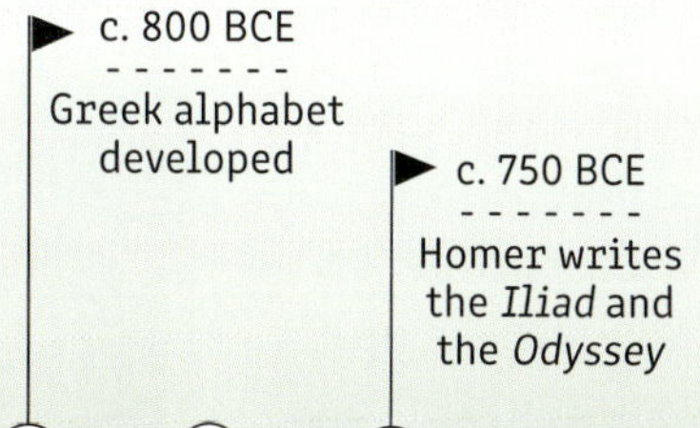

- c. 3200 BCE – Start of Bronze Age
- 2600 BCE – Minoan Civilisation
- 1400 BCE – Great Age of Mycenae begins
- 1300 BCE – Start of the Iron Age
- c. 800 BCE – Greek alphabet developed
- 776 BCE – First Olympic Games
- c. 750 BCE – Homer writes the *Iliad* and the *Odyssey*

Bronze Age c. 3200–1200 BCE

Minoan Civilisation 2600–1100 BCE

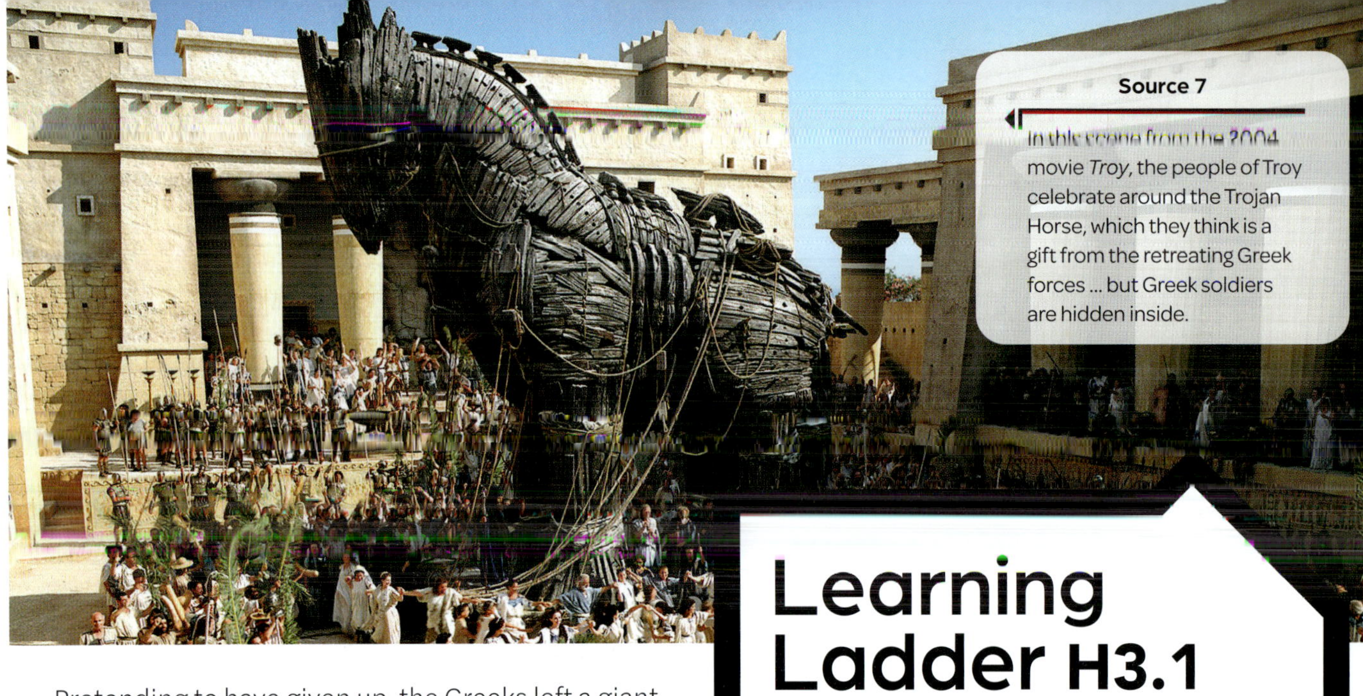

Source 7

In this scene from the 2004 movie *Troy*, the people of Troy celebrate around the Trojan Horse, which they think is a gift from the retreating Greek forces ... but Greek soldiers are hidden inside.

Pretending to have given up, the Greeks left a giant wooden horse outside Troy's city gates and sailed away. Hidden inside the horse was Odysseus the king of Ithaca, and 20 other warriors.

The Trojans (as people of Troy were called) saw the Greeks retreating and thought they had left the horse as a gift. The Trojans dragged the horse inside the city walls and Odysseus and his men snuck out at night and opened the city gates allowing the returning Greek soldiers to capture Troy.

In 1870, an amateur German archaeologist, Heinrich Schliemann, used information in Homer's *Iliad* and other ancient writings to locate what he thought was the site of ancient Troy, at Hisarlik in modern-day Turkey.

Excavations at Hisarlik have uncovered sections of a defensive wall and gates thought to be the site of ancient Troy. Several monuments including the temple of Athena, a concert hall and a meeting place (a *bouleuterion*; see page 103) have been uncovered. The site is classified and protected as a World Heritage Site.

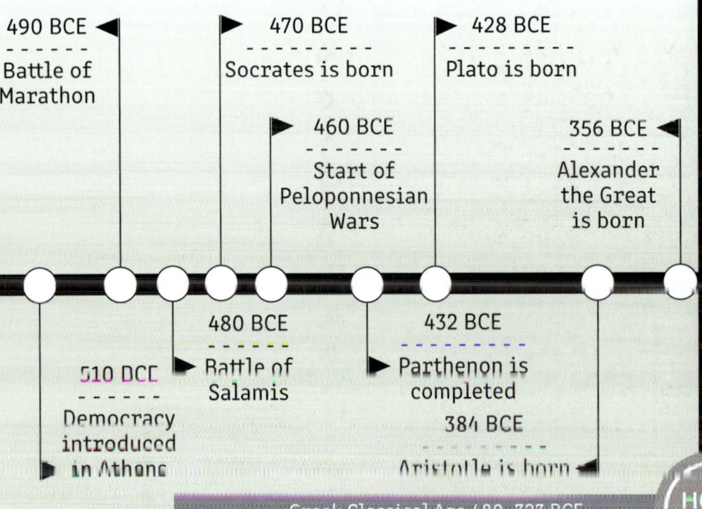

Learning Ladder H3.1

Show what you know

1 How did the geography of Greece affect where people lived and what they did?

2 Sources 2 and 4: What can we learn about Minoan civilisation by looking at the ruins of the palace at Knossos?

3 Source 3: Describe the location of the Minoan civilisation.

4 What was used to help locate the site of Troy?

Chronology

Step 1: I can read a timeline

5 Use the timeline to answer these questions.
   a How many years before the Great Age of Mycenae did the Minoan civilisation begin?
   b Which civilisation lasted longer: the Minoan or the Mycenaean?

Step 2: I can place events on a timeline

6 Create your own timeline of the Minoan and Mycenaean civilisations using the dates and information on these pages.

Step 3: I can create a timeline using historical conventions

7 The timeline shows two different scales – from 356 to 800 BCE and from 1300 to 3200 BCE. Estimate how much one centimetre represents in years for each scale.

Step 4: I can distinguish causes, effects, continuity and change from looking at timelines

8 Minoan civilisation lasted for around 1500 years. Why do you think we still don't know much about this civilisation?

HOW TO

Chronology, page 206

H3.2

Why did the ancient Greeks become seafarers?

Greece is mountainous and is surrounded by seas. Smaller peninsulas jut out from the main peninsula, creating more coastline and many natural harbours. There are hundreds of small Greek islands in the Aegean and Mediterranean seas. The ancient Greeks became expert shipbuilders and developed different craft for pleasure, transport, trading and warfare. Much of the communication and trading between ancient Greek city-states was done by sea.

Life by the sea

The ancient Greeks took advantage of their access to the sea and protected harbours by becoming experienced fishermen, sailors and shipbuilders. Greek ships were powered by oars or sails – or both. Small trading ships stayed in sight of the shore and sailors on longer voyages used the stars to navigate when no land was in sight. Before a voyage, the sailors prayed to the sea god Poseidon for a safe journey (see pages 98–99).

Skilled seafarers

Larger ships were designed to carry cargo or to fight at sea. They were powered by sails and oarsmen and built to turn quickly and move rapidly through the water. By the third century BCE, the Greeks had started to use the Little Bear (Ursa Minor) constellation of stars in the northern sky to navigate at night.

Ships were used for trading with Greek city-states and other civilisations around the Mediterranean Sea. Through trade, the Greeks acquired new ideas and skills in navigation and shipbuilding: from Egypt they gained new ideas about astronomy, shipbuilding and mathematics; from the Syrians they learnt metalworking techniques.

In times of war, the specially designed ships, with their strong sails and well-trained oarsmen, helped Athens win many sea battles.

Greek warships

Greek warships had oars as well as sails. The largest warships were called **triremes**, and they had three banks of oars on each side. A trireme was about 35 metres long and needed 170 men to row it. The large number of oarsmen made the trireme very fast in battle.

The oarsmen sat under a long narrow deck. Soldiers ran along the deck and fought from it. In a battle, the captain ordered the ship to steer straight at an enemy ship. The bronze prow at the front of the ship was used like a battering ram to smash a hole in the side of an enemy ship. The damaged ship either sank or had to be beached. Soldiers on the triremes sometimes jumped onto a damaged ship to capture it.

Source 1

Ancient Greeks used the Little Bear (Ursa Minor) constellation for navigation.

Source 2

A cutaway view of a trireme. A trireme was steered by long oars at its stern (or back). The front of the trireme had a metal battering ram for smashing holes in the sides of enemy ships.

The Battle of Salamis

The greatest sea battle in Greek history was the battle of Salamis in 480 BCE. It was fought between the Greeks and the invading Persians, and took place at the island of Salamis, near the port of Piraeus.

Greek historian Herodotus recorded advice that Artemisia (the female ruler of Halicarnassus) gave to Persian king Xerxes: 'Spare thy ships, and do not risk a battle; for these people are as much superior to your people in seamanship ...'.

Xerxes ignored Artemisia's advice and sent a massive fleet of 500 Persian ships into the narrow strait at Salamis. The Greeks had small, fast moving ships. They rammed the huge Persian ships with their triremes and tossed aboard burning wood to incinerate them. About 300 Persian ships were sunk or damaged in a great Greek victory.

Learning Ladder H3.2

Show what you know

1 How did Greeks become expert shipbuilders and navigators?

2 Describe three types of ships used by ancient Greeks.

3 What features of a trireme would be intimidating to foreign navies?

4 Draw a trireme diagram. Label at least two attacking features, two defensive features, and two movement features.

Cause and effect

Step 1: I can recognise a cause and an effect

5 Which of these was a cause of Greeks becoming skilled seafarers: good timber, Persian traders, olive farming, mountainous geography, military threats, being surrounded by sea.

Step 2: I can determine causes and effects

6 What caused the Greeks to win the Battle of Salamis?

Step 3: I can explain why something is a cause or an effect

7 Explain what effect the development of the trireme had for Greece in warfare.

Step 4: I can analyse cause and effect

8 How did trading by sea influence new innovations in navigation and shipbuilding in ancient Greece?

HOW TO Cause and effect, page 315

Source 1

Mount Olympus is the highest mountain in Greece. Ancient Greeks believed it was the home of the gods because the summit was usually hidden by clouds.

How did gods influence daily life?

The people of ancient Greece believed that the gods of Mount Olympus controlled everything in their lives. To honour the gods, they would pray daily at a shrine in their home, or they would visit the temple of a specific god if help were required in a particular area of their lives.

Source 2

Many modern superheroes are based on Greek gods from mythology. These two superheroes have specific links to Greek mythology. Wonder Woman is the biological daughter of Zeus, king of the gods. Aquaman is heir to the throne of Atlantis, the mythical underwater ancient Greek city.

HADES

God of Death and the Underworld

Hades was Zeus and Poseidon's elder brother. Hades rarely left the Underworld, where he ruled over the dead.

ZEUS

King of the gods and God of the Sky and Thunder

Zeus was the king of the gods and the god of the sky and thunder. Zeus controlled the weather and hurled lightning bolts at those who displeased him.

HERA

Goddess of Marriage

Hera was the wife of Zeus and queen of the gods. She was the goddess of women, marriage and family.

POSEIDON

God of the Sea

Poseidon was Zeus' bad-tempered brother and god of the sea and earthquakes. His weapon was a trident that could shake the Earth.

HERMES

Messenger of the gods

Hermes was a son of Zeus. He was the fastest god and had winged sandals and hat. Hermes was the messenger of the gods and responsible for guiding the dead to the Underworld.

ATHENA

Goddess of Wisdom and Warfare

Athena was one of Zeus' daughters and the Greek goddess of wisdom and war. She would protect soldiers during war.

Gods and goddesses

Ancient Greeks believed in many different gods and goddesses that controlled everything in their lives and their environment. People thought it was important to please the gods, as happy gods helped you and unhappy gods punished you.

Ancient Greeks believed that a family of the most important gods and goddesses lived at the top of Mount Olympus in northern Greece.

Source 3

The Statue of Zeus at Olympia was one of the Seven Wonders of the Ancient World. The 13-metre-tall figure was completed around 435 BCE and destroyed in the fifth century CE.

Honouring the gods

To honour the gods, ancient Greeks built temples in every town, each one dedicated to a specific god or goddess.

Temples were usually built on a hill called an acropolis (see pages 106–109). Temples were elaborately decorated, inside and out. A statue of the god for whom the temple was built was erected inside.

Temples were cared for by priests who had the power to talk to the gods. People would visit different temples according to what they were praying for. For example, those wanting help in their love life would attend a temple dedicated to Aphrodite, the goddess of love.

PERESEPHONE

Queen of the Underworld

ARES

God of War

ARTEMIS

Goddess of the Hunt

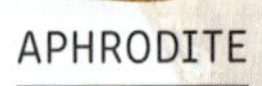

APHRODITE

Goddess of Love and Beauty

Aphrodite was the goddess of love and beauty. She had a girdle that had the power to cause people to fall in love with the wearer.

EROS

God of Love and Attraction

APOLLO

God of the Sun

DIONYSUS

God of Wine

Source 4

A sixteenth-century bronze statue entitled *Perseus with the Head of Medusa*. [Benvenuto Cellini, *Perseus with the Head of Medusa* (16th century CE), bronze, Piazza della Signoria, Florence, Italy.]

Medusa and Perseus

One of the most popular stories in Greek mythology is that of Medusa and Perseus.

Medusa was a beautiful maiden who was one of the priestesses of Athena, the goddess of warfare. Medusa had vowed to Athena that she would remain a virgin forever and never marry ... but then she became pregnant to Poseidon, god of the sea.

Athena punished Medusa for breaking her vow. She turned Medusa into a horrible creature. The strands of her beautiful hair morphed into poisonous snakes. Her skin turned green and her eyes became bloodshot – and anyone she looked at turned to stone. Medusa was shunned and hated by everyone, and went off to a secret hiding place.

Meanwhile, King Polydectes had fallen in love with Perseus' mother, and wanted Perseus out of the way, preferably dead. Polydectes gave Perseus a challenge: to slay Medusa and bring back her head.

Athena gave Perseus a shiny bronze shield. Hermes the messenger god gave him a sword. He was also given winged shoes so that he could fly, a cap of invisibility, and a bag in which to carry Medusa's head.

Perseus flew to the home of the aged ones: three women who shared a single eye and tooth between them. Wearing his cap of invisibility, Perseus snatched their eye and threatened to keep it unless they revealed Medusa's secret hiding place. In return for their eye, the aged ones told Perseus where Medusa lived.

Perseus arrived at Medusa's cave and found her asleep, along with her two hideous sisters. Perseus could not look directly at Medusa's face for fear of turning to stone. Instead, he looked at her reflection on his shiny shield. He chopped off Medusa's head with his sword and dropped it into his bag.

When Perseus returned, Polydectes was surprised to see him still alive and asked what was in his bag. Perseus pulled out Medusa's head – and one look turned Polydectes to stone.

Source 5

Medusa was either beautiful or hideous, depending on which version of the myth you read. [Michelangelo Merisi da Caravaggio, *Medusa* (1597 CE), oil on canvas, Uffizi Gallery, Florence, Italy.]

Practising religion

Each morning a Greek family would pray at their household shrine and leave offerings to the gods, such as wine or food. At temples, people said their prayers and made offerings to the temple god, such as incense, flowers and food. Those who could afford it also organised to sacrifice an animal at a temple. People visited temples to ask Asclepius, the god of medicine, to help overcome sickness.

Death customs

Death was believed to be the beginning of a long journey to the Underworld, ruled by Hades, the god of the dead. The souls of the dead were led to the River Styx by Hermes, the messenger god.

The mythical River Styx separated the Underworld from the living world. Charon the ferryman rowed the dead souls across the River Styx to the Underworld. The fare was paid with a coin that had been placed in the mouth of the dead person.

Source 6

An illustration showing ancient Greek people going to a temple to make offerings to the gods.

Learning Ladder H3.3

Show what you know

1 Name the gods and goddesses in Zeus' family mentioned on pages 98–99.

2 Source 2: Which modern-day comic book superheroes have links to Greek mythology?

3 Describe the things a wealthy Athenian might do to worship the gods.

4 What evidence might historians have found to help them learn about ancient Greek religious customs?

Source analysis

Step 1: I can determine the origin of a source

5 When was the statue of Zeus (Source 3) built, where was it, and why do you think it was built?

Step 2: I can list specific features of a source

6 Describe in detail what you can see in the statue in Source 4, *Perseus with the Head of Medusa*.

Step 3: I can find themes in a source

7 What lesson (or lessons) can we learn from the story of Medusa and Perseus?

Step 4: I can use my outside knowledge to help explain a source

8 Source 2: Research the history of the comic book hero Wonder Woman. What links does she have to Greek mythology?

HOW TO

Source analysis, page 209

How were cities planned in ancient Greece?

Travel over mountainous land was difficult, so small farming villages developed independently into what was called a city-state (*polis* in Greek, or *poleis* if talking about more than one). Each polis stayed independent rather than growing into a single Greek nation. City-states spent much time fighting one another, but sometimes banded together to fight invaders.

City-states of Greece

The ancient Greeks lived in about 100 separate **city-states** (or ***poleis***). City-states were independent, self-governing urban centres. They all spoke dialects of the Greek language and worshipped the same gods, but each *polis* had its own laws and governing body. Often, a *polis* was ruled by a council of rich men, known as an **oligarchy**.

City-states often fought against one other; at other times they banded together to fight against an invading army such as Persia (see page 116). Athens and Sparta were the two most powerful *poleis*.

Source 1

Ancient Greece and some of its city-states

Ancient Greek houses were built along narrow streets. Homes were built around a central courtyard. The walls were often made from mud bricks. They had small window openings to keep out the sun.

Merchants paid a fee for their space in the marketplace. The citizens looked down on the merchants, and some people thought that merchants sold goods stolen by pirates.

People gathered in the agora to listen to philosophers.

Slaves were traded in the slave market at the agora. Slaves laboured in mines and on farms, or did household chores. It is estimated that slaves made up about 30 per cent of Athens' population.

The councils met at a round building called the *tholos*.

Military headquarters

The *stoa* (a covered walkway) contained shops and offices, and provided a shaded place to meet and escape from the sun.

The Council of Citizens (called the *boule*) met at the *bouleuterion*.

Important public buildings such as temples were built on raised ground near the agora. Each temple was built for a specific god.

Source 2

A *polis* had a central place where most activities took place, called the **agora**. People met there to do business, listen to philosophers or be entertained. Markets were held at the agora, along with slave auctions. Surrounding the agora were public buildings where governing councils met. Important buildings such as temples and palaces were usually built on raised ground near the agora. People's homes were built around the city centre and these residential areas were surrounded by farmland to provide the *polis* with food.

Learning Ladder H3.4

Show what you know

1 Define all the history key words describing parts of an ancient Greek city.

2 What is an oligarchy and where might they meet in the *polis*?

3 Summarise what might happen on a typical day in the agora.

4 Explain how city-states were different from each other.

5 Explain why city-states sometimes fought against each other or teamed together.

6 Source 1: Describe where the two most powerful city-states were located.

7 Looking at Source 2, describe in detail the Greek architectural style.

Continuity and change

Step 1: I can recognise continuity and change

8 What is the modern equivalent of the busy market in the agora that attracted many shoppers?

Step 2: I can describe continuity and change

9 Compare city-states to modern countries. How are they different? How are they the same?

Step 3: I can explain why something did or did not change

10 Why did *poleis* develop separately and what did they have in common?

Step 4: I can analyse patterns of continuity and change

11 Many people mixed in the agora and many events took place. How could this mixture lead to changes in ancient Greece? Give one example.

HOW TO

Continuity and change, page 212

Source 1

Ostraka are sherds of broken pots that were re-used as voting 'ballots' cast by Athenians.

What is democracy?

Democracy is where people control the government, rather than having a single ruler such as a queen or a dictator. Democracy began in ancient Athens, where select citizens voted on important issues. A different form of democracy continues today in countries like Australia.

Source 2

Greek leader Pericles shown giving a speech to the *ecclesia*. Athens developed the first democracy in the fifth century BCE. [Phillip von Foltz, *Pericles' Funeral Oration* (1852 CE).]

Government in ancient Greece

An ancient Greek city-state was made up of a major city and the surrounding area. Each city–state had its own type of government, and these varied over time. These were:

- **democracy**: rule by the people (male citizens), who elect officials and leaders
- **monarchy**: rule by a single ruler who inherited the role
- **oligarchy**: rule by a small group of individuals
- **tyranny**: rule by an individual who seized power illegally.

The birth of democracy

The word *democracy* comes from two Greek words; *demos* meaning 'people' and *kratos* meaning 'rule'. Around the fifth century BCE, Athens developed the first democracy, where citizens decided what to do. But in order to have a say, you had to be a **citizen**, and only men who had completed their military training were citizens. Each citizen was expected to vote for every law.

Citizens would come together to vote as an ***ecclesia*** or assembly. All Athenian citizens could participate. The ecclesia was responsible for deciding new laws and making important decisions, such as declaring war.

Beginning as early as 507 BCE, the Athenian citizens would hold assemblies once a month on a hill called the Pnyx. A typical meeting of the ecclesia could attract 6000 citizens. Votes were taken by a show of hands, or by counting stones or pieces of broken pottery called ***ostraka***.

Democratic government in Australia

Australia is a **representative democracy**. Voting for representatives in Australia's federal and state parliaments is a democratic right, and is compulsory for all Australian citizens over the age of 18. At elections, each voter can select candidates that most closely represent their views about how their country or state should be governed.

Source 3

All Australian citizens aged 18 and over can vote for their representatives in parliament.

Federal parliament

Australia's federal **parliament** consists of two houses: the **House of Representatives** and the **Senate**. The House of Representatives has 151 members, each one representing one local area of Australia. They are elected for a term of three years.

The government is formed by the political party with the most members in the House of Representatives. At the 2019 federal election, the Liberal and National parties had 77 members (just over half) so the coalition of these two parties formed government. The leader of the Liberal Party, Scott Morrison, became the prime minister.

The Senate has 76 members elected for a term of six years. Half of the senators face election every three years. Each of the six states has 12 senators, and the Australian Capital Territory and Northern Territory each have two senators.

Representing electorates

Members of Parliament (MPs) provide a direct link between the people in their **electorate** and the parliament. On average, each House of Representatives electoral division has 100 000 voters.

Each MP has an office in their electorate where they can hear the concerns of the voters. Some groups of voters may try to influence (or lobby) the local member to present their interests about a special issue.

In parliament, MPs represent their voters on matters of special interest to their electorate, such as a major road construction or the closure of a local industry.

Members of Parliament must represent their electorate well in parliament, for at the end of their term they must face an election where voters decide whether to vote for them or not.

Learning Ladder H3.5

Civics and citizenship

Step 1: I can identify topics about society

1 State where the word democracy comes from.

2 Who could and could not become citizens in Athens?

Step 2: I can describe societal issues

3 What is different about how members of the Australian Senate and the House of Representatives are elected? Why do you think there is a difference?

Step 3: I can explain issues in society

4 Outline how modern Australia is a representative democracy.

5 Sources 2 and 3: Looking at the images of ancient Greece (Source 2) and modern Australian democracy (Source 3), what similarities and differences can you see?

Step 4: I can explain different points of view

6 Do you think Australian democracy is better than ancient Greek democracy? Why or why not? Use evidence in your response.

Step 5: I can analyse issues in society

7 List the four main forms of government in ancient Greece and how they ruled. Rank the four forms of government from most democratic to least democratic.

8 Australia is a constitutional monarchy where all laws need to be signed off by the King's representative – the governor-general. Can Australia ever be a true democracy while this is the case?

key individual

How did Pericles forge the Golden Age of Athens?

During its golden age in the fifth century BCE, Athens was a thriving city and a centre for learning and philosophy. The Greek leader Pericles built many splendid public buildings – such as the Parthenon, which still stands today.

The Golden Age of Athens

Athens was the largest, wealthiest and most powerful of the Greek city-states. It was very different from the warrior city-state of Sparta (see pages 110–111). The Golden Age of Athenian culture was 480 BCE–431 BCE, the years of peace between the Persian and Peloponnesian wars, when Athens was a centre for the arts, learning and philosophy.

Athens was the home of key philosophers, politicians and writers, such as Socrates, Plato, Aristotle (see pages 122–123) and Pericles. Ancient Athens is recognised as the birthplace of democracy (see pages 104–105).

How did Pericles forge the Golden Age of Athens?

Pericles grew up in one of Athens' wealthiest families during the Persian wars. He benefited from having some of the best teachers in Athens and grew to appreciate art and philosophy. Pericles became a brilliant general and a clever politician. He quickly rose through the ranks to lead Athens in 461 BC.

Athens flourished under the leadership of Pericles (who lived from 495 BCE to 429 BCE). He gave Athens great splendour by rebuilding many temples on the city's hilltop **Acropolis**.

Source 1

The Acropolis imagined in an 1846 CE painting by Leo von Klenze. The Acropolis is the large rocky hill where elegant buildings such as the Parthenon were erected. Below the Acropolis was the agora, [illegible] markets. [Leo von Klenze, *The Acropolis at Athens* (1846 CE), oil on canvas, 147.7 x 102.8 cm, Neue Pinakothek, Munich, Germany.]

Source 2

Statue of Athena, after whom Athens is named. Athena was the goddess of warfare and wisdom. [Leonidas Drosis, Statue of Athena, marble, Academy of Athens, Athens, Greece.]

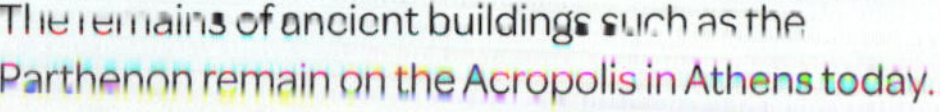

Source 3

The remains of ancient buildings such as the Parthenon remain on the Acropolis in Athens today.

The Acropolis had been destroyed earlier, during the war with Persia. The buildings were made of white marble and decorated with statues carved by Athens' greatest sculptors. The most famous of these buildings is the Parthenon, a temple built to honour the goddess Athena (see pages 98–99).

The Parthenon was built between 447 and 432 BCE. It was the centrepiece of Pericles' ambitious building program to celebrate Athens' victory in the 50-year war with Persia and project Athenian power and culture. Its enormous size and sophisticated architecture showed advances in mathematics and technology, and the resources used to build it showed Athens' wealth. The Parthenon was built to house a statue of the city's patron, Athena. She was the goddess of wisdom – a characteristic celebrated by Athenians.

Democracy

Pericles was not the founder of democracy, but he played a major role in involving common people in the running of government. Pericles encouraged all Athenians to be educated in philosophy and introduced payment for jury service so that poor people could afford to take time off work to be involved in politics. Although democracy was still limited to adult male citizens, Athenian citizenship gave full participation in all decisions, regardless of wealth or class.

Pericles was a great speaker. He delivered a famous speech called the Funeral Oration to honour soldiers who had died in the Peloponnesian War (see page 117) with Sparta. Pericles also spoke passionately about the greatness of Athens and the democratic rights of its citizens. The speech was written down and today it is a key source for historians to understand this period of history.

Source 4

Pericles defending his partner, who had been accused of corrupting other women. [Margaret Dovaston, *Pericles pleading for Aspasia*, in [illegible] by Walter Hutchinson, 1915.]

'Our constitution does not copy the laws of neighbouring states; we are rather a pattern to others than imitators ourselves. Its administration favours the many instead of the few; this is why it is called a democracy. If we look to the laws, they afford equal justice to all in their private differences; if no social standing, advancement in public life falls to reputation for capacity, class considerations not being allowed to interfere with merit; nor again does poverty bar the way, if a man is able to serve the state, he is not hindered by the obscurity of his condition.'

Source 5

Pericles, Funeral Oration, 431 BCE, from Thucydides, *History of the Peloponnesian War*, Modern Library, translated by Richard Crawley, 1951

Source 6

This is a Roman copy of a bust of Pericles. The original bust was created in 430 BCE. [Replica bust of Pericles, marble, British Museum, London, England.]

Daily life in Athens

Like most other Greek city–states, Athens had important temples and buildings protected on a high acropolis. Below the Acropolis in Athens was the centre of activity – the agora.

The agora was a place where plays were performed and men gathered to hear the ideas of philosophers such as Socrates, Plato and Aristotle. It was also where public meetings were held and political speeches were made.

The *agora* was a busy marketplace where markets stalls traded goods such as animals, food, pots and jewellery. Slaves were also bought and sold there.

Gender roles in Athens

In Athenian homes an area called the *andron* was set aside for men to relax and entertain other male guests. Women were not allowed to be seen. At the back of the homes was a room for women only, called the *gynaikeion*.

Women spent most of their lives in the home, running the household with the help of daughters and slaves. Men spent most of their time away from the house with work, at the gymnasium, meeting friends in the agora or attending dinner parties with other male friends.

The social divisions between men and women in ancient Athens were very clear. Men made all of the decisions, including whether a baby born into the family would live or die. A child might be abandoned by the father if it were a girl or appeared weak. The father also decided who his daughters would marry. Most girls were married at 13 or 14 to older men.

Source 7

The agora was a large square below the Acropolis. It was the centre of democratic government and also a place to buy goods at market stalls and shops. Men visited the agora to listen to philosophers and chat with their friends.

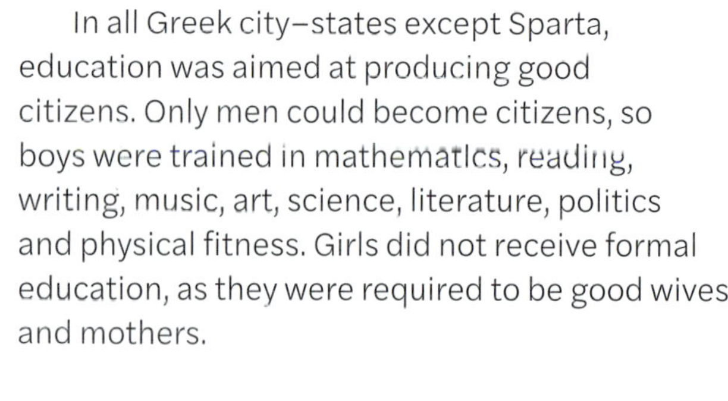

In all Greek city–states except Sparta, education was aimed at producing good citizens. Only men could become citizens, so boys were trained in mathematics, reading, writing, music, art, science, literature, politics and physical fitness. Girls did not receive formal education, as they were required to be good wives and mothers.

Learning Ladder H3.6

Show what you know

1 Summarise the way that Pericles led Athens.

2 Source 1: Use the painting in Source 1 on pages 106–107 to help you describe what ancient Athens looked like during its golden age.

3 List the activities that took place in the agora.

4 Compare the lives of males and females in Athens.

5 Why do you think only men could become citizens in ancient Athens?

6 Why do you think the Parthenon was built on a hill?

7 Why do you think Pericles rebuilt the Parthenon on such a grand scale?

Historical significance

Step 1: I can recognise historical significance

8 Which of these is more significant?
- a Athenian art or Athenian political theories?
- b Athenian gender roles or Athenian architecture?

Step 2: I can explain historical significance

9 Give three reasons why Pericles is considered to be such an important leader.

Step 3: I can apply a theory of significance

10 How did Pericles help the common people of the time become involved in running Athens?

Step 4: I can analyse historical significance

11 Source 5: From reading Pericles' Funeral Oration, what did he value?

HOW TO

Historical significance, page 219

What was life like in Sparta?

Sparta was a warrior state, where men were trained from the age of seven to be soldiers, and women were expected to be physically active to produce courageous and fit young soldiers. Culture and the arts had little value in Sparta.

A military state

Unlike Athens, which was a centre for arts and learning, Sparta was a military state with a permanent army. Sparta reached the height of its power when it defeated Athens in the Peloponnesian Wars (460–404 BCE).

Spartan society was geared to protecting the state. Luxuries, culture and art had little value. The key role of a Spartan man was to be a brave soldier, and the main role of a Spartan woman was to produce strong children. Spartan society consisted of three main groups:

1 Spartans, who were full citizens
2 **Perioikoi**, traders and craftsmen, who were not citizens
3 **Helots**, slaves who undertook manual labour.

Boys left home at the age of seven to begin their military training at the army barracks. Known as the **agoge**, the training system emphasised commitment, discipline and endurance. The Greek writer Plutarch wrote about the training of boys to become brave warriors.

'Of reading and writing, they learned only enough to serve their turn; all the rest of their training was calculated to make them obey commands well, endure hardships, and conquer in battle. Therefore, as they grew in age, their bodily exercise was increased; their heads were close-clipped, and they were accustomed to going bare-foot, and to playing for the most part without clothes. When they were twelve years old, they no longer had tunics to wear, received one cloak a year, had hard, dry flesh, and knew little of baths and ointments; only on certain days of the year, and few at that, did they indulge in such amenities.'

Source 1

Plutarch, *Lives: Volume VII*, Loeb Classical Library, translated by Bernadotte Perrin, 1919

Source 2

At 20 years of age, Spartan males became full-time soldiers – a role they kept until age 60. In battle, they wore a red cloak, bronze helmet, breastplate and ankle guards. They carried a round bronze shield and a long spear and sword.

Source 3

From the age of seven, young Spartan boys trained for military life. The boys slept outside and were given small rations of food. Boys were flogged to teach them to be brave and strong. [Dromos (racecourse) in ancient Sparta, engraving later colourised.]

Life in the barracks was very harsh. Boys were flogged to teach them to become courageous. They slept outside and made their own beds from reeds. Boys were given small rations of food and encouraged to become skilled at stealing food to survive – but if they were caught stealing, they were flogged. The agoge training had the young men continually exercising, playing war games and learning about Sparta's rules of conduct. Men lived in military camps until the age of 30, when they could become a citizen and marry.

Spartan women

In the warrior city–state of Sparta, women were taught to be courageous and outspoken. They exercised in competitions such as javelin throwing and wrestling to stay fit. Spartan women also sang and danced competitively. They cut their hair short, wore plain clothing and no perfume or jewellery.

A Spartan woman's key role was to bear strong, healthy children. The Spartan leaders pressured couples to have boys, so they could grow to become warriors and replace those who died in battle. Domestic chores such as cooking, cleaning and sewing were undertaken by helots.

Spartan women could not be citizens, vote or hold public office. However, they could own and manage property. So many Spartan men died in battle that Spartan women came to own a third of Sparta's land.

Source 4

A 520–500 BCE bronze statue from Sparta of a girl running. [Figurine of a girl running (520–500 BCE), bronze, British Museum, London, UK.]

Learning Ladder H3.7

Show what you know

1 How was life in Sparta different to life in Athens?

2 Describe how women contributed to Spartan life.

3 Source 1: How does Plutarch describe life in the agoge?

Continuity and change

Step 1: I can recognise continuity and change

4 Use the information on this page to identify three things that have changed between Spartan society and modern society. Can you identify any examples of things that remain relatively similar?

Step 2: I can describe continuity and change

5 Why did Spartan boys leave home at the age of seven? Where did they go and what training did they receive?

Step 3: I can explain why something did or did not change

6 Explain why Spartan men underwent such punishing training.

Step 4: I can analyse patterns of continuity and change

7 Re-read the paragraph about Spartan women. Describe how Spartan women's lives compare to contemporary Australian women's lives and suggest reasons for similarities and differences.

HOW TO

Continuity and change, page 212

key event

Why was the Battle of Thermopylae important?

In 480 BCE Xerxes, the leader of the Persian Empire, assembled a massive army and navy and set out to conquer Greece. His father Darius had tried to conquer Greece 10 years earlier but had lost at the Battle of Marathon (see page 116).

Ancient Greece was made up of hundreds of city–states, of which Athens and Sparta were the most powerful. Although the city–states fought with one another, they also joined together to defend themselves against foreign invasion.

To reach their destination at Attica, the region in which Athens is located, the Persians needed to go through the narrow coastal pass of Thermopylae. The Athenian general Themistocles planned for the allied Greek forces to block the Persian army at Thermopylae. The allied Greek force of approximately 7000 men led by King Leonidas of Sparta marched north to block the pass. Like all Spartan men, Leonidas had been trained since childhood to become a warrior (see pages 110–111).

The Persian army of between 100 000 and 150 000 soldiers descended on Thermopylae. Xerxes demanded that the Greeks surrender their weapons. King Leonidas replied 'Come and get them!'. The Greek army was greatly outnumbered, but managed to hold off the attacking Persian forces for three days.

During those three days of battle, the small force led by Leonidas blocked the only road by which the huge Persian army could pass. The small Greek force suffered light losses and imposed heavy casualties on the Persian army. A Greek traitor showed the Persians a secret mountain route that would allow them to surround the Greeks from the front and the rear. When Leonardis found out that he had been betrayed, he ordered most of the Greek army to flee.

Then Leonidas hand-picked 300 Spartans to join him in a fight to the death, to delay the Persian advance for as long as possible. They were joined by 700 Thespians and 400 Thebans.

Fight to the death

Why were the small number of Spartan and Greek soldiers prepared to fight to the death against the huge Persian army?

Herodotus (c. 480–429 BCE) is often called the world's first historian. In his books *The Histories*, which were written between 450 and 420 BCE, he describes how the Persian King Xerxes is surprised that such a small force would try to stand in his way.

If his date of birth is correct, Herodotus was a child during the Persian Wars and could not have been an eyewitness to most of the events he wrote about in *The Histories*. So, we cannot consider his text to be primary evidence. Instead, Herodotus asked people for their recollections of events. This type of historical investigation is called **oral history**. The Greek word *historia* means 'to question and investigate'.

According to Herodotus, Xerxes did not believe the small Greek army seriously considered facing the Persian forces. Xerxes tried to find out what the Greeks were doing. He had found out that the Spartans were spending a lot of time combing and grooming their long hair, which was curious. He sent Demaratus to find out what was going on. Demaratus was a former Spartan king, who was now working for the Persians.

Source 1

King Leonidas of Sparta as depicted in the 2006 film *300*. King Leonidas led a tiny force of Spartan soldiers against a huge Persian invading force.

Demaratus told Xerxes:

> ‘These men have come to fight with us for the [mountain] passage, and this is it that they are preparing to do; for they have a custom which is as follows: whenever they are about to put their lives in peril, they attend to the arrangement of their hair.’

Demaratus tried to convince Xerxes that the Spartans were serious warriors and if the Persians defeated them, no other forces would stand in the Persians' way.

> ‘Be assured however, that if thou shalt subdue these and the rest of them which remain behind in Sparta, there is no other race of men which will await thy onset ... or will raise hands against thee: for now thou art about to fight against the noblest kingdom and city of those which are among the Greeks, and the best men.’

Xerxes still didn't believe Demaratus and asked him again why so few Spartan warriors were prepared to fight to the death against the Persians.

Demaratus said:

> ‘O king, deal with me as with a liar, if thou find not that these things come to pass as I say.’

Source 2

All three quotes from Herodotus, *The History of Herodotus,* Dutton & Co. translated by George Rawlinson, 1862

Source 3

The Battle of Thermopylae

The last stand

King Leonidas rallied his small force on the third day of the battle to defend the pass in the hope of delaying the Persians' progress. This time Xerxes could attack from both front and rear. Leonidas moved his troops to the widest part of the pass to use all of his men at once. They were quickly defeated by the vast Persian army. The Spartan king was killed and the remaining troops were slaughtered with a barrage of Persian arrows.

After the Battle of Thermopylae, the Persian King Xerxes ordered that the head of Leonidas was to be put on a stake and displayed on the battlefield. Although the Battle of Thermopylae resulted in defeat for the Greeks, it remains a symbol of heroic resistance against overwhelming enemy numbers.

Source 4

King Leonidas of Sparta and his 300 hoplites face the huge Persian army at a narrow pass at Thermopylae, near Athens. [George Mamos, Battle of the Brave (2010), oil on linen, 91.4 x 121.9 cm.]

Learning Ladder H3.8

Show what you know

1 Source 4: Who were the different groups and leaders on each side of the battle?

2 Source 3: Describe the movements of the two forces by looking at the map.

3 How did the Persians win the battle?

4 Why is the Battle of Thermopylae so famous?

5 Think of a military tactic that could have been applied at the time that might have helped the Greeks win the battle.

Source analysis

Step 1: I can determine the origin of a source

6 Source 2: What type of source is Herodotus' *The Histories*? How do you know that? How might Herodotus have got the material for his book if he was just a child at the time of the battle?

Step 2: I can list specific features of a source

7 Why did Xerxes not believe Demaratus? What words did Demaratus use to persuade Xerxes he was telling the truth? Use examples from the text.

Step 3: I can find themes in a source

8 In Source 4, what has the artist done to show what the confrontation was like for those involved?

Step 4: I can use my outside knowledge to help explain a source

9 What knowledge do you have that can explain these parts of the image in Source 4?

a the mountains

b the two sides facing each other

c the clothing and equipment of the people in the left foreground

HOW TO

Source analysis, page 209

How did ancient Greeks fight?

The Greek city-states spent a great deal of time fighting one another, including the Peloponnesian Wars that spanned almost 60 years. Hoplite soldiers from city-states would sometimes band together to ward off foreign enemies, such as the Persians.

Greece at war

Greek city-states each had their own armies, which went back to their regular jobs when wars ended. The warrior city-state of Sparta was the only one with a full-time army that was always on duty (see pages 110–111).

Greek soldiers were called **hoplites**. They wore protective helmets, chest plates and leg protectors, and each carried a large shield. Hoplites were armed with a double-edged sword and a two-metre-long spear with an iron blade at one end and a spike at the other.

Hoplites used a battle formation called a **phalanx** to protect themselves while advancing on enemy troops. They packed together tightly and overlapped their shields to provide a protective outer barrier to deflect arrows and spears. Spears in the front row were held forward to penetrate the enemy and those in the other rows were held skyward to deflect incoming missiles.

Greek warships were called triremes. They were powered by sails and oarsmen, and designed to turn quickly and move rapidly through the water. In times of war, these specially designed **triremes** and their well-trained oarsmen helped the Greek navy defeat larger Persian warships (see pages 96–97).

The city-states of Greece often fought one other, but most of them united when the Persian Empire (modern-day Iran, Palestine, Syria and Egypt) attacked Greece in 490 BCE and again in 480 BCE. Then in the fourth century BCE, Greece was invaded by their northern neighbours – the Macedonians.

The Battle of Marathon

In 490 BCE, King Darius I of Persia sent a large fleet of ships to invade Greece. Darius wanted to punish Athens for supporting a rebellion against the Persian Empire. The Persian forces landed on the plain of Marathon, about 40 kilometres from Athens.

About 10 000 Athenian hoplites marched to Marathon to meet the invading Persians. The Athenians blocked both exits from the plain of Marathon and after five days they sprung a surprise attack and defeated the 50 000 strong Persian army.

The Battle of Marathon is now more famous as the inspiration for the 42-kilometre marathon race, which was supposedly the distance run by Greek messenger Pheidippides from Marathon to Athens with news of the Greek victory. The first marathon race was held between Marathon and Athens at the 1896 Athens Olympic Games (see pages 120–121).

Further Persian wars

King Darius I died in 486 BCE but his son Xerxes organised an even larger invasion of Greece. In 480 BCE, a massive Persian army of more than 100 000 soldiers marched into Greece. They were supported by a large naval fleet sailing down the coast.

The feuding Greek city-states joined together to defend themselves against the invading Persian forces. Spartan King Leonidas and his troops held up the huge Persian invasion force at a narrow mountain pass at Thermopylae (see pages 112–115). The Greek forces delayed the Persian advance, but were eventually defeated.

Shortly after the Battle of Thermopylae, the Greek fleet, led by Athens, fought three battles against the much larger Persian fleet. When the nimble Athenian ships attacked the larger Persian navy at the Battle of Salamis (see pages 96–97), the Persians were completely defeated.

The Peloponnesian War

Cooperation among the Greek city–states did not last long and soon they were again fighting one another. In the Peloponnesian Wars (which went from 460–404 BCE with a truce in the middle), Greece was divided into the city–states and colonies that supported Athens or supported Sparta. In 431 BCE, Sparta attacked the walled city of Athens and destroyed its farms. Sparta repeatedly invaded Athens for the next decade, wearing down resources of both sides.

The Peloponnesian Wars finally ended when Sparta gained Persian support to force Athens to surrender. The Spartans set up their oligarchy-style of government in Athens. Sparta then dominated Greece until the city-state of Thebes defeated the Spartans in 371 BCE.

Source 1

Greek hoplites in a phalanx formation. They marched tightly together, overlapping shields to protect themselves from arrows and spears. Two-metre-long spears were held skyward to deflect incoming missiles and in the front row they were held forward to attack the enemy. [Line of armed soldiers, decorative detail from an Attic vase (6th century BCE), black-figure pottery, Museo Archeologico Nazionale, Naples, Italy.]

Learning Ladder H3.9

Show what you know

1 What weapons and armour did hoplites use?

2 How did the Athenians win the Battle of Marathon?

3 Why was warfare so common in ancient Greece?

Cause and effect

Step 1: I can recognise a cause and an effect

4 What was the effect of the more agile Athenian ships at the Battle of Salamis?

Step 2: I can determine causes and effects

5 Why did the Persian attacks cause Greek city–states to unite?

Step 3: I can explain why something is a cause or an effect

6 Source 1: The phalanx formation caused Greek soldiers to be successful. How?

Step 4: I can analyse cause and effect

7 Here are some effects of Greece's warlike nature. Rank them in order from least important to most important. Then write a paragraph explaining the order you have put them in.

- Conflict with Persia
- Development of military technology
- Frequent civil war

HOW TO Cause and effect, page 215

key individual

How did Alexander the Great spread Greek culture?

The Macedonian king Alexander the Great established the largest empire of the ancient world through a bloodthirsty rampage across the Middle East and Asia.

Alexander the Great

The continual conflict between Greek city-states left them too weak to withstand a new invasion from the north: Macedonia. By 338 BCE, Macedonian king Phillip II had taken control of the weakened Greek city-states and set his eyes on the Persian Empire. When Phillip II was assassinated in 336 BCE, his son Alexander became king. He was 20 years old and would come to be known as Alexander the Great.

Alexander proved to be a ferocious, inspirational and capable leader. When the Greek city-state of Thebes staged a revolt against him, Alexander crushed the rebellion, killed 6000 people and sold the rest into slavery.

Alexander established the largest **empire** the ancient world had ever seen – and he never lost a battle. His love of Greek culture saw him introduce Greek customs, practices and leaders in his conquered lands. The new lands in Alexander's empire soon had Greek temples, as well as public buildings, amphitheatres and houses based on Greek designs.

Alexander and his loyal army conquered the Persian Empire and Egypt. By 323 BCE, Alexander was head of an enormous empire, established through a military campaign that spanned 11 years and a journey of 34 000 kilometres.

Alexander died of a fever in 323 BCE at the age of just 32. He had no direct heirs, and his empire was quickly divided up by military generals after his death.

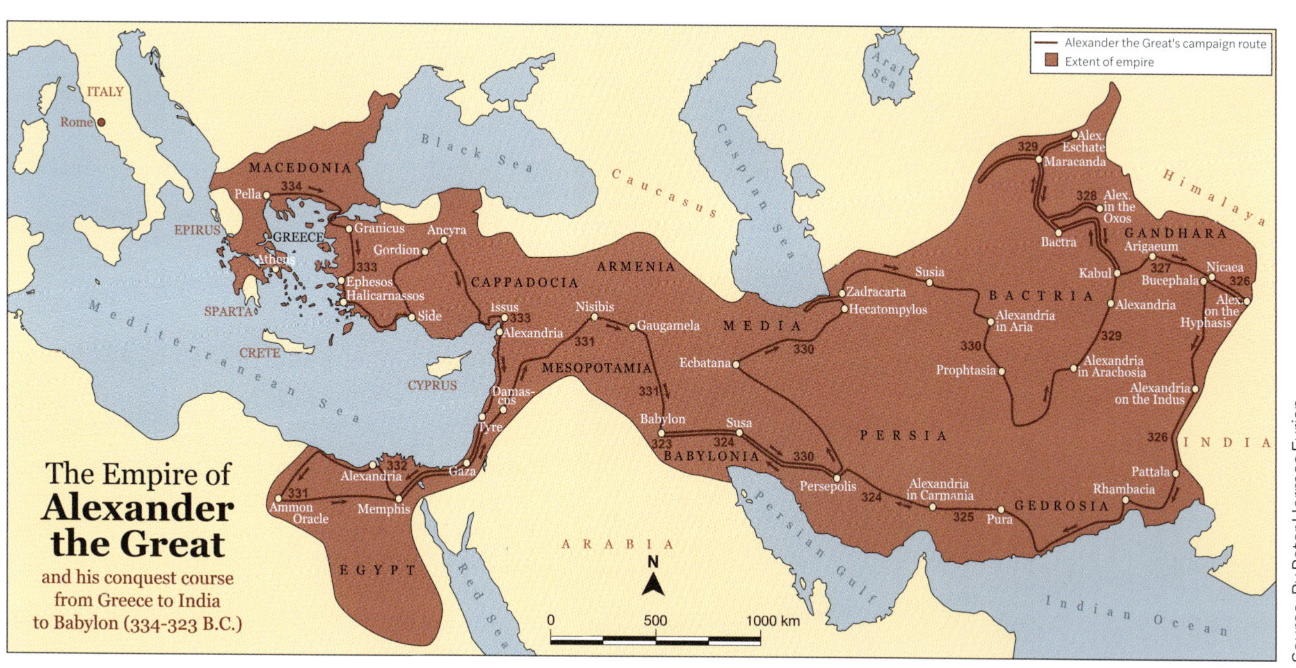

Source 1

The empire of Alexander the Great

Learning Ladder H3.10

Show what you know

1 What kind of person was Alexander the Great?

2 How did he deal with challenges to his authority?

3 Source 2: Describe the image of Alexander from the 2004 movie. How does the photo show Alexander's character?

4 Do you think Alexander deserves to be called 'the Great'?

Continuity and change

Step 1: I can recognise continuity and change

5 What changes did Alexander the Great bring to conquered lands?

Step 2: I can describe continuity and change

6 Look carefully at Source 1.

a Use the scale on the map to estimate how far Alexander travelled east to conquer lands.

b Which was the largest territory that came under the control of Alexander?

c List the places that Alexander named for himself during his conquests.

Step 3: I can explain why something did or did not change

7 Source 1: What does the map suggest about what did and did not change under Alexander the Great's rule?

Step 4: I can analyse patterns of continuity and change

8 Is it fair to say that Alexander the Great influenced great change but not continuity? Explain your response.

Continuity and change, page 212 HOW TO

Source 2

This scene from the 2004 movie *Alexander the Great*, shows Alexander in battle on his horse Bucephalus. Alexander tamed the enormous stallion when he was just 12 years old and Bucephalus became his battle companion for most of his life.

How did the Olympic Games originate?

The ancient Olympic Games were festivals held every four years in honour of Zeus (see pages 98–99), king of the gods. Only freeborn men were allowed to compete at the games – and in most events they competed in the nude.

Ancient Olympic Games

The idea for our modern Olympic Games came from the ancient Olympic Games – a religious festival held every four years from at least 776 BCE at Olympia, in the city–state of Elis. The five-day festival attracted competitors from all Greek city–states. Only freeborn Greek men could participate, and married women were banned from attending – although they could train horses for events.

A truce was declared between warring city–states for each Olympics, but the athletic events became a political tool for city–states to show their dominance. The city–states gave many rewards to their athletic heroes, although the only official Olympic prize was a crown of olive leaves.

Olympic events

The main events in the ancient Olympics were running and fighting. Running races included a sprint called the *stade*, a long-distance race the *dolichos* and an event called *hoplitodromos*, where competitors ran in their hoplite armour.

Source 1

This is a *stamnos*, a type of pottery used to store liquids. It shows two fighters training for the *pankration*. [Attic black-figure pottery stamnos depicting an athletics scene (510–500 BCE), Ashmolean Museum, Oxford, England.]

Source 2

The site of the ancient Olympic Games at Olympia in Greece.

1 The Temple of Zeus housed a 13-metre-tall statue of Zeus made from gold and ivory. The statue was one of the Seven Wonders of the Ancient World. It was completed around 435 BCE and destroyed in the 5th century CE.

2 The gymnasium was used by athletes to train before the games began.

3 Running races were held at the stadion. Athletes competed naked to honour the gods, except for the *hoplitodromos* where competitors wore hoplite armour.

4 The Temple of Hera, wife of Zeus. Like the modern Olympic Games, the sacred flame used to light the altar was kept burning throughout the Olympic festival. Today, the Olympic flame is lit at Olympia.

Fighting events included wrestling, boxing and the *pankration*. The *pankration* combined boxing, wrestling and kicking. Crowds found the *pankration* exciting, with fights sometimes leading to severe injuries – even death.

The festival highlights were chariot racing and the pentathlon. The pentathlon included discus throwing, long jump, javelin throwing, running and wrestling. Chariot races had up to 40 chariots racing at high speeds.

Source 3

Stefano Baldini of Italy won the gold medal in the men's marathon at the 2004 Athens Olympic Games.

Learning Ladder H3.11

Show what you know

1 Why were the ancient Olympics held?

2 How are the ancient and modern Olympics different? What is still the same?

3 Source 1: What sport is shown here? Why did crowds enjoy this sport?

4 What benefits might there be for city-states to compete in the Olympics?

Chronology

Step 1: I can read a timeline

5 Look at the timeline on page 94. How many years before the Battle of Marathon (after which the marathon race is named) was the first Olympic Games held?

Step 2: I can place events on a timeline

6 List these modern Olympic Games in order, from oldest to most recent.

2004 Athens
2020 Tokyo
1896 Athens
1956 Melbourne

Step 3: I can create a timeline using historical conventions

7 Using correct conventions, create a timeline with these events on it.

c. 776 CE the first ancient Olympics
490 CE the Battle of Marathon
Four decades after the Battle of Marathon, Athens is a powerful city
Three centuries before the event in 8c, Homer writes the *Iliad* and *Odyssey*

Step 4: I can distinguish causes, effects, continuity and change from looking at timelines

8 Look at the timeline on page 94.

a Which events are linked by cause and effect?

b Which events, if any, show things that are continuous with the modern world?

c Which events show things that are a change from the modern world?

HOW TO

Chronology, page 206

How did ancient Greeks change how we think?

Ancient Greek philosophers introduced a fresh outlook to help people view life from a different perspective, rather than solely relying on a belief that the gods controlled their lives.

Source 1

Pythagoras lived from about 570 BCE to about 500 BCE. He thought that the entire world could be explained with mathematics. Pythagoras developed an advanced system of geometry that we still use today to find the length of sides of right-angled triangles. It is called the Pythagorean theorem.

Philosophers

Greek philosophers began a new type of thinking. Rather than simply believing in myths and the actions of gods, they tried to make sense of the world in a non-religious way. The Greek word *philo* means 'love' and *sophia* means 'wisdom', so philosophy means 'love of wisdom'.

Early Greek philosophers questioned the world around them to try and make sense out of what they saw. Philosophers discussed their ideas in the agora, where they spoke to crowds of people and instructed pupils. Many philosophers studied mathematics and physics, and some opened their own schools.

The ancient Greek philosophers were so influential that their ideas are still used today.

Socrates

Socrates lived c. 470–399 BCE. He was the first key Greek philosopher and one of the first to explore the idea of ethics: what is right and wrong. Socrates developed a method of studying issues and problems by continually asking questions. This became known as the Socratic method.

Socrates welcomed all students into his lectures, including enemies of the state. He was convicted by a jury of 'corrupting young people', and sentenced to death by poisoning. He died age 70.

Plato

Plato was born around 428 BCE and was a student of Socrates. Following Socrates' death, Plato travelled to Italy, Sicily and Egypt to study with other philosophers, including followers of the mathematician Pythagoras.

Plato returned to Athens and established his own school of philosophy, where he passed on learnings from Socrates and developed his own ideas. The following quotations are from Plato:

- Wise men speak because they have something to say; fools because they have to say something.
- Ignorance is the root and stem of all evil.
- Better a little which is well done, than a great deal imperfectly.

Aristotle

Aristotle (384–322 BCE) was a student of Plato's and a tutor for Alexander the Great (see pages 118–119). Aristotle took a more practical approach to learning. He made detailed observations and even dissected animals to learn more about their anatomy, thus laying the foundation of science today. Aristotle also studied astronomy. He was the first to know that Earth was round.

Archimedes

Archimedes was born in 287 BCE. He was a mathematician and inventor. Archimedes lived in the city of Syracuse on the island of Sicily.

Archimedes worked out, while taking a bath, that the further down he sank, the higher the water rose. From this observation, he developed Archimedes' principle – a scientific law that states that the volume of an object in water is equal to the weight of the water that it displaces.

Archimedes made a machine that farmers could use to lift water from a river or lake. It was called Archimedes' screw. In 213 BCE, when Romans attacked Syracuse, Archimedes invented the 'claw of Archimedes' to defend the city from the attacking ships.

Source 2

Archimedes' screw transferred water from a lake or river into irrigation ditches on farms. Water is pumped by turning a screw inside a cylinder.

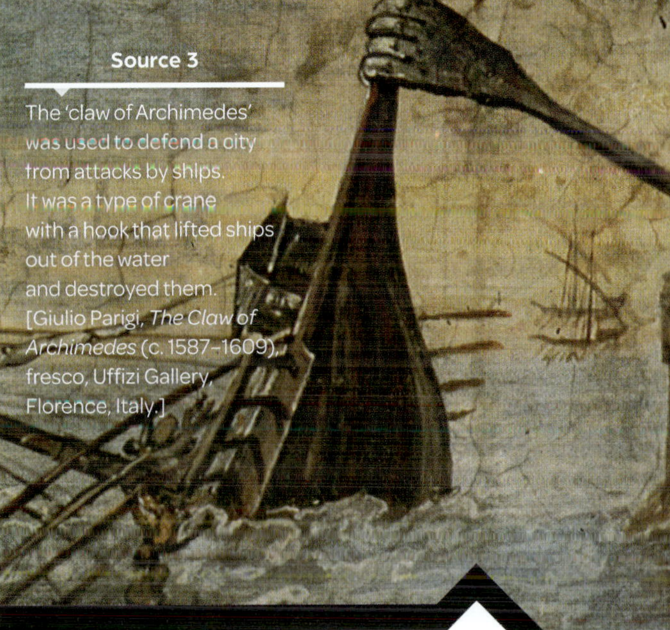

Source 3

The 'claw of Archimedes' was used to defend a city from attacks by ships. It was a type of crane with a hook that lifted ships out of the water and destroyed them. [Giulio Parigi, *The Claw of Archimedes* (c. 1587–1609), fresco, Uffizi Gallery, Florence, Italy.]

Learning Ladder H3.12

Show what you know

1 Which word means 'love of wisdom'?

2 Source 1: What in this image suggests that learning is taking place?

3 List the ancient Greek philosophers, in order of who was the teacher of whom.

4 Summarise Plato's beliefs.

5 Why did Aristotle think it important to make observations about the natural world?

Historical significance

Step 1: I can recognise historical significance

6 Which of the thinkers do you think is the most important, and why?

Step 2: I can explain historical significance

7 How has Greek philosophy improved human life?

Step 3: I can apply a theory of significance

8 Apply Partington's theory of significance. Show your response to the philosophy of ancient Greece in a table, looking at importance, depth, number, time and relevance.

Step 4: I can analyse historical significance

9 The philosophies of Socrates, Plato and Aristotle were significant. Break down why this is so, listing at least six reasons. Back up your answer with evidence from the text, or from independent research

HOW TO Historical significance, page 219

Masterclass

Learning Ladder

Work at the level that is right for you or level-up for a learning challenge!

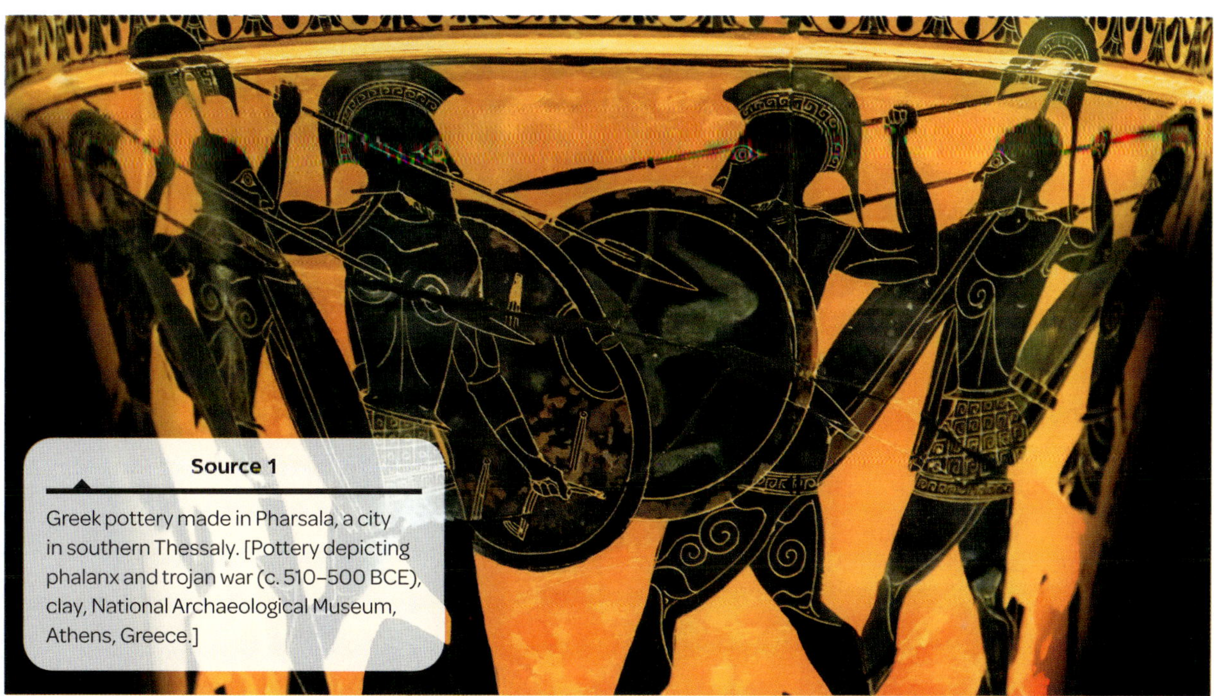

Source 1

Greek pottery made in Pharsala, a city in southern Thessaly. [Pottery depicting phalanx and trojan war (c. 510–500 BCE), clay, National Archaeological Museum, Athens, Greece.]

Step 1

a **I can read a timeline**

Look at the timeline on page 94. How long was it between the start of the Bronze Age and the start of the Iron Age?

b **I can determine the origin of a source**

Look at Source 1. When and where is this piece of pottery from?

c **I can recognise continuity and change**

Rank these elements of ancient Greek society on a scale in terms of how different they are to today. One end is 'not different', the other end is 'very different'.

- religion
- sport
- clothing
- warfare
- food and drink
- gender roles
- politics

d **I can recognise a cause and an effect**

Which of these was a cause of the Greek city-states falling to Alexander the Great?

- they were great seafarers
- they ate oil and grapes
- they were divided
- they despised the Persians

e **I can recognise historical significance**

Put these events on a scale, from least important to most important.

- the death of Socrates
- the death of Alexander the Great
- the first Olympics
- the first democracy

Source 2

The Parthenon, Athens

Step 2

a I can place events on a timeline

Put these events in the correct order.

460–404 BCE Peloponnesian War

447–432 BCE Parthenon built

1600–1200 BCE Mycenean civilisation

461–429 BCE Pericles leading statesman in Athens

776 BCE First Olympic Games

479–431 BCE Golden Age of Athens

334–326 BCE Alexander the Great's conquests

b I can list specific features of a source

Look at Source 1. Describe in detail everything you see on this piece of pottery.

c I can describe continuity and change

Copy and complete the table below.

d I can determine causes and effects

What caused the Greeks to win the Battle of Marathon?

e I can explain historical significance

Source 2: Explain why the building of the Parthenon was (or was not) historically significant.

Step 3

a I can create a timeline using historical conventions

Use the events in the list that follows to create a historical timeline with correct conventions, including equal spacing for equal years.

460–404 BCE Peloponnesian War

447–432 BCE Parthenon built

1600–1200 BCE Mycenean civilisation

461–429 BCE Pericles is leading statesman in Athens

776 BCE First Olympic Games

479–431 BCE Golden Age of Athens

334–326 BCE Alexander the Great's conquests

b I can find themes in a source

Look at Source 1. What do you think the theme on the piece of pottery is?

c I can explain why something did or did not change

Explain why seafaring became more important in Greek culture, but religion stayed the same.

d I can explain why something is a cause or an effect

How did Alexander the Great conquer so much territory so easily?

e I can apply a theory of significance

Using Partington's model of significance, explain why the Peloponnesian Wars were or were not significant. Give your response in table format.

Element	Golden Age of Athens	Modern Greece	Continuity or change?	How?
Transport				
Military power				
Technology				
Jobs people have				

Masterclass

Step 4

a **I can distinguish causes, effects, continuity and change from looking at timelines**

Look at the timeline on page 94. Which events are linked by cause and effect? What things have changed between 3200 BCE and 356 BCE?

b **I can use my outside knowledge to help explain a source**

Look at Source 1. What do you know about ancient Greece that would help you explain what the image on this piece of pottery means?

c **I can analyse patterns of continuity and change**

Compared to Minoan and Mycenaean culture, what was similar and what was different about Greek culture?

d **I can analyse cause and effect**

Alexander the Great brought Greek culture to areas that he conquered. Describe at least three effects of the spread of Greek culture.

e **I can analyse historical significance**

Greek culture and art was significant. Describe at least four reasons why.

Step 5

a **I can describe patterns of change**

Look at the timeline on page 94. Can you see a pattern that relates to war and conflict? Write a paragraph explaining the connection between these events.

b **I can use the origin of a source to explain its creator's purpose**

The piece of pottery in Source 1 was created just prior to the Greek Classical Age. Knowing this, why do you think the pottery was created with an image like this?

c **I can evaluate patterns of continuity and change**

Democracy was first developed in ancient Greece. Is it good or bad that we still have democracy in Australia? Why or why not?

d **I can evaluate cause and effect**

Sparta won the Peloponnesian Wars and Athens declined afterwards. Was this a good thing? Why or why not? Use historical evidence in your response.

e **I can evaluate historical significance**

Greece has been called the 'cradle of Western civilisation'. Do you agree? Why or why not? Use historical evidence in your response.

Historical writing

1 Structure

Imagine you are answering an essay question: 'Geography is destiny. The geography of Greece determined how their civilisation developed. Discuss.' Write an essay plan for [illegible] 3 main paragraphs.

2 Draft

Using the drafting and vocabulary suggestions on page 226, draft a 400–600 word essay responding to the topic.

3 Edit and proofread

Using the editing and proofreading guidelines on page 227, edit your draft, then proofread it.

Historical research

4 Organise and present information

Assume you are going to produce a research assignment on the technological achievements of ancient Greece. Create a graphic organiser with the different sub-topics that you would take notes about.

Capstone

How can I understand ancient Greece?

In this chapter, you have learnt a lot about ancient Greece. Now you can put your new knowledge and understanding together for the capstone project to show what you know and what you think.

In the world of building, a capstone is an element that finishes off an arch or tops off a building or wall. That is what the capstone project will offer you, too: a chance to top off and bring together your learning in interesting, critical and creative ways. You can complete this project yourself, or your teacher can make it a class task or a homework task.

Scan this QR code to find the capstone project online.

mea.digital/GHV7_H3

H4

Ancient Rome

WHAT WAS THE COLOSSEUM?

How can I understand ancient Rome?

The Roman Empire was highly significant in history, and still impacts the world deeply today. Ancient Roman civilisation lasted for 1200 years and, at its peak, built a huge empire that united Europe. Ancient Romans developed the idea of the balance of power between three branches of government, and created legal ideas such as trial by jury, civil rights, contracts and personal property that we still use today.

learning ladder

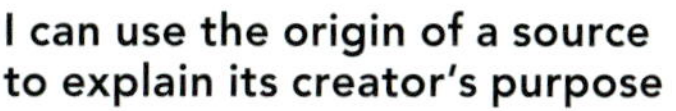

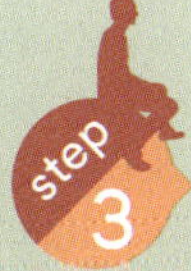

Step	Chronology	Source analysis	Continuity and change
step 5	**I can describe patterns of change** I read timelines and see the 'big picture'. I group timeline events and see if they show patterns of change. I know typical historical patterns to look for.	**I can use the origin of a source to explain its creator's purpose** I combine knowledge of when and where a source was created to answer the question, 'Why was it created?'	**I can evaluate patterns of continuity and change** I answer the question, 'So what?' about patterns of continuity and change. I weigh up different things and debate the importance of a continuity or a change.
step 4	**I can distinguish causes, effects, continuity and change from looking at timelines** I read timelines and find events that are linked by cause and effect. I find things that are the same or different from then until later times.	**I can use my outside knowledge to help explain a source** I have enough outside knowledge about ancient Rome to help me explain a source.	**I can analyse patterns of continuity and change** I see beyond individual examples of continuity and change in ancient Rome and identify broader patterns, and I explain why they exist.
step 3	**I can create a timeline using historical conventions** When given a set of ancient Roman events, I construct a historical timeline, making sure I use correct terminology, spacing and layout.	**I can find themes in a source** I look a bit closer into a source and find more than just features. I find themes or patterns in the source.	**I can explain why something did or did not change** I answer the question 'why?' something changed or stayed the same between historical periods.
step 2	**I can place events on a timeline** When given a list of ancient Roman events, I put them in order from earliest to latest, the simplest kind of timeline.	**I can list specific features of a source** I look at an ancient Roman source and list detailed things I can see in it.	**I can describe continuity and change** I have enough content knowledge about two different historical periods to recognise what is similar or different about them, and can describe it.
step 1	**I can read a timeline** I read timelines with ancient Roman events on them and answer questions about them.	**I can determine the origin of a source** I can work out when and where an ancient Roman source was made by looking for clues.	**I can recognise continuity and change** I recognise things that have stayed the same and things that have changed from ancient Rome until now.

Source 1

The Gladiator Mosaic is a famous mosaic depicting gladiators. Measuring approximately 28 metres, it was discovered in 1834 on the Borghese estate at Torrenova, on the Via Casilina outside Rome. [*Gladiator Mosaic* (4th century CE), Galleria Borghese, Rome, Italy.]

Warm up

Cause and effect	Historical significance
I can evaluate cause and effect I answer the question 'So what?' about cause and effect. I weigh up different things and debate the importance of a cause or an effect.	**I can evaluate historical significance** I answer the question 'So what?' about things that are supposedly historically important. I weigh up events against each other and cast doubt on how important things are.
I can analyse cause and effect I don't just see a cause or effect as one thing. I determine the factors that make up causes and effects.	**I can analyse historical significance** I separate out the various factors that make something historically important in ancient Roman history.
I can explain why something is a cause or an effect I can answer 'how?' or 'why?' a cause led to an effect in ancient Rome.	**I can apply a theory of significance** I know a theory of significance. I use it to rank the importance of ancient Roman events.
I can determine causes and effects Applying what I have learnt about ancient Rome, I can decide what the cause or effect of something was.	**I can explain historical significance** I answer the question 'why?' about things that were important in ancient Rome.
I can recognise a cause and an effect From a supplied list, I recognise things that were causes or effects of each other in ancient Rome.	**I can recognise historical significance** When shown a list of things from ancient Roman history, I can work out which are important.

Chronology

1 If the Roman civilisation fell in 476 CE and lasted about 1200 years, roughly when did it start?

Source analysis

2 Source 1: When was this source created? When was it discovered?

3 Source 1: What does this source reveal about ancient Rome in the 4th century CE?

4 Source 1: The mosaic depicts animals from modern-day Africa. What does this tell us about the ancient Roman Empire?

Continuity and change

5 Using your 21st century perspective, what adjectives could be used to describe ancient Roman sport and entertainment?

Cause and effect

6 Up until c. 404 CE, the ruling class of the Roman Empire allowed the public to view gladiatorial battles for free. During the same period, the ruling class was constantly worried that the lower classes would rebel against them. Can you identify a cause and effect between these two things (a causal relationship)? Explain your answer.

Historical significance

7 How are Roman politics still important today?

How did ancient Roman civilisation develop?

Backed by a strong permanent army and innovations in engineering, ancient Rome grew from a small city to become a powerhouse in the ancient world. As the Romans conquered new lands they absorbed new technologies and beliefs and used their captives as slaves to build their empire. Rome experimented with different forms of government and ultimately created a form of government by the people that is still in use today.

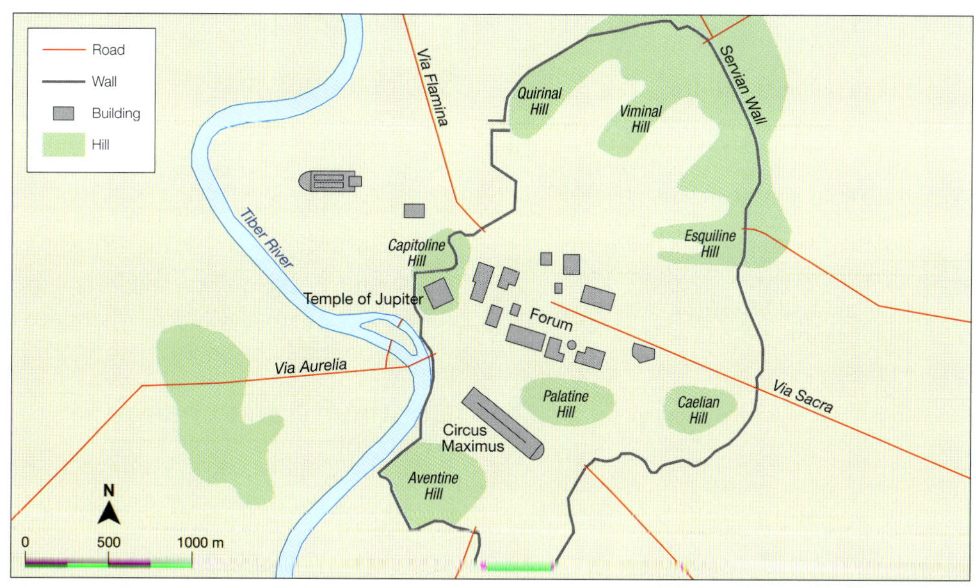

Source: Matilda Education Australia

Source 1

Rome was built on the Tiber River, which supplied fresh water and a trading route to the Mediterranean Sea. The first settlers built homes on the seven hills of Rome, which could be easily defended. A large wall, called the Servian Wall, was built to protect Rome from invaders.

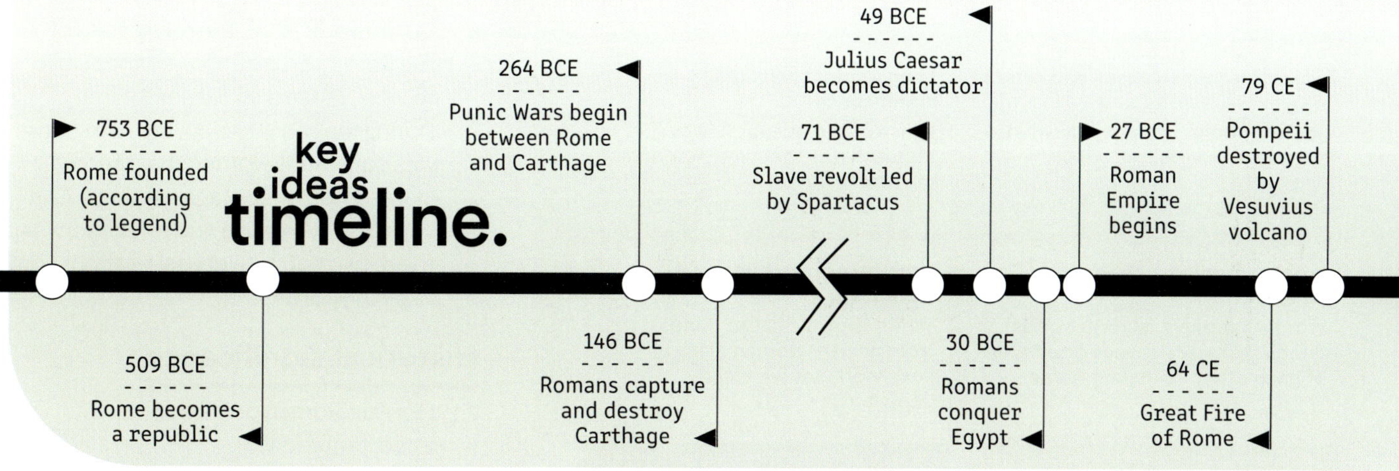

Source 2

In 509 BCE, the king was overthrown and the Roman Republic was established. Rule by the people has become the basis of most governments in the world today. [Cesare Maccari, *Cicero Denounces Catiline* (c. 1882–88 CE), fresco, Palazzo Madama, Rome, Italy.]

Source 3

The Punic Wars (264–146 BCE) were fought between Rome and the north African city of Carthage. The armies of Carthage were led by Hannibal, who defeated larger Roman armies with his use of war elephants. [Jacopo Ripanda, *Hannibal Crossing the Alps* (c. 1505 CE), fresco, Palazzo dei Conservatori, Musei Capitolini, Rome, Italy.]

Source 4

In 27 BCE, the Roman Empire began. Emperor Augustus – the adopted son of Julius Caesar – becomes Rome's first emperor. Augustus' long reign was an age of prosperity, with Rome expanding its territory to become one of the greatest empires in world history.

Source 5

The Colosseum hosted large crowds that came to watch gladiators, slaves and animals fight to the death. The events were put on to entertain the Roman people and gain their support.

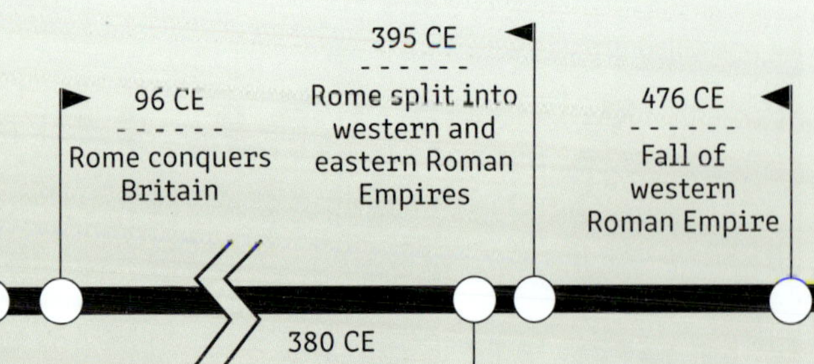

Learning Ladder H4.1

Show what you know

1 Source 5: How did gladiator contests help leaders continue their rule?

2 What significant event happened in 476 CE?

3 Source 4: Why was Emperor Augustus significant?

4 Which image is a primary source and why is it different to other images here?

Cause and effect

Step 1: I can recognise a cause and an effect

5 As the Romans conquered new lands, what advantages did they gain?

Step 2: I can determine causes and effects

6 The Roman Empire split into two in 395 CE. Research online to identify a cause of this event.

Step 3: I can explain why something is a cause or an effect

7 A volcano destroyed Pompeii in 79 CE. What effect did this event have on its residents? (See pages 162–165.)

Step 4: I can analyse cause and effect

8 Trace the map of ancient Rome (Source 1).

a Label the Tiber River. Why was Rome located here?

b Label the hills where the first settlers built homes. Why did they build here?

c Label the Servian wall. Why was it built?

HOW TO

Cause and effect, page 215

How did Rome grow?

Ancient Rome grew from a collection of villages to become the superpower of the ancient world. The ancient Romans revolutionised engineering, warfare and government, and built an empire that covered much of modern Europe and northern Africa.

Beginnings of an empire

The ancient world was not an easy place to live – it was full of disease and warring tribes. According to legend, the strongest tribes united under general Romulus to take control of a region centred on the Tiber River. This small settlement grew into the city of Rome, which was named after Romulus.

The Tiber River supplied fresh water for the settlement, with good access to farmland and the Mediterranean Sea. The early settlements were built on hilltops so they were safe from flooding and easy to defend.

Recent archaeological digs in Rome have uncovered the foundations of huts built around the time that Romulus took the throne in 753 BCE. However, historians continue to look for evidence to prove that Romulus actually existed.

The Roman Kingdom lasted until 509 BCE. It was overthrown during the reign of Lucius Tarquinius Superbus. Superbus was a deeply unpopular ruler. The Roman people rioted and rose up against him. This led to the abolition of the **monarchy** and the beginning of the Roman Republic.

The powers previously held by the king were passed to two elected officials called **consuls**, and new powers were given to the **senate**. The people had more power to create laws and govern Rome. Rome expanded greatly through the Roman Republic period, which lasted 480 years and ended in 27 BCE.

Growth of an empire

After a period of instability following Caesar's death, Rome became an **empire**. The people had decided they did not want to be ruled by kings or dictators. The long reign of its first **emperor**, Augustus, began a golden age of peace and prosperity for Rome that lasted more than 500 years – until 476 CE.

The Roman Empire expanded to become one of the largest empires in world history. It was at its peak in 117 CE when it ruled 70 million inhabitants – then 20 per cent of the world's population. The empire covered 5 million square kilometres and included most of Europe, western Asia and northern Africa, as well as the Mediterranean islands.

The Roman Empire was a very powerful military, economic, cultural and political force in the ancient world. The length of its rule and vast area it covered ensured the lasting influence of the Latin-based languages of Italian, French and Spanish. Both the modern calendar and the popularity of **Christianity** began in ancient Rome.

The design of government in the Roman Empire has become the blueprint for modern **democracies** (see pages 148–151).

Source 1

The leader of the Gauls, Vercingetorix, surrenders to Caesar after the Battle of Alesia in 52 BCE. Vercingetorix is shown throwing down his weapons. Under Caesar's expansion, Roman territory grew to include Gaul (modern-day France). [Lionel-Noël Royer, *Vercingetorix Throws Down his Arms at the Feet of Julius Caesar* (1899 CE), 321 x 482 cm, Musee Crozatier, Le Puy-en-Velay, France.]

THE GROWTH OF THE ROMAN EMPIRE
- 270 BCE
- 130 BCE (after the Punic Wars)
- 44 BCE (Death of Julius Caeser)
- 14 CE (Death of Augustus)
- 70 CE (Accession of Vespasian)
- 116 CE (Reign of Trajan)

Source: Matilda Education Australia

Source 2

The growth of the Roman Empire

Learning Ladder H4.2

Show what you know

1 What made the site of Rome a great place for a city?

2 Look at Source 2 and describe the growth of the Roman Empire over time.

3 Why did ancient Rome change from a monarchy to a republic, and then to an empire?

4 Source 1: What clues are there in the painting that tell who the different sides are?

5 Make a table listing at least five challenges of running an empire as large as the Roman Empire. Then suggest five possible ways of dealing with those challenges.

Historical significance

Step 1: I can recognise historical significance

6 Put these events in order, from most important to least important:
- The assassination of Julius Caesar
- Roman Empire at its largest in 117 CE
- The foundation of Rome
- Rome goes from a monarchy to a republic
- Rome goes from a republic to an empire.

Step 2: I can explain historical significance

7 Why was the growth of the Roman Empire important in history?

Step 3: I can apply a theory of significance

8 Apply Partington's theory of significance (page 219). Show your response to the Roman Empire in a table, looking at importance, depth, number, time and relevance.

Step 4: I can analyse historical significance

9 Source 2: Compare the map of the Roman Empire with a modern map of the same region. At its peak, how many modern countries would have been in the Roman Empire?

Historical significance, page 219

Leaders wore tall plumes on their helmets to show their rank, and so their soldiers could see them.

How did the ancient Romans fight?

The ancient Romans understood the importance of a strong army. Disciplined training and clever military tactics enabled the army to increase Roman territory, and to then defend it against dangerous threats such as Carthaginian commander Hannibal during the Punic Wars.

Upper body armour protected the chest from attack. Some soldiers decorated the leather belts that held their sword and dagger. The short sword was used in close battle for stabbing the enemy.

Source 1

Standard-bearers known as *aquilifers* were selected for their leadership. Each legion in the Roman army marched into battle behind a battle flag, or a standard. The Roman standard was a silver eagle (or aquila) mounted on a pole. Aquilifers led legions into battle, often while wearing an animal headdress.

Source 2

Film still from the 2011 film *The Eagle*. The film is set in Roman Britain in 140 CE and tells the fictional story of a Roman soldier, Marcus Aquila, who sets out to find out what happened to the Ninth Legion of the Roman army that disappeared in the north of Britain. The ancient Roman army was one of the most disciplined and successful armies in history.

Soldiers stood side by side and overlapped their shields, forming a single protective shield. This was called the *testudo* formation: *testudo* is the Latin word for 'tortoise'.

The Roman army

In the early days of the Roman Republic, only landowners could join the army. Those landowners who were wealthy enough to own a horse became part of the **cavalry**.

As the Roman Republic grew, a more permanent and professional army was needed. In about 406 BCE Roman soldiers began to be paid, with their food and equipment expenses taken from their wages. Around 107 BCE, the restriction on owning land before being able to join the army was removed, and thousands of young men aged 17–22 enlisted.

The Roman army was one of the most disciplined and well organised military forces of all time. Roman soldiers were known as **legionaries** and the army consisted of around 30 **legions**.

- Each legion had around 6000 soldiers led by a general known as a *legatus*.
- Every legion was broken down into 10 **cohorts**.
- Each cohort contained six **centuries** of around 100 soldiers, led by a soldier known as a *centurion*.
- Extra soldiers called **auxiliaries** were recruited across the empire and used as specialist forces, such as archers. They were expected to fight on the frontline alongside Roman soldiers.

Life of a Roman soldier

The Roman army became very strong through harsh and disciplined training. The ancient Roman historian Polybius described the punishment given to soldiers for breaking military codes.

'A court martial composed of the tribunes is convened at once to try [a soldier being careless on patrol duty]. If he is found guilty he is punished by the bastinado [foot-whipping]. This is inflicted as follows: the tribune takes a cudgel [short stick] and just touches the condemned man with it, after which all in the camp beat or stone him, in most cases dispatching [killing] him in the camp itself.'

Source 3

Polybius, *The Histories, Vol. III*, Loeb Classical Library, translated by WR Paton, 1922–7.

Roman soldiers had to be fit for marching and fighting in heavy body armour. The chest armour of a typical Roman soldier was made from overlapping iron plates. They also wore a metal helmet and a belt with studded leather strips. Each soldier carried a dagger, a sword, a spear and a large shield.

The army also carried out building works to construct military camps, walls, roads and aqueducts (see pages 160–161). Camps were always laid out the same way so that soldiers could find their way around.

Each legion carried a **standard** (or battle flag) into battle. The Roman standard used to identify each legion was a silver eagle (or **aquila**) mounted on a pole. The aquila was held by a soldier chosen for his ability to lead men, called an *aquilifer*. The standard was a symbol of the legion's strength, and it became a matter of great shame if the aquila was lost in battle.

Soldiers were given incentives to ensure their support for their leaders. They could often keep the plunder from a battle and were given land when they retired from the army. The length of military service varied depending on the era, but could be up to 25 years.

Hannibal threatens Rome

The Punic Wars were fought between Rome and Carthage from 264 BCE to 146 BCE. Carthage was a powerful north African city with colonies and a large trading empire right around the Mediterranean. Rome did not have any warships, but the Carthaginians (as people from Carthage were called) had a strong navy and were considered a threat to Roman dominance.

The Romans quickly built a navy and, despite their inexperience at sea, triumphed over Carthage in the First Punic War (264 BCE–241 BCE), taking control of Sardinia, Corsica and Sicily.

The leader of Carthage's forces in the First Punic War was Hamilcar Barca. Barca had instilled hatred for the Romans in his young son, Hannibal. In the Second Punic War (218 BCE–202 BCE), Hannibal took over command of the Carthaginian army. At just 26 years old, he was a brilliant military leader.

In 218 BCE Hannibal launched a daring attack on Rome. He went 'the long way around', marching 31 000 soldiers and 37 war elephants over the Alps and south into Italy. Hannibal's superior strategies led to decisive victories over much larger Roman army legions in major battles at Trebbia, Lake Trasimene and Cannae.

The Romans decided to change tactics. They avoided Hannibal's army, attacking Carthaginian-held territories in Spain and north Africa instead. Hannibal's army was forced to leave Italy and return to defend their territory in north Africa.

Hannibal met his match in 202 BCE, in a battle at the Carthaginian-held town of Zama, when Roman general Scipio Africanus proved his strategic equal. Both sides had roughly the same number of troops (around 40 000 men), although the Romans had more cavalry and Hannibal had 80 elephants. Faced with a wall of soldiers on elephants, Scipio's troops blew trumpets to frighten the creatures. The elephants scattered among the Carthaginian soldiers, putting Hannibal's army into disarray.

The Roman cavalry defeated the Carthaginian cavalry and then joined in the battle to slaughter 20 000 of Hannibal's troops and capture a similar number. Hannibal escaped, but later poisoned himself rather than surrender to the Romans.

Following a long siege at the end of the Third Punic War (149 BCE–146 BCE), the Roman army eventually captured Carthage and burnt it to the ground. The people of Carthage were either killed or sold into slavery.

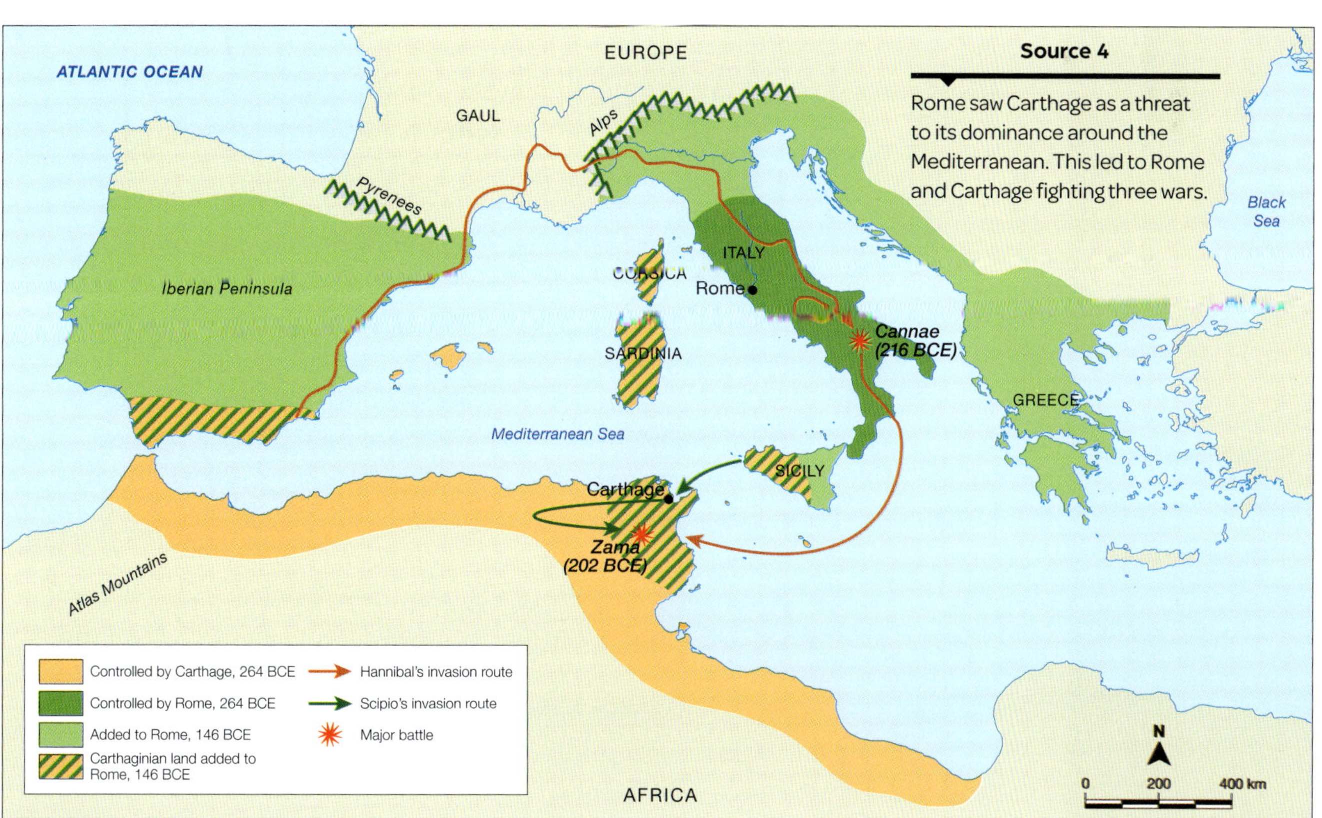

Source 4

Rome saw Carthage as a threat to its dominance around the Mediterranean. This led to Rome and Carthage fighting three wars.

Source: Matilda Education Australia

The fall of Rome

The Roman Empire eventually grew too large to be effectively governed. There was growing greed and corruption among soldiers, and Roman armies began fighting each other.

This instability allowed Germanic tribes – such as the Franks, Saxons, Huns, Goths and Vandals – to attack Roman territories. The invasions continued until the Goths finally brought about 'the fall of Rome' in 476 CE.

Source 5

Hannibal of Carthage marched 31 000 soldiers and 37 war elephants over the Alps and into Italy to attack the Romans in the Second Punic War. Tancredi Scarpelli, *Hannibal Crossing the Alps* (20th Century).

Learning Ladder H4.3

Show what you know

1 Source 1 (page 134): What was a standard-bearer? Why were they important?

2 Source 2: Describe what a typical Roman soldier would have looked like as he charged into battle.

3 Which of the leaders in the Punic Wars had the best tactics? Justify your answer.

4 Draw a diagram showing how an army is divided up into legions, cohorts and centuries.

5 Research the structure of the Australian army. Compare the Australian army and the Roman army.

Chronology

Step 1: I can read a timeline

6 Use the timeline on pages 130–131 to answer these questions.
- How long was it between the founding of Rome and the fall of the western Roman Empire?
- For how long was Rome a republic?
- How long did the Punic Wars last?

Step 2: I can place events on a timeline

7 Put these events in chronological order.

Roman soldiers first got paid in 396 BCE.

It took a further 386 years for the restriction that soldiers had to own land to be removed.

Four years before the first payment for soldiers, the Roman cavalry was expanded to include those who could pay for their own horse.

The First Punic War began in 264 BCE and lasted 23 years.

Caesar set out to conquer Gaul in 58 BCE. It took him seven years.

Step 3: I can create a timeline using historical conventions

8 Using correct conventions, create a timeline with the events in question 7 on it.

Step 4: I can distinguish causes, effects, continuity and change from looking at timelines

9 Look at Source 4. Use the dates on the map to describe the change of control of the land during the Punic Wars.

HOW TO

Chronology, page 206

What was the religion of ancient Rome?

Romans tolerated most religions and adopted many gods of the people they conquered. Many leaders, such as Julius Caesar, were declared gods after their death (see page 145). However, people who followed Christianity and Judaism believed there was only one god, and they refused to make offerings to former Roman leaders. This led to the persecution of Christians, in particular.

Romulus and Remus

According to Roman myth, Rome was founded by twin boys named Romulus and Remus. Their mother Rhea was a mortal woman who fell pregnant to Mars, the god of war. Rhea's brother Amulius wanted Romulus and Remus out of the way, and ordered they be drowned in the Tiber River. But Amulius's men put the twins in a basket instead and floated them down the river, hoping they would be found.

Romulus and Remus were discovered by a female wolf who raised them as her own cubs. When the boys became men, they decided to build a city and rule over it as kings. Romulus wanted the city to be built on the Palatine Hill and Remus wanted to build it on the Aventine Hill.

They had a fight, and Romulus killed Remus. Romulus then founded the city of Rome, named after himself.

Source 2

The image of Romulus and Remus being suckled by a she-wolf has been a symbol of the city of Rome since ancient times. [*Capitoline Wolf* (c. 11th century CE), bronze, 75 × 114 cm, Palazzo dei Conservatori, Capitoline Museums, Rome, Italy.]

Source 1

Romans believed that their former leaders were gods and regularly made offerings to them. However, Christians believed that there was only one god and refused to make sacrifices to past Roman leaders or for the good of the empire. Defying the Roman expectations went against the social order and led to Christians being violently punished. They were beheaded, nailed to crosses and burnt, as well as being torn apart by animals for entertainment. [Jean-Léon Gérôme, *The Christian Martyrs' Last Prayer* (1883 CE), oil on canvas, 150 x 88 cm, Walters Art Museum, Maryland, United States.]

Gods of ancient Rome

The people of ancient Rome believed that gods controlled everything that happened, and their actions were used to explain the good and bad events that occurred. Every god had at least one job to do, such as watching over the crops, or helping soldiers in battle. Ancient Romans prayed to their gods at least once a day.

As Rome expanded, Romans came into contact with people who worshipped other gods. From the Greeks they heard about powerful Greek gods such as Zeus and Athena. The Romans adopted the Greek gods, giving them Roman names and changing their personalities to fit Roman culture.

The most important Roman god was Jupiter. Jupiter was king of the Roman gods – and modelled on the Greek god Zeus. Most of the planets in our solar system are named after Roman gods: Mercury, Venus, Mars, Jupiter, Saturn and Neptune, as well as the dwarf planet Pluto.

Source 3

Some of the ancient Roman gods and goddesses

JUPITER
King of the Gods

JUNO
Goddess of Marriage and Childbirth

VESTA
Goddess of the Hearth

NEPTUNE
God of the Sea

VULCAN
God of Fire

MARS
God of war

VENUS
Goddess of Love and Beauty

DIANA
Goddess of the Moon and Hunting

APOLLO
God of the Sun

MINERVA
Goddess of Wisdom and War

Worshipping the gods

Religion was part of everyday life for Romans, from slaves through to the emperor. There were hundreds of gods, and a lot of work went into making sure that all of them remained happy.

Each home had a shrine dedicated to the gods that were important to that family. Small statues of gods were placed on the shrine.

Every town had a temple dedicated to a specific god or goddess. People would visit different temples, depending on what they were praying for. For example, people wanting help in their love life would visit a temple dedicated to Venus, the goddess of love.

PLUTO
God of the Underworld

CERES
Goddess of Agriculture

MERCURY
Messenger of the gods

BACCHUS
God of Wine

Learning Ladder H4.4

Show what you know

1. According to Roman myth, how was Rome founded?
2. Why were Christians persecuted? How were they treated?
3. If you wanted to grow successful crops, which gods might you pray to? Why?
4. Why do you think the Romans named the planets after gods?
5. How is the way the Romans worshipped their gods different to a religion you know about?
6. Look at Source 1. How has the painter represented the Christians' fate?

Cause and effect

Step 1: I can recognise a cause and an effect

7. Why did people wanting help in their love life pray at a temple dedicated to Venus?

Step 2: I can determine causes and effects

8. The ancient Romans worshipped hundreds of gods. As the Roman Empire expanded, they discovered new gods that were worshipped by conquered people. Is there a causal relationship between these two things?

Step 3: I can explain why something is a cause or an effect

9. Source 1: Romans tolerated most religions, so why did they persecute Christians and Jews?

Step 4: I can analyse cause and effect

10. Following a famine in Rome in 496 BCE, a new temple was built on the slope of Aventine Hill in Rome (see page 130), dedicated to the goddess Ceres. Is there a causal relationship between these two events? What kind of people would most likely pray there?

HOW TO
Cause and effect, page 215

How was Roman society organised?

The structure of Roman society was based on heredity (who your parents were), property, wealth and citizenship. Women and slaves had no voice at all.

Citizens

Roman **citizens** had to be males born to parents who were citizens. **Slaves** could not be citizens and had no rights at all – their master decided what they could and could not do. Women had limited citizenship, and could not vote. A woman's key role was to raise children and run the household.

Roman citizens enjoyed much greater power than people who were not citizens. They could vote when 25 years of age, be elected as a magistrate and have protection under the law.

Roman citizens were divided into two groups:

- **patricians**: wealthy landowners whose families had held influential positions in society
- **plebeians**: poorer farmers and craftspeople who made up the bulk of the population.

Governing the people

In 509 BCE, Rome introduced a system of government called a republic where citizens could have a say in how Rome was run. Each year citizens elected more than 50 **magistrates** to run the city.

Learning Ladder H4.5

Show what you know

1 Who could and could not vote in ancient Rome?

2 How were ordinary people represented in government?

3 Source 1: What was the difference between a senator, a consul and a tribune?

4 Do you think the Romans had the right balance between ordinary people and powerful people in government?

Continuity and change

Step 1: I can recognise continuity and change

5 Rank these things in order of how much they have changed from ancient Roman times until today, from least changed to most changed:
- voting once you are an adult
- two people are joint leaders
- only men can vote
- slavery was common.

Step 2: I can describe continuity and change

6 What important change occurred in 471 BCE? What impact did it have?

Step 3: I can explain why something did or did not change

7 Gender roles of men and women have evolved from ancient to modern times. What has changed? What has stayed the same?

Step 4: I can analyse patterns of continuity and change

8 What rights did slaves have, if any? How did Spartacus try to change the situation and what was the result (page 157)?

Continuity and change, page 212

HOW TO

Once a man had served as a magistrate, he became a member of the Senate.

The Senate advised the two most senior leaders: the consuls. To make sure no consul could become a dictator or a king, two new consuls were elected each year. The consuls served only one year, and had the power to veto each other (see pages 148–9).

Key positions in government were held by patricians. From 471 BCE, plebeians were given a say in their own assembly, and they had 10 elected tribunes who could veto laws made by the Senate.

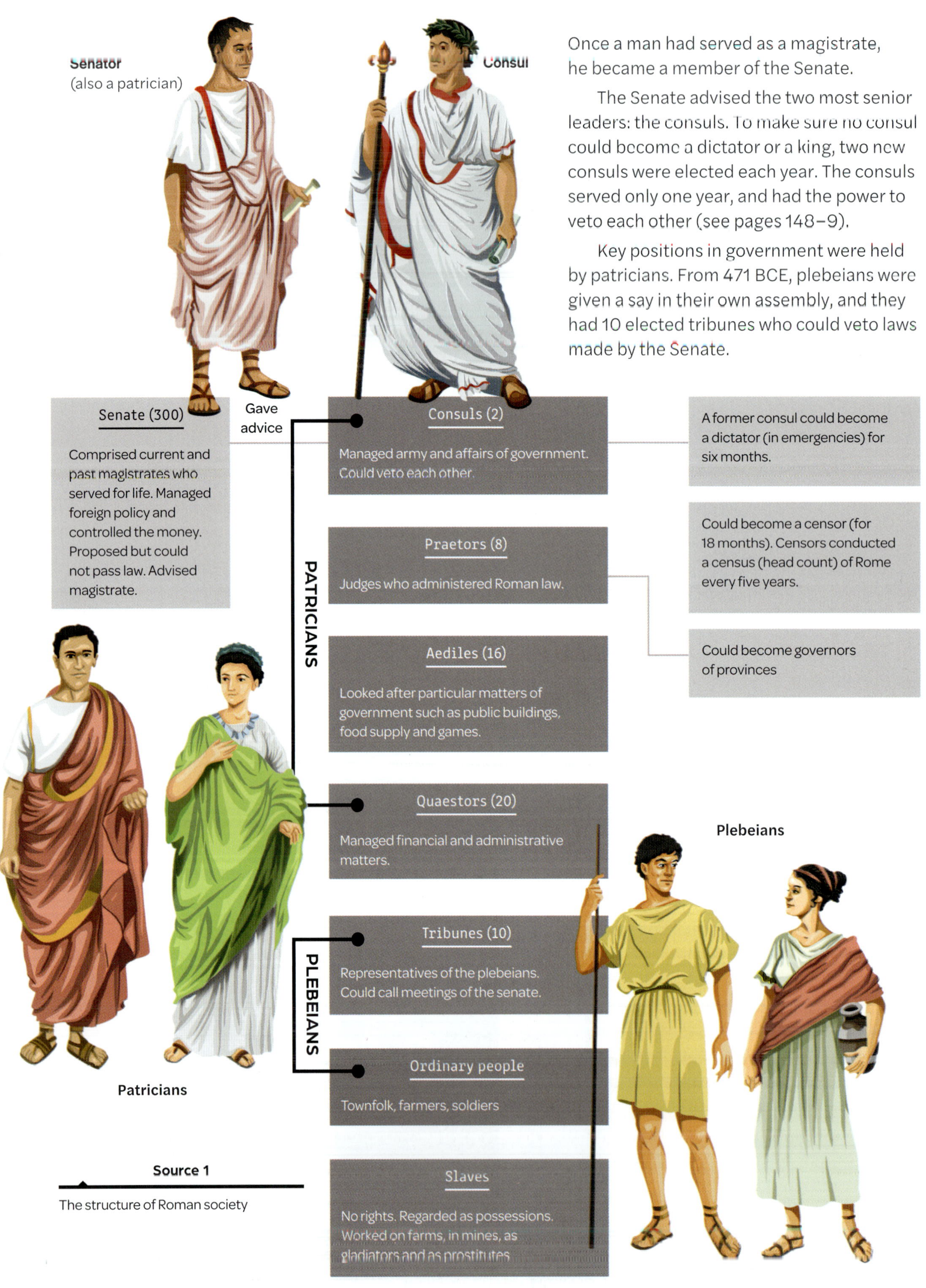

Source 1

The structure of Roman society

How did Julius Caesar rise to power?

Julius Caesar was a powerful politician and military commander. With the support of the people and the army, Caesar swept to power in 49 BCE and laid the foundations for the Roman Empire.

The early years

Julius Caesar was born into a wealthy patrician family in 100 BCE. He received a good education and followed his father into a career in politics. When his father died, Caesar became the head of his family, aged just 16. At the age of 18, he married Cornelia, the daughter of the powerful patrician Lucius Cornelius Cinna.

Caesar and his family fled Rome when Sulla, a general, became a dictator. Caesar joined the army where he showed he was a very capable soldier.

Caesar's rise to the top

Caesar returned to Rome after the death of Sulla in 78 BCE. He quickly gained popularity and power through his excellent leadership and successful military campaigns. By 68 BCE, Caesar was elected as a **quaestor** (financial officer), and in 62 BCE he became a *praetor* (assistant to the consul) and governor of the province of Spain.

Caesar returned to Rome from Spain in 60 BCE. He made a deal with two key politicians, Pompey and Crassus, to become elected as a consul, which was one of two top governing positions in the Roman Republic. In 59 BCE Caesar was appointed a consul for the one-year term, and he took the position of governor of the province of Gaul (most of modern-day France).

As governor and military commander of Gaul, Caesar seized new lands and extended Roman territory. He was a brilliant military strategist and very popular with his men.

But back in Rome, Pompey – who was now the leader of the **Senate** – was worried that Caesar was starting to act without consulting the Senate. In 49 BCE, the Senate ordered Caesar to give up his governorship of Gaul. Caesar refused to give it up. Instead, he and his loyal army attacked Pompey's troops.

In the **civil war** that followed, Caesar won control of Rome and Pompey fled to Egypt, where he was killed by King Ptolemy – who wanted to show his loyalty to Caesar.

King Ptolemy's co-ruler (and sister) was a woman named Cleopatra. Cleopatra and Caesar became lovers, and he helped her gain the Egyptian throne (see pages 82–85).

Source 2

Julius Caesar became the most powerful man in the world. The Senate granted him the title 'Dictator for Life' and he ruled like a king. Following his death in 44 BCE, Caesar was declared a god and his adopted son, Octavian (under the name Augustus), became the first emperor of Rome. In 42 BCE, Julius Caesar was declared a god, and a temple was dedicated to him in Rome. [Andrea Ferrucci, bust of Julius Caesar (c. 1512–14 CE), marble, 68.6 cm, The Met Fifth Avenue, New York, United States.]

Caesar returned to Rome in 47 BCE, and began many reforms. Among them were:

- several new Roman colonies in Africa and Gaul
- grants of land to about 15000 of his soldiers
- establishment of a police force
- a new Julian calendar that added a leap day at the end of February every fourth year, setting the length of the year to 365.25 days.

Caesar's downfall

In 44 BCE Caesar was **assassinated**, leading to 15 years of civil war in Rome. Caesar's adopted son, Octavian, eventually gained control and in 27 BCE Octavian (given the name Augustus) became Rome's first emperor. It was the beginning of the Roman Empire and the end of the Roman Republic.

Source 1

The Death of Julius Caesar, by Vincenzo Camuccini, 1804–1805. Up to 60 senators attacked Caesar in the Senate, stabbing him with knives they had hidden under their togas. Caesar was stabbed 23 times and died on 15 March, 44 BCE. [Vincenzo Camuccini, *The Death of Julius Caesar* (c. 1805 CE), oil on canvas, 195 × 112 cm, La Galleria Nazionale, Rome, Italy.]

Learning Ladder H4.6

Show what you know

1 How do you think Caesar's early life made him a better leader?

2 Why was Caesar popular? Why do you think he was declared a god?

3 Describe Caesar's personality, using lots of adjectives.

4 Source 1: What do the looks on the people's faces tell you about what was happening?

Historical significance

Step 1: I can recognise historical significance

5 List the events that led to Caesar becoming 'Dictator for Life'.

Step 2: I can explain historical significance

6 Rank the dot-point list of Caesar's reforms in order, from most important to least important.

Step 3: I can apply a theory of significance

7 Write five sentences about the importance of Julius Caesar to history, one of each about:
- his importance at the time
- how deeply people's lives were affected by him
- how many people were affected
- how long they were affected for
- his relevance today.

Step 4: I can analyse historical significance

8 Source 1: Julius Caesar was murdered. List four different causes of this event. Rank them from least to most crucial. Write this up in a short paragraph to answer this question: 'What causes resulted in the murder of Caesar? How important was each cause?'

HOW TO

Historical significance, page 219

Who ruled the Roman Empire?

The emperors of ancient Rome were supreme rulers. They were often chosen by the Senate, but they ruled like kings. Some were brave soldiers or bold leaders who were celebrated for their achievements. Others abused their power and were despised for their cruelty.

Augustus

There was a period of civil war after Caesar's assassination. In 27 BCE Caesar's adopted son, Octavian, became the first emperor of Rome. He was granted the name Augustus, meaning 'the revered one'. During his 40-year reign, Augustus used Rome's military might to enlarge the Roman Empire to include Egypt and Spain.

Augustus embarked on a program of road-building, and developed trade with new partners such as India. He expanded the Roman system of censuses and taxation to help integrate newly conquered lands into the Roman Empire. Augustus founded the Roman postal service, and introduced firefighters and police (called *vigiles*).

Augustus laid the foundations for a long period of peace. The western Roman empire lasted almost 450 years (until 476 CE) and the eastern empire lasted nearly 1500 years (until 1453 CE). Augustus renamed a month on the calendar after himself (August), just as Julius Caesar had done earlier with the month of July.

Source 1

Nero ruled Rome as emperor 54 CE–68 CE. Historians believe his mother had Emperor Claudius killed to give Nero the throne. Nero continued on this murderous path when in power, killing his wives, his mother and anyone who stood in his way. He is infamous for the mass slaughter of Christians (see page 139). [Pieter Fransz de Grebber, Roman Emperor (c. 1625–30), oil on canvas.]

Nero

Nero was a cruel tyrant who abused his power as emperor. He became emperor in 54 CE when he was just 17 years of age. He had his influential mother killed, along with two of his wives. To fund his excessive parties and public spectacles, Nero raised money through taxes and by confiscating property.

The ancient Roman writer Suetonius (c. 69 CE–140 CE) gave his account of Nero's passion for entertainment.

'He gave many entertainments of different kinds: ... chariot races in the Circus, stage-plays, and a gladiatorial show ... For the games in the Circus he ... even matched chariots drawn by four camels. At the plays which he gave for the 'Eternity of the Empire', which by his order were called the Ludi Maximi [Great Games] ... Every day all kinds of presents were thrown to the people ... [Nero] devised a kind of game, in which, covered with the skin of some wild animal, he was let loose from a cage and attacked the private parts of men and women, who were bound to stakes.'

Source 2

Suetonius, *The Lives of the Caesars*, Loeb Classical Library, translated by JC Rolfe, 1914.

Source 3

Nero's Torches is a painting by Henryk Siemiradzki from 1876 CE. It depicts Christians about to be burned alive. Emperor Nero blamed the 64 CE Great Fire of Rome on the Christians and, for revenge, burned them as torches to light his gardens. [Henryk Siemiradzki, *Nero's Torches* (1876 CE), oil on canvas, 705 × 385 cm, National Museum, Kraków, Poland.]

In 64 CE a great fire almost destroyed Rome. Some Romans believed that Nero had started the fire so that he could rebuild the whole city to his taste. In return, Nero blamed Christians for the fire. He ordered some Christians to be fed to lions (see page 139) and had others painted with tar and set alight in the gardens of his palace.

In 65 CE the Senate unsuccessfully plotted to remove Nero from power. Later, in 68 CE, his personal guard abandoned him and he was forced to flee Rome. Nero was so humiliated that he committed suicide.

Marcus Aurelius

Marcus Aurelius became emperor of Rome in 161 CE, at 40 years of age. He was one of the most respected emperors in history. Aurelius consistently placed the needs of the people before his own desires and passed laws that benefited all classes of Romans. He increased the size of the army and gave more rights to women and slaves.

His reign was marred by a famine that ravaged Rome after the Tiber River flooded, and an outbreak of plague that spread through the empire, killing thousands of people.

Learning Ladder H4.7

Show what you know

1 What actions did Augustus take as emperor?

2 Source 2: How does Suetonius show that Nero was reckless?

3 How do we know that Suetonius's account is a primary source? Does that make it more reliable? Explain your answer.

4 Describe the traits that make an emperor either good or bad.

Chronology

Step 1: I can read a timeline

5 Look at the timeline on pages 130–131. What event occurred in 64 CE and which emperor was rumoured to be behind it?

Step 2: I can place events on a timeline

6 List these connected events in chronological order. How are they connected?

49 BCE Julius Caesar becomes dictator

27 BCE Augustus becomes first emperor

44 BCE Caesar is assassinated

Step 3: I can create a timeline using historical conventions

7 There are seven events or developments listed in this section. Create a timeline that includes them all.

Step 4: I can distinguish causes, effects, continuity and change from looking at timelines

8 Look at the timeline on pages 130–131. Which events relate to 'great individuals'?

Chronology, page 206

How does Roman government compare to ours?

The design of government in the Roman Republic has become the blueprint for many modern democracies. The power of government is in the hands of the people, with safeguards built in to stop leaders abusing their power.

Roman Republic

The first government of ancient Rome was a **monarchy**, ruled by kings. In 509 BCE the king was overthrown by the people – and the Roman Republic was established.

Government in the Roman Republic was led by two **consuls**, who were elected for a one-year term. To prevent a single consul from becoming a king or dictator, each consul could veto the decision of the other. The consuls made key decisions, such as declaring war, collecting taxes and making laws.

The consuls were advised by the **Senate**. The Senate was a group of 300 wealthy landowners known as **patricians**. Senators were appointed by consuls and, once appointed, were senators for life.

The Senate made important decisions about key issues, such as finances and relationships with other countries. Consuls usually followed what the Senate recommended.

At first, ordinary Romans, the **plebeians**, had no say in the government, and no one to represent them. In 471 BCE, a Plebeian Council was formed to represent common people. This allowed plebeians to elect their own leaders, pass laws, and try legal cases.

The Roman Republic ruled Rome until 27 BCE, when a single **emperor** ruled the Roman Empire.

Source 1

The formal function of the Roman Senate was to debate issues and advise consuls with proposed actions. They had no legal power, but their recommendations were usually followed by the consuls. [Cesare Maccari, *Cicero Denounces Catiline* (c. 1882–88 CE), fresco, Palazzo Madama, Rome, Italy.]

Constitution

The Constitution of the Roman Republic was a set of unwritten guidelines. It created three separate branches of government:

- the **Assemblies**, where plebeians democratically elected magistrates and passed laws
- the Senate, which advised the magistrates and the state, and managed Rome's day-to-day affairs
- the Magistrates, who were elected by the people, and had religious, military and legal powers. Consuls were the two highest Magistrates.

The constitution ensured rule by the people and reduced the risk of corruption. It became the basis for modern democracies.

Government in Australia

After Great Britain claimed the east coast of Australia in 1788, the Westminster parliamentary system was set up. Parliaments were set up in the colonies, each with a lower house and an upper house – with final approval from the king or queen of England. There was no acknowledgement of the laws that Aboriginal peoples had put in place.

The six separate colonies of Queensland, New South Wales, Victoria, Tasmania, South Australia and Western Australia agreed to unite and form one Australian government. On 1 January 1901, the *Commonwealth of Australia Constitution Act 1900* became law, setting up the framework for how Australia is governed.

The Commonwealth **Parliament** has been in Canberra since 1927. It makes laws that concern the whole country, such as defence, foreign affairs, trade and immigration.

Australia's Constitution

The **Constitution** established a system where the responsibility to make or change laws is shared between a federal parliament and six state parliaments.

The Commonwealth Parliament consists of two houses and the queen or king of Great Britain, who is represented in Australia by the **Governor–General**.

The Lower House is the **House of Representatives**, and the Upper House is the **Senate**. All laws made by the Commonwealth Parliament must be passed by a majority of members in both houses of parliament and approved by the Governor-General.

Each of Australia's six states and two territories also has a parliament to make laws on state matters, such as health, education and transport.

Source 2

The formal function of Australia's Parliament is to propose, debate and decide on laws.

The Constitution separates the powers the government has. This is to prevent one group having power over both the law-making and law-judging systems. The powers are broken into three branches:

- **executive**: the prime minister and senior ministers of the government and the governor-general have the power to administer or implement the law
- **legislature**: parliament has the power to make the law
- **judiciary**: the courts have the power to interpret and apply the law.

Changes to the Constitution can only be made by a **referendum**, where a majority of Australian voters and a majority of states vote to approve the changes proposed by parliament.

Learning Ladder H4.8

Civics and citizenship

Step 1: I can identify topics about society

1. List the various groups and roles in the ancient Roman government and describe the responsibilities of each.
2. List the various groups and roles in the Australian Commonwealth or Federal Parliament and describe the responsibilities of each.

Step 2: I can describe societal issues

3. Who had the most power in the Roman system of government: wealthy people or plebeians? What makes you say that?
4. Which of these is a cause of Rome being governed by two consuls?
   - It gave more rights to plebeians.
   - One consul alone might have too much power.
   - One consul would be better than two consuls.

Step 3: I can explain issues in society

5. What is the Australian Constitution for?
6. Is the separation of powers in the Australian Constitution important? Explain your response.
7. Create a class constitution. List four rights that everyone has and four rules everyone must obey. How could you enforce this constitution?

Step 4: I can explain different points of view

8. Compare the images in Source 1 and Source 2. What is similar about the images? What is different? Explain whether you think this shows similarity or difference between the two systems.

Step 5: I can analyse issues in society

9. How is Australia's separation of powers similar to Rome's requirement to have two consuls? Do you think we need checks on the power of rulers in society?

What was daily life like in ancient Rome?

How people lived in ancient Rome depended on which class they belonged to. Plebeians lived in rented rooms without kitchens, toilets, heating or running water. Patricians had comfortable city homes with slaves to attend to their needs – and they also had large country homes, called villas.

Housing for plebeians

In Rome, the vast majority of people – the plebeians – lived in cramped multi-storey apartments called *insulae* (literally 'islands'). The ground floor of the insulae had shops, called *tabernae*. The streets were skinny and dirty.

Often whole families lived in one room of an *insulae*, with up to 40 people in each housing block sharing bathing and cooking facilities in a central courtyard.

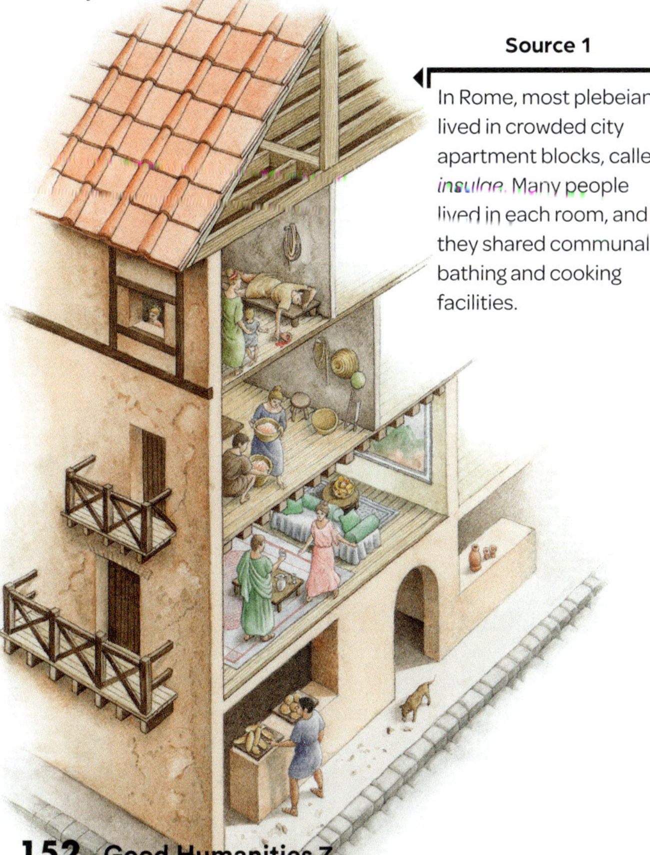

Source 1

In Rome, most plebeians lived in crowded city apartment blocks, called *insulae*. Many people lived in each room, and they shared communal bathing and cooking facilities.

People had to walk to a public toilet where they sat over large holes and chatted with others as they did their business. They wiped their bottoms with wet sponges attached to sticks, which they then dropped into the sewer system.

Housing for patricians

Wealthy Romans – patricians – owned homes in Rome, as well as large country estates called villas.

A patrician's townhouse was called a **domus**, and had multiple rooms. The entrance hall led to the atrium, which contained an altar and statues of the household gods (see pages 140–141). Positioned off the atrium were bedrooms, and a dining room, living room and kitchen.

A **villa** was larger than a *domus*, and included courtyards, baths, pools, storage rooms, exercise rooms, servants' quarters and gardens.

A well-appointed *domus* had glass windows, but most homes used animal skins stretched across the window openings instead.

Family life

Ancient Roman families were headed by the oldest male, who was called the **paterfamilias**, or 'father of the family'. The *paterfamilias* had control over all decisions and had the right to:

- punish any family members
- decide whether a baby should live or die
- disown his children

Source 2

The *domus* was the city house for wealthy Roman patricians, who also owned large villas in the countryside.

- sell his children as slaves
- choose a husband for his daughter (or sister).

Sons were important for continuing the family name. If a father had no sons, he could adopt one. Only the *paterfamilias* could own property. His sons received an allowance to manage their households.

Education was a privilege, and it was for wealthy boys only. Boys generally finished their schooling at 16 years of age, at which point they registered as a full Roman citizen. Rich girls learnt to read and write, but much of their education related to skills such as spinning and weaving.

Women were expected to be good wives and mothers. Any money that women earnt automatically belonged to their husbands.

Girls in ancient Rome were married young, often around 13 or 14 years old. The girl's father always arranged the marriage – and the girl had no say in it.

Learning Ladder H4.9

Show what you know

1 Sources 1 and 2: Why did patricians and plebeians have different living arrangements?

2 Source 1: What basics do you have in your home that plebeians did not have?

3 Source 1: Choose three figures from the image of the *insulae*. What are they doing?

4 List the differences between the lives of males and females.

5 Why did girls learn spinning and weaving?

Continuity and change

Step 1: I can recognise continuity and change

6 What was expected of women in ancient Rome? How does this compare with expectations of women today?

Step 2: I can describe continuity and change

7 Do we have divisions in our society such as patricians and plebeians? Explain your answer, giving examples.

Step 3: I can explain why something did or did not change

8 Education for boys and girls used to be very different. Why do boys and girls learn the same things now?

Step 4: I can analyse patterns of continuity and change

9 In ancient times there was inequality between rich and poor, and between men and women. Is there more or less inequality now? Why? Explain your answer using historical and modern examples.

Continuity and change, page 212

How did ancient Romans entertain themselves?

Life was tough for ordinary Romans. To prevent poor people from rebelling, the ruling classes provided entertainment and free grain. Large arenas hosted bloody battles between gladiators and wild animals, as well as chariot races where death and injury were common.

Public entertainment

The number of days of free entertainment provided by the Roman government varied, depending on the era, but at the end of the first century BCE, there was free entertainment for nearly half the year. The free entertainment had one purpose: to make sure the plebeians (see pages 142–143) did not become restless and rebel. Entertainment also helped to keep the rulers popular. The ancient Roman writer Juvenal said that Romans were kept happy and peaceful by two things: *panem et circenses* (bread and circuses).

There was a special arena in Rome for chariot races. It was called the Circus Maximus, and it could seat close to 250 000 people. Death and injury were common in chariot races – but that was part of the entertainment.

Source 1

Chariot racing at the Circus Maximus was a very popular form of entertainment in ancient Roman times. This is a reconstruction of a Roman chariot with the inclusion of original metallic parts, made c. 2003 CE.

Source 2

Wild animals such as tigers and lions were captured and forced to fight gladiators. The animals were kept in cages below the surface of the arena, then released. Animals were often forced to fight one another for the crowd's amusement. The battle between a lion and a bull was said to be very popular. [Alexander von Wagner, *A Roman Bull Fight* (19th century CE), colour lithograph.]

Source 3

Each type of gladiator wore distinctive armour and used different weapons.

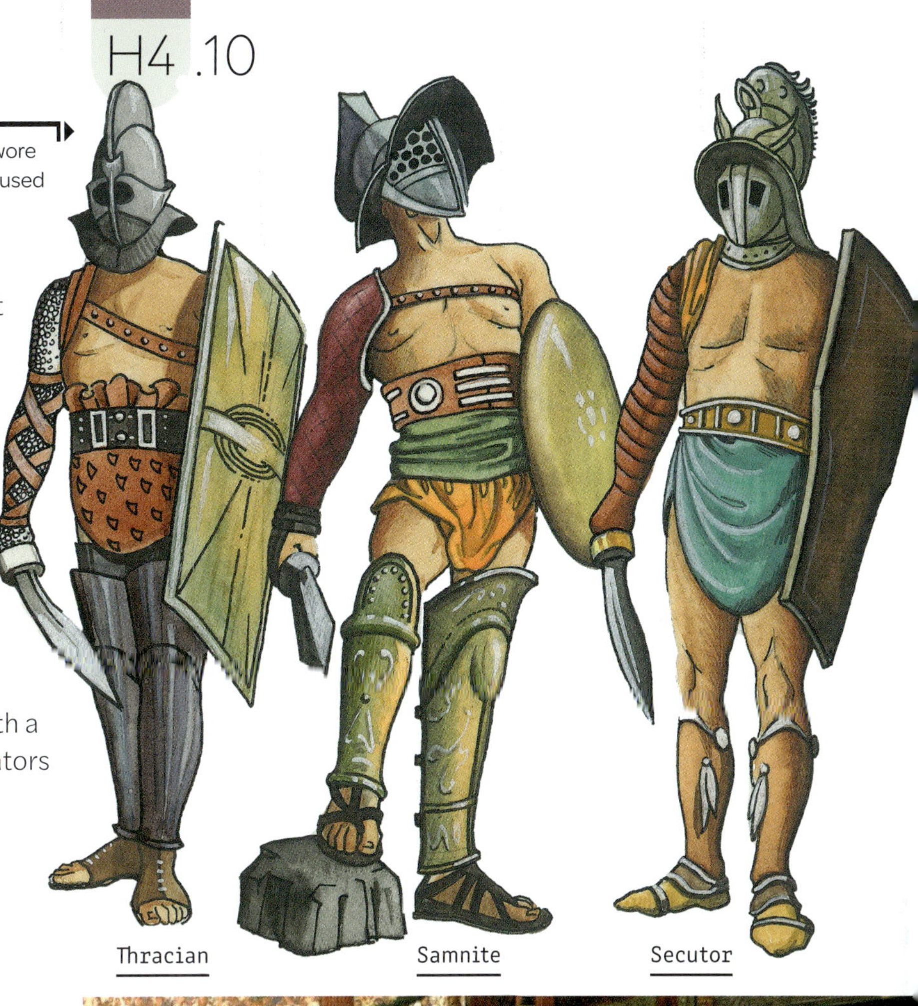

Gladiators

A very popular form of public entertainment in ancient Rome was watching **gladiators** fight to the death in huge arenas such as the Colosseum (pages 158–159). Gladiators were first introduced in 264 BCE. They were generally slaves, criminals or prisoners.

Tens of thousands of Romans went to the Colosseum to watch gladiators fight and kill each other. According to legend, the emperor would decide whether a gladiator would live or die by signalling with a thumb up or thumb down. Successful gladiators were admired like sporting stars today and, if they were lucky, they would be granted their freedom.

Types of gladiators

To add interest to fights, there were many different types of gladiators.

- An *andabata* wore helmets without eyeholes. They charged blindly at each other for the crowd's amusement.
- A *bestiarius* fought beasts using a spear and a knife. Many exotic imported animals such as lions and tigers were used.
- A *murmillo* wore a helmet and armguards, and carried a short sword and shield. They often fought against *Thracians*.
- A *retiarius* wore only an armguard for protection and carried a trident (a spear with three prongs), a dagger, and a net to catch their opponent's weapon.
- A *samnite* wore a high-crested helmet and used a short sword. Their right arm and left leg were protected with leather strapping, and greaves were worn on the legs (normally the left leg only).
- A *secutor* used the same armour and weapons as a murmillo. Their helmets had only two small eyeholes to prevent a *retiarius'* trident from piercing their face.
- A *Thracian* used a shield and a curved sword known as a *sica*.

Spartacus

Spartacus served in the Roman army before he was captured and sold into slavery. He attended gladiatorial training school near Capua, where he trained as a *murmillo* gladiator and survived a number of gladiatorial contests.

In 73 BCE, Spartacus escaped to Mount Vesuvius (see pages 162–165) where, over the next two years, about 90 000 other runaway slaves joined him. As one of the leaders of the slave army, Spartacus defeated several Roman legions over the next two years, before his death on the battlefield in 71 BC. His followers who survived the final battle were crucified, and their bodies kept on display as a warning to anyone else who wanted to challenge the Roman state.

Retiarius

Source 4

Pollice Verso is a painting created by French artist Jean-Léon Gérôme in 1872. The title refers to the Latin term for the thumbs up/thumbs down hand gesture the crowd used to decide the fate of the fallen gladiator. The gesture on the painting is given by the spectators to the victorious *murmillo* who has just defeated the *retiarius* at the Colosseum. We assume today that thumbs up meant the fallen gladiator lived and thumbs down meant he died. However, thumbs down may have meant 'swords down', which meant the losing gladiator would live. Its exact meaning is unknown. [Jean-Léon Gérôme, *Pollice Verso* (1872 CE), oil on canvas, 149.2 × 96.5 cm, Phoenix Art Museum, Phoenix, United States.]

Learning Ladder H4.10

Show what you know

1 What problems could there be with almost half a year of free public entertainment? Why was it provided?

2 Describe which features of Roman sport made it brutal.

3 What was the object in Source 1 used for and where could you have seen this activity in Rome?

4 What can we learn about Roman culture from looking at Source 1 and Source 2?

5 What did people do to end up becoming gladiators?

6 Why do you think Spartacus was such a legend?

7 Rank the seven types of gladiator, from most deadly to least deadly.

Source analysis

Step 1: I can determine the origin of a source

8 Are the images in this section primary sources or secondary sources? How do you know?

Step 2: I can list specific features of a source

9 Source 3: Look at the images of gladiators. How do they show that gladiators were fierce competitors?

Step 3: I can find themes in a source

10 Source 4: Why is the *murmillo* looking to the crowd and how are they responding? What did the signal from the crowd mean?

Step 4: I can use my outside knowledge to help explain a source

11 Read this account of Spartacus, written by ancient Roman historian Appian (95 CE–165 CE):

'[Spartacus] took refuge on Mount Vesuvius, where he allowed many runaway domestic slaves and some free farm hands to join him ... he plundered the nearby areas, and because he divided the spoils in equal shares his numbers quickly swelled.'

a Who joined Spartacus' army?
b Why was Spartacus so successful in attracting recruits?
c How did the Romans respond and why?

HOW TO

Source analysis, page 209

What was the Colosseum?

The Colosseum was larger than the Melbourne Cricket Ground, and could hold up to 50 000 people. In ancient Roman times, it hosted events to entertain people and gain support for their leaders, including having gladiators fight to the death.

The building

The Colosseum is a massive freestanding stone stadium. Work began on it began around 70–72 CE and it was completed in 80 CE. Emperor Titus opened the Colosseum, known as the Flavian Amphitheatre, with 100 days of games for the Roman people to enjoy, including gladiator fights.

Much of the original stadium has fallen into disrepair, caused by weathering, natural disasters, theft and vandalism. However, the Colosseum is still popular, attracting tourists from all over the world.

The events

The arena hosted bloody battles between gladiators, and between gladiators and wild animals (see pages 154–157). Many gladiators were slaves or criminals, but some fighters volunteered because they wanted fame and fortune. Some emperors even showed off their combat skills in the arena to gain the support of the people.

The arena could be decorated to look like a natural landscape, where the gladiators hunted animals. Sometimes the arena was flooded and mock sea battles were held. Crocodiles were released to grab those who fell overboard.

There were 76 public entrances for plebeians, and four entrances for wealthy patricians, senators and the emperor. It could accommodate 45 000 people seated, with standing room for an extra 5000 people.

Animals were kept in cages under the Colosseum and hoisted up to ground level by a pulley system, where they entered the arena through trapdoors.

Source 2

The remains of the Colosseum are a major tourist attraction. Restoration work on the stadium, which is almost 2000 years old, was completed in 2016.

Awnings were pulled out from the top storey to protect the audience from the sun.

The emperor had his own private box. Other important people sat in the seats surrounding the arena. Women, poor men and slaves sat or stood in the top tier.

Gladiators entered the stadium through gates at the arena level.

Gladiators, animals and props were held in cells and rooms beneath the arena.

Source 1

This cutaway illustration shows how spectators accessed the Colosseum seating via the 80 entrances. Gladiators, soldiers and animals were housed in rooms, cells and corridors beneath the arena.

Learning Ladder H4.11

Show what you know

1 What kinds of things took place in the Colosseum?

2 How were people, especially wealthy people, made to feel comfortable while at the Colosseum?

3 List five interesting adjectives to describe the Colosseum.

4 Source 1: Where were gladiators, animals and props held before appearing on the arena? How did gladiators and animals enter the arena?

Source analysis

Step 1: I can determine the origin of a source

5 Sources 1 and 2: Which of these is a primary source? Make a sketch of the primary source and use labels to show its features using information from the secondary source.

Step 2: I can list specific features of a source

6 Sources 1 and 2: Describe the Colosseum physically, including its size, shape, areas and artistic features.

Step 3: I can find themes in a source

7 Do you think it is right to host fighting competitions, pretend sea battles and conflicts between people and animals? Did people think differently about such competitions in ancient times? Explain your response.

Step 4: I can use my outside knowledge to help explain a source

8 Sources 1 and 2: What differences can you see between these sources? What has happened to the Colosseum since it was built? Why do you think it is still standing?

Source analysis, page 209

What did the Romans build?

The Romans developed sophisticated materials and tools, and were the leaders in engineering in the ancient world. They built 84 000 kilometres of roads to connect the cities of their growing empire. They also built protective walls and fortresses, as well as reservoirs and aqueducts to provide a constant water supply.

Master builders

As the ancient Roman Empire grew, the Romans needed to become increasingly skilful builders. They revolutionised the construction of buildings and roads with the invention of concrete, which they made by mixing volcanic dust with lime and water.

Their engineering and construction skills were so accurate and sturdy that many ancient Roman structures are still in use today. Most of the construction was done by the Roman army, aided by slaves.

Water supply

Reservoirs were built to store water required for daily needs and public baths. Water from rivers was pumped into settling ponds, where any dirt would settle before being pumped into reservoirs. Purer mountain water was directed to reservoirs via **aqueducts** that could run for up to 100 km.

The Romans used pipes to connect reservoirs to public baths, fountains and toilets – and also to the homes of the wealthy. The advanced plumbing systems of the ancient Romans also carried away wastewater and sewage.

Public baths were very popular in ancient Rome, where people went to wash and socialise. The water was heated by large fires kept burning by slaves in the boiler room. Hot air was fed under the floors through a series of tunnels to heat the pools.

Source 1

Roads and bridges were vital technology in ancient Rome.

Source 2

Romans developed different ways of using the arch, and eventually invented the dome. Domes became a feature of many public buildings. The most spectacular Roman dome is at the Pantheon, a temple rebuilt around 120 CE. The dome has a 43-metre diameter – and it is still the second-biggest dome in the world.

Learning Ladder H4.12

Show what you know

1 Why were arches and domes so important?

2 Why did the Romans build so many roads?

3 What made public baths so popular? Include some of your own ideas in your answer.

4 What challenges do you think there would have been in running public baths in ancient times? Select one challenge and suggest how it could have been overcome.

5 Source 2: What artistic features were used in building the Pantheon to make the dome appear large and imposing?

Cause and effect

Step 1: I can recognise a cause and an effect

6 If the cause was the need to supply settlements with water, what was the effect created by ancient Roman technology?

Step 2: I can determine causes and effects

7 Sources 1 and 2: What did the development of the arch enable the ancient Romans to do?

Step 3: I can explain why something is a cause or an effect

8 How did improved building tools affect Roman achievements?

Step 4: I can analyse cause and effect

9 Consider how important roads were to the Roman Empire. List three effects of these roads. How do these three effects combined make roads important? How does each effect support and strengthen the others?

HOW TO

Cause and effect, page 215

key event

What happened at Pompeii?

The thriving Roman city of Pompeii disappeared under volcanic ash in just 24 hours.

In 1748, explorers looking for ancient artefacts rediscovered Pompeii, which had been buried by a volcanic eruption in 79 CE. The volcanic ash that covered Pompeii had acted as a blanket, preserving Pompeii almost exactly as it had been nearly 1700 years earlier.

The buildings were still intact, and everyday objects littered the streets. **Archaeologists** even uncovered loaves of bread and jars of preserved fruit. Skeletons of the people lay where they'd fallen. The city of Pompeii was frozen in time, which has allowed us to find out a lot about everyday life in ancient times.

The town of Pompeii was just nine kilometres from Mount Vesuvius, an active volcano. Pompeii was a resort for Rome's wealthiest citizens. Fancy homes lined its paved streets and people enjoyed cafes, taverns and bathhouses, and watched performances in the local arena. The volcano had not erupted in living memory, so the local people did not see it as a threat.

On 24 August in 79 CE, Mount Vesuvius exploded with a huge eruption that continued for a whole day. Vesuvius spewed massive clouds of ash and rock into the air. The wind carried the ash towards Pompeii and deposited a layer of ash over the town. Following the initial blast came an avalanche of ash, rock and gas (called a pyroclastic flow) travelling at more than 100 kilometres per hour.

Source 1

Eruption of Vesuvius, by Pierre Henri de Valenciennes, 1813, oil on canvas

Ancient historian Pliny the Younger watched the eruption from across the bay:

'Ashes were already falling on the ships, hotter and thicker the nearer they approached; and even pumice and other stones, black and scorched, and cracked by the fire. There had been a sudden retreat by the sea, and the debris from the mountain made the shore unapproachable ... Meanwhile, from many points of Mount Vesuvius, vast sheets of flame and tall columns of fire were blazing, the flashes and brightness of which were heightened by the darkness of the night ... for the walls nodded under the repeated and tremendous shocks, and seems, as though dislodged from their foundations, to be swaying to and fro, first in one direction and then in another. On the other hand, in the open air, there was the fall of the pumice stones (though they were light and burnt out) to be apprehended.

Source 2

TA Schneer, *The History of Vesuvius from AD 79 to AD 1907*, 1907.

What was the eruption of Vesuvius like for the people of Pompeii?

Most people living in Pompeii had time to flee, but those who remained were instantly killed by the heat of the pyroclastic flow, which swallowed everything in its path. The residents of nearby Herculaneum suffered the same fate. The superheated blast of gas would have boiled their brains and incinerated their soft tissue into ash.

By the end of the eruption, 1500 people in Pompeii and Herculaneum had died, and Pompeii was covered in about four to six metres of ash and pumice.

Giuseppe Fiorelli was an Italian archaeologist who directed excavations at Pompeii between 1860 and 1875. His radical new methods helped uncover what the final moments of life were like for the people of Pompeii.

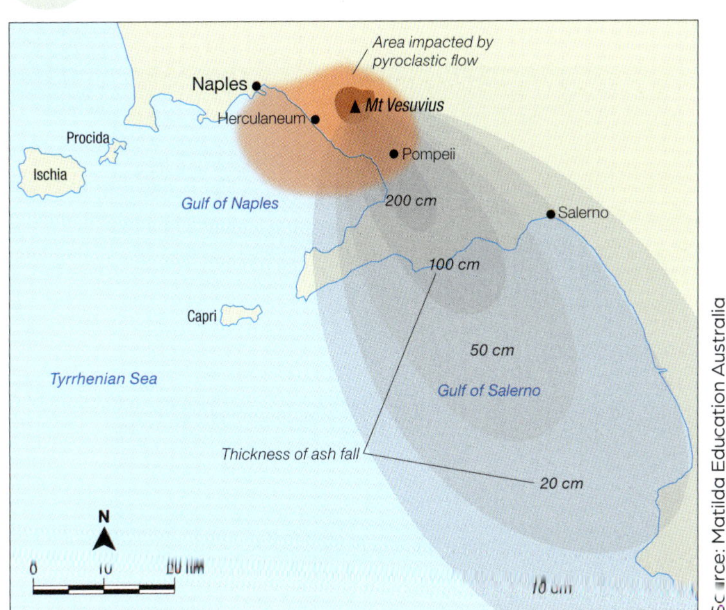

Source: Matilda Education Australia

Source 4

Areas impacted by the pyroclastic flow and ash fall from the eruption of Mount Vesuvius in 79 CE

Source 3

Italian archeologist Giuseppe Fiorelli

Source 5

A plaster cast of a dog. This cast was made by pumping plaster of Paris into cavities in the hardened ash at Pompeii.

Fiorelli realised that the holes he found in the hardened ash were actually the cavities left after the bodies of victims at Pompeii had rotted over time. He developed the use of plaster casts to recreate the shapes of bodies, both human and animal. Whenever a cavity was discovered at the excavation site, plaster of Paris was poured in and left to harden. The hardened ash around the plaster was then carefully removed, revealing a plaster replica of that person or animal at the moment of their death.

Fiorelli's new method of plaster casts gave accurate information about how people had died from the eruption of Mount Vesuvius.

Source 6

Plaster casts of Pompeii victims who died from the superheated blast of ash, rocks and gas

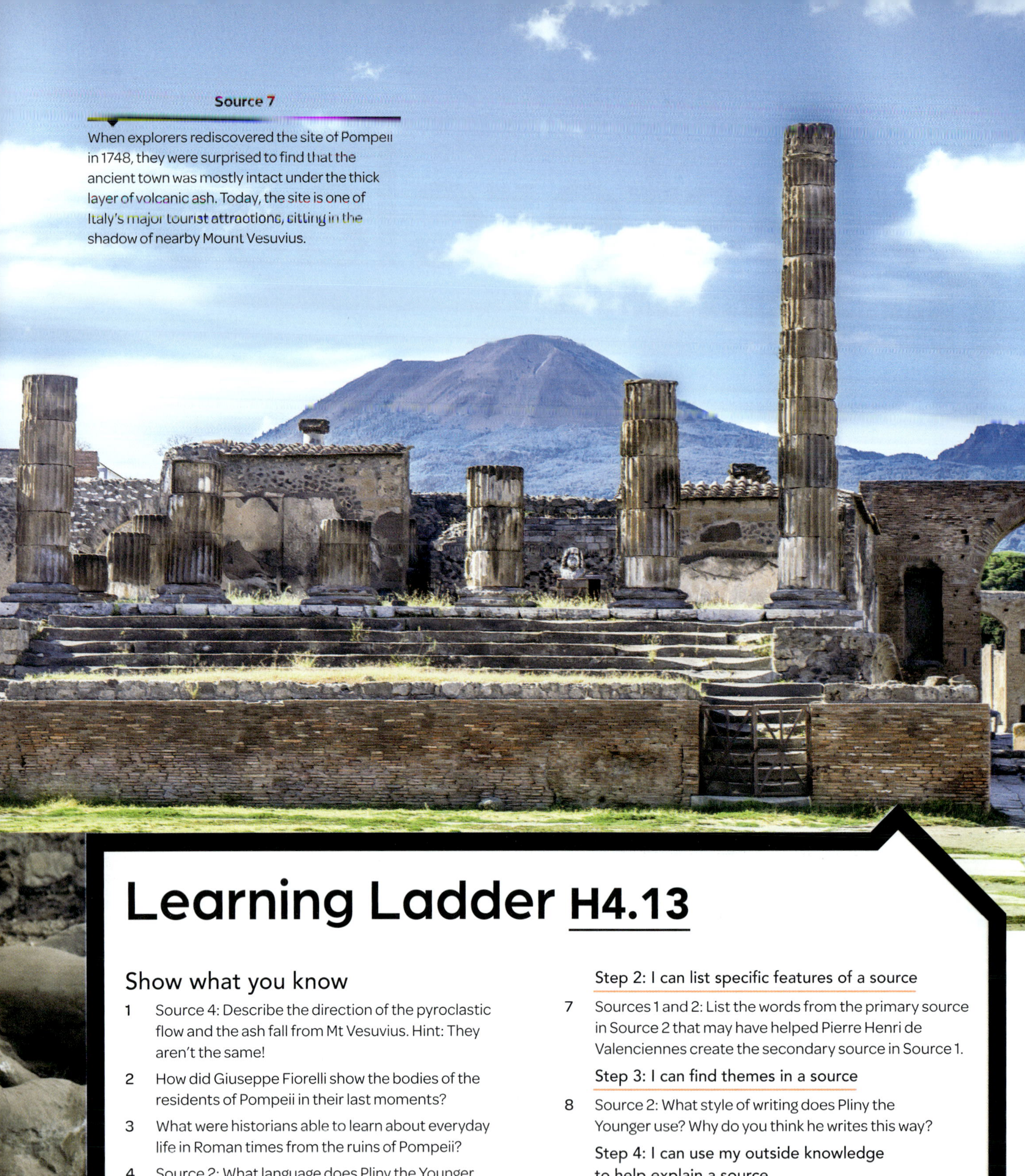

Source 7

When explorers rediscovered the site of Pompeii in 1748, they were surprised to find that the ancient town was mostly intact under the thick layer of volcanic ash. Today, the site is one of Italy's major tourist attractions, sitting in the shadow of nearby Mount Vesuvius.

Learning Ladder H4.13

Show what you know

1 Source 4: Describe the direction of the pyroclastic flow and the ash fall from Mt Vesuvius. Hint: They aren't the same!

2 How did Giuseppe Fiorelli show the bodies of the residents of Pompeii in their last moments?

3 What were historians able to learn about everyday life in Roman times from the ruins of Pompeii?

4 Source 2: What language does Pliny the Younger use to describe the eruption?

5 How could the leaders of Pompeii have protected their people from volcanic eruptions?

Source analysis

Step 1: I can determine the origin of a source

6 Source 6: When was this source created and how was it made?

Step 2: I can list specific features of a source

7 Sources 1 and 2: List the words from the primary source in Source 2 that may have helped Pierre Henri de Valenciennes create the secondary source in Source 1.

Step 3: I can find themes in a source

8 Source 2: What style of writing does Pliny the Younger use? Why do you think he writes this way?

Step 4: I can use my outside knowledge to help explain a source

9 Source 6: Look at the plaster casts of volcano victims. Based on the information you have read in this section, is it more likely these were wealthy people or poor people? Back up your answer.

Source analysis, page 209

HOW TO

Masterclass

Learning Ladder

Caesar (c. 100 BCE–44 BCE)

By Plutarch, 75 CE

‘This love of honour and passion for distinction were inspired into them and cherished in them by Caesar himself, who, by his [generous] distribution of money and honours, showed them that he did not heap up wealth from the wars for his own luxury, [...] all he gave to deserving soldiers as so much increase to his own riches. Added to this also, there was no danger to which he did not willingly expose himself, no labour from which he pleaded an exemption. His contempt of danger was not so much wondered at by his soldiers because they knew how much he [wanted] honour. But his enduring so much hardship, which he did to all appearance beyond his natural strength, very much astonished them. For he was a spare man, had a soft and white skin, was distempered [emotionally disturbed] in the head.’

Source 1

Plutarch, *Cicero*, translated by John Dryden, 1906

Step 1

a I can read a timeline

Use the timeline on pages 130–131 to answer these questions.

- How long after Caesar’s death was the Great Fire of Rome?
- For how many years did Spartacus lead a slave revolt?
- How long did the Punic Wars last?

b I can determine the origin of a source

Source 1: When was this written and who was it written about?

c I can recognise continuity and change

Rank these features of daily life in ancient Rome:

- clothing
- politics
- technology
- religion

d I can recognise a cause and an effect

Copy this table and match each cause with its effect.

Christians refuse to pray to Roman gods or make offerings to previous leaders	Leader is assassinated by senators and civil war follows
Mount Vesuvius erupts	People overthrow leader and start a republic
Popular Julius Caesar is made dictator for life	Army overthrows leader and he commits suicide
Unpopular Emperor Nero rules as a tyrant	Christians were persecuted and killed
Unpopular King Superbus angers the population	Towns of Pompeii and Herculaneum are destroyed

e I can recognise historical significance

Put these events in order, from most important to least important:

- Christianity becomes official religion of Roman Empire
- Great Fire of Rome
- Hadrian’s Wall built to protect against Scots
- Roman Empire split into east and west
- Spartacus slave uprising
- Germanic tribes attack Rome

Work at the level that is right for you or level-up for a learning challenge!

Step 2

a **I can place events on a timeline**

Place these events in order, from earliest to most recent. Note: there are eight different points in time mentioned.

Emperor Hadrian built a wall to defend against the Scots in 121 CE

Rome was invaded by Germanic tribes in 410 CE

66 years after Germanic tribes invaded Rome, the western Roman Empire fell

202 BCE was when Hannibal fought Roman troops at the Battle of Zama

Hannibal first invaded Rome 16 years before the Battle of Zama

The Great Fire of Rome was in 64 CE, when Nero was emperor. He stopped being emperor four years later. He was emperor for 14 years altogether.

b **I can list specific features of a source**

Source 1: What did Caesar do to help his soldiers?

c **I can describe continuity and change**

Do we have divisions in our society such as patricians and plebeians? Explain your answer, giving examples.

d **I can determine causes and effects**

Sort these effects into types of things that caused them: social conditions, political conditions, economic conditions, individual actors or group actors.

- the murder of Caesar
- women don't receive the same education as men
- Colosseum is built
- slavery is widespread
- soldiers overthrow Nero

e **I can explain historical significance**

Why is the murder of Caesar such a big deal?

Step 3

a **I can create a timeline using historical conventions**

Turn the chronological list from Step 2a into a historical timeline using correct conventions.

b **I can find themes in a source**

Source 1: What language does the author use to show what he thinks about Caesar?

c **I can explain why something did or did not change**

What standards did the Romans bring with them to places they conquered, and why were they enforced?

d **I can explain why something is a cause or an effect**

Use the statements about cause and effect to write a short paragraph explaining how the cause led to the effect.

Cause	Effect
Emperor Marcus Aurelius made laws to benefit all, and improved rights for women and slaves	Aurelius was popular and well respected
Roman soldiers were well paid and most young male citizens could join	The Roman army was large and well trained

e **I can apply a theory of significance**

How important was the eruption of Mount Vesuvius at the time? How many people were affected by the eruption? How deeply were their lives affected?

Masterclass

Step 4

a I can distinguish causes, effects, continuity and change from looking at timelines

Which of the events listed on the timeline on pages 130–1 are linked by cause and effect and which show change over time?

b I can use my outside knowledge to help explain a source

What do you know about Caesar that backs up what Plutarch is saying in Source 1?

c I can analyse patterns of continuity and change

Which large one-off events impacted ancient Roman civilisation? How did they affect it?

d I can analyse cause and effect

The western Roman Empire fell in 476 CE. List four causes that contributed to its fall. Suggest a cause for each of those causes.

e I can analyse historical significance

Roman civil engineering, such as the building of arches, domes and roads, was historically significant. Give five reasons why.

Step 5

a I can describe patterns of change

Look at the timeline on pages 130–131 and list one pattern that you can see. What events are part of this pattern? What makes it a pattern of change?

b I can use the origin of a source to explain its creator's purpose

Plutarch wrote biographies of Roman leaders. Why do you think he wrote the one cited in Source 1? Justify your answer using your historical knowledge and quotes from the text.

c I can evaluate patterns of continuity and change

i Was the change from monarchy to republic a good thing or a bad thing? Justify your answer.

ii Was the change from republic to empire a good thing or a bad thing? Justify your answer.

d I can evaluate cause and effect

Was the fall of the Roman Empire a good thing or a bad thing? Back up your answer with evidence from ancient and modern times.

e I can evaluate historical significance

Historians debate the importance of the death of Caesar, the civil war that came after it, and the change from republic to empire. Are these events really that important? Explain your response.

Historical writing

1 Structure

Imagine you are answering an essay question: 'The downfall of Rome was inevitable. Discuss.'

Write an essay plan for this topic. Have at least three main paragraphs.

2 Draft

Using the drafting and vocabulary suggestions on page 226, draft a 400–600 word essay responding to the topic.

3 Edit and proofread

Use the editing and proofreading tips on page 227 to help edit and proofread your draft.

Historical research

4 Organise and present information

Imagine you are doing a large research project: 'The Importance of Geography to Rome'. Write a contents page for this project. There should be an introduction, conclusion, at least four main sections and many subsections. Number your chapters.

Capstone

How can I understand ancient Rome?

In this chapter, you have learnt a lot about ancient Rome. Now you can put your new knowledge and understanding together for the capstone project to show what you know and what you think.

In the world of building, a capstone is an element that finishes off an arch or tops off a building or wall. That is what the capstone project will offer you, too: a chance to top off and bring together your learning in interesting, critical and creative ways. You can complete this project yourself, or your teacher can make it a class task or a homework task.

Scan this QR code to find the capstone project online.

mea.digital/GHV7_H4

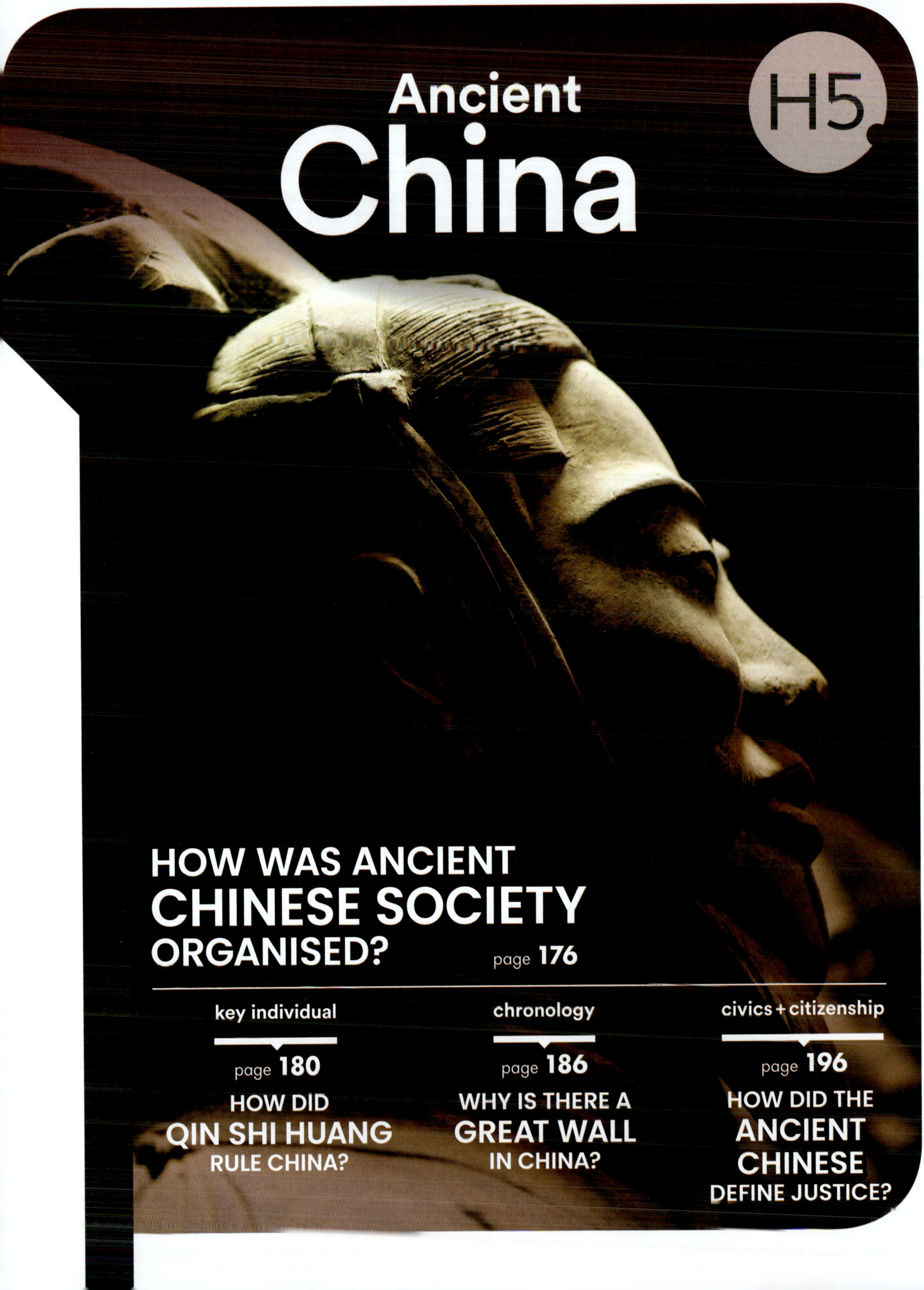

Ancient China

HOW WAS ANCIENT CHINESE SOCIETY ORGANISED?

page 176

key individual

page 180

HOW DID QIN SHI HUANG RULE CHINA?

chronology

page 186

WHY IS THERE A GREAT WALL IN CHINA?

civics + citizenship

page 196

HOW DID THE ANCIENT CHINESE DEFINE JUSTICE?

How can I understand ancient China?

China has made impressive cultural, philosophical, technological and political achievements over thousands of years of history. Chinese values, such as the Confucian ideals of authority, respect and the importance of education are the founding beliefs of over a billion people. With an imposing past and emerging future, understanding this culture is crucial to grasping the modern world

learning ladder

Step	Chronology	Source analysis	Continuity and change
step 5	**I can describe patterns of change** I read timelines and see the 'big picture'. I group timeline events and see if they show patterns of change. I know typical historical patterns to look for.	**I can use the origin of a source to explain its creator's purpose** I combine knowledge of when and where a source was created to answer the question, 'Why was it created?'	**I can evaluate patterns of continuity and change** I answer the question, 'So what?' about patterns of continuity and change. I weigh up different things and debate the importance of a continuity or a change.
step 4	**I can distinguish causes, effects, continuity and change from looking at timelines** I read timelines and find events that are linked by cause and effect. I find things that are the same or different from then until later times.	**I can use my outside knowledge to help explain a source** I have enough outside knowledge about ancient China to help me explain a source.	**I can analyse patterns of continuity and change** I see beyond individual examples of continuity and change in ancient China and identify broader patterns, and I explain why they exist.
step 3	**I can create a timeline using historical conventions** When given a set of ancient Chinese events, I construct a historical timeline, making sure I use correct terminology, spacing and layout.	**I can find themes in a source** I look a bit closer into a source and find more than just features. I find themes or patterns in the source.	**I can explain why something did or did not change** I answer the question 'why?' something changed or stayed the same between historical periods.
step 2	**I can place events on a timeline** When given a list of ancient Chinese events, I put them in order from earliest to latest, the simplest kind of timeline.	**I can list specific features of a source** I look at an ancient Chinese source and list detailed things I can see in it.	**I can describe continuity and change** I have enough content knowledge about two different historical periods to recognise what is similar or different about them, and can describe it.
step 1	**I can read a timeline** I read timelines with ancient Chinese events on them and answer questions about them.	**I can determine the origin of a source** I can work out when and where an ancient Chinese source was made by looking for clues.	**I can recognise continuity and change** I recognise things that have stayed the same and things that have changed from ancient China until now.

Source 1

Yu the Great, c. 2200–2100 BCE. A legendary ruler in ancient China famed for his introduction of flood control, he is said to have protected the Chinese empire from the effects of a mighty deluge by using tortoise shells as drain pipes. He also made a system of irrigation canals that relieved floodwater into fields, and he spent great effort dredging the riverbeds.

Warm up

Chronology

1 Source 1: Roughly how long ago did Yu the Great start his reign?

Source analysis

2 Source 1: Describe the people's clothing in this image.

Continuity and change

3 Do you think flood control is still managed in the same way in China now? How might it be different?

Cause and effect

4 Why did Yu create canals?

Historical significance

5 Source 2: Why do you think Yu was called 'the Great'?

Source 2

Yu the Great

Cause and effect	Historical significance
I can evaluate cause and effect I answer the question 'So what?' about cause and effect. I weigh up different things and debate the importance of a cause or an effect.	**I can evaluate historical significance** I answer the question 'So what?' about things that are supposedly historically important. I weigh up events against each other and cast doubt on how important things are.
I can analyse cause and effect I don't just see a cause or effect as one thing. I determine the factors that make up causes and effects.	**I can analyse historical significance** I separate out the various factors that make something historically important in ancient Chinese history.
I can explain why something is a cause or an effect I can answer 'how?' or 'why?' a cause led to an effect in ancient China.	**I can apply a theory of significance** I know a theory of significance. I use it to rank the importance of ancient Chinese events.
I can determine causes and effects Applying what I have learnt about ancient China, I can decide what the cause or effect of something was.	**I can explain historical significance** I answer the question 'why?' about things that were important in ancient China.
I can recognise a cause and an effect From a supplied list, I recognise things that were causes or effects of each other in ancient China.	**I can recognise historical significance** When shown a list of things from ancient Chinese history, I can work out which are important.

How did geography determine China's destiny?

Because of China's physical isolation, ancient China developed its own culture, and a belief that it was the centre of the world – the Middle Kingdom. Ancient Chinese society was highly organised, with a common written language and cultural values, many of which continue today.

A key reason why China has had a continuous culture is because it is geographically isolated. China is separated from the rest of the world by the Pacific Ocean to the east, thick rainforest in the south, rugged and high mountain ranges to the west and deserts to the north.

key ideas timeline.

- c. 6500 BCE – Agriculture begins along the Yangtze River
- c. 2205 BCE – Chinese start to make bronze
- c. 1600 BCE – Chinese priests engrave signs on bones – earliest example of Chinese writing
- Shang dynasty begins
- 566 BCE – Buddha is born
- 551 BCE – Confucius is born
- c. 513 BCE – Cast iron invented in China

Zhou dynasty 1046–256 BCE

Source 1

Emperor Qin Shi Huang was a significant figure in ancient China. He had terracotta warriors buried near his tomb to protect him in the afterlife. The Terracotta Army contained more than 8000 soldiers, 130 chariots and 670 horses. [Terracotta warrior from the Terracotta Army (3rd century BCE), terracotta, Lintong District, Shaanxi, China.]

Source 2

Geography is one of the key reasons why Chinese culture has survived continuously. China is bounded by an ocean to the east, rainforest in the south, mountains to the west and desert to the north. The Great Wall of China – the world's longest wall – was built to protect China's northern border from invaders.

Natural barriers

Ancient China was isolated from the rest of the ancient world because of its geography, and the earliest recorded evidence of trade between China and Rome is 166 CE.

To the east of China is the world's largest ocean: the Pacific. Thick tropical rainforests are found in border regions to the south (modern-day Myanmar, Laos and Vietnam). Invasion from the west was almost impossible because of the world's highest mountain range: the Himalayas.

Most of China's northern border with Mongolia is covered by the Gobi Desert. For extra protection along its northern border, the Chinese built the world's longest defensive structure – the Great Wall of China – to protect against invasions by **nomadic** tribes (see page 186).

Great river systems

China has two major river systems: the Yangtze and the Yellow (Huang He) rivers. The first villages and towns were built along the banks of these rivers. The rivers helped people to move around the country. Then canals were built to join up rivers, and boats were used to help merchants carry goods between towns, and to help farmers sell their produce. Soldiers were moved to posts along rivers and the riverbanks were the sites of major battles in ancient China.

The Yellow River is also called 'China's Sorrow', as floods in ancient times led to destruction of crops and the deaths of millions of people. Irrigation systems were developed along the Yangtze River during the Han dynasty (206 BCE–220 CE). Irrigation made farms more productive, and helped the Yangtze region became one of the wealthiest regions in China.

Source 3
The Gobi desert

Source 4
Himalayan mountains

Source 5
Laos rainforest

Source 6

Map of east Asia

Learning Ladder H5.1

Show what you know

1 What modern countries border China to the south?

2 Source 6: Match the numbers 1, 2 and 3 on the map with Sources 3, 4 and 5 on page 174.

3 What were the two main river systems used for?

4 Was it a good thing that China was geographically isolated? Give evidence in your response.

Cause and effect

Step 1: I can recognise a cause and an effect

5 Match the following causes with their effects.

Causes	Effects
Himalayas	Impede movement from north
Pacific	Impede movement from south
Jungle	Impede movement from east
Deserts	Impede movement from west

Step 2: I can determine causes and effects

6 Rank the four causes in question 5 from most effective barrier to least effective barrier. Justify your selection.

Step 3: I can explain why something is a cause or an effect

7 Look carefully at Source 6.

a Suggest a positive and negative effect of locating villages along rivers.

b Where is the Yellow River located? Why is it referred to as China's Sorrow?

8 Look carefully at the structure shown on pages 186 and 187.

a What is it?

b What did the builders hope the effect of this structure would be?

c What does the construction of this structure suggest about the effectiveness of deserts as a barrier to movement?

Step 4: I can analyse cause and effect

9 Compare the development of China's distinct culture to the development of city-states in Ancient Greece (pages 92 and 102). What are the common causes and effects in the two examples?

Cause and effect, page 215

H5.2

How was ancient Chinese society organised?

The society of ancient China was a strict hierarchy. At the top was the imperial family: the emperor and his family. Below the imperial family were four social classes – nobles and officials, farmers, artisans (skilled tradesmen) and merchants.

Dynastic rule

The rule of Chinese kings and **emperors** in **dynasties** lasted from the Xia dynasty in 2100 BCE right up until 1912 CE. A dynasty was when one family maintained power, handing down the throne to an heir. Dynasties usually began after a battle for power between rival kingdoms – the victor became the new emperor and started a new dynasty.

In 221 BCE, Qin Shi Huang, leader of the Qin state, defeated six warring states that were fighting for power and territory. He brought all of China under his rule.

Before the Qin dynasty came to rule in 221 BCE, Chinese rulers were known as kings. From the Qin dynasty onwards, rulers were known as emperors. Only one Chinese emperor was a woman – Wu Zetian, who ruled China 690–705 CE.

For nearly 4000 years, ancient Chinese society was ruled by a succession of dynasties. The average length of a dynasty was about 200 years, and this brought great stability and continuity to China. The longest dynasty was the Zhou dynasty, which lasted for 789 years between 1045 BCE and 256 BCE.

Middle Kingdom

People in China referred to their country as the **Middle Kingdom**, because they believed they were living in the centre of the world.

During the Xia dynasty, rulers developed the idea of the **Mandate of Heaven**, which meant that a king could rule as long as his actions were wise and fair. But if he were selfish or cruel, it was a sign the gods had withdrawn his mandate (or authority) to rule. Natural disasters such as famines and floods were also signs of heaven's displeasure with the king. People would often revolt against their rulers after major disasters, which they saw as signs that the gods had withdrawn the king's mandate to rule.

Dixin was the last king of the Shang dynasty (1766 BCE–1045 BCE). He inflicted torture and hardship on his people. In 1045 BCE, King Wen and his allies claimed that Dixin had lost the Mandate of Heaven – and thus lost the right to rule. King Wen overthrew the Shang dynasty and established the Zhou dynasty.

The emperor

Emperors of ancient China had tremendous power and responsibility. The emperor was the 'Son of Heaven', and had a Mandate of Heaven to rule. The emperor lived apart from ordinary people, secluded in the imperial palace.

No one was allowed to speak directly to the emperor – or to even look at him. When anyone approached the emperor they had to perform a ***kowtow*** – kneeling and bowing so low that their head touched the ground. Everyone had to keep their head bowed, even if they were addressing the emperor. When the emperor left the imperial palace, the roads were cleared of people and he was carried in a screened carriage so that no one would see him.

Source 1

Empress Wu Zetian was the only female emperor in Chinese history. At the age of 14 she entered the imperial palace as a concubine (secondary wife) of Emperor Taizong of the Tang dynasty. After Taizong died, Wu Zetian became concubine to the new emperor, Gaozong. After Gaozong died, she had any rivals to the throne removed and took control of the government. Wu Zetian was empress of China 690–705 CE.

The only other males allowed in the imperial palace were castrated males called **eunuchs**, who had had their sexual organs removed. This was to make sure that any children born to the empress – or to any of the **concubines** – were fathered by the emperor.

Social order in ancient China

In ancient China, the social order was a hierarchy, with clear divisions between the social classes. Beneath the emperor and imperial family there were four social groups: nobles and officials, farmers, artisans and merchants.

- Nobles and officials were known as the *shi*. This class included the emperor's relatives, commanders of the army, lords of conquered kingdoms and government officials.
- Farmers were known as the *nong*. They were not wealthy, but were highly respected for their ability to feed the population. Farmers could live on the land in return for working on it. They used food to pay their tax, and had to serve as soldiers or labour on building projects when required.
- Artisans were known as the *gong*. Artisans earned more than farmers. Their skills as carpenters, metal workers, painters, potters and jewellers were highly regarded, as they created essential goods that people needed. An artisan's skills were usually handed down from father to son.
- The merchant class were known as the *shang*. They were traders and moneylenders. They were on the lowest level of the social hierarchy because they did not produce anything, and only worked for their own gain. Some merchants bought land to farm in a bid to improve their social status.

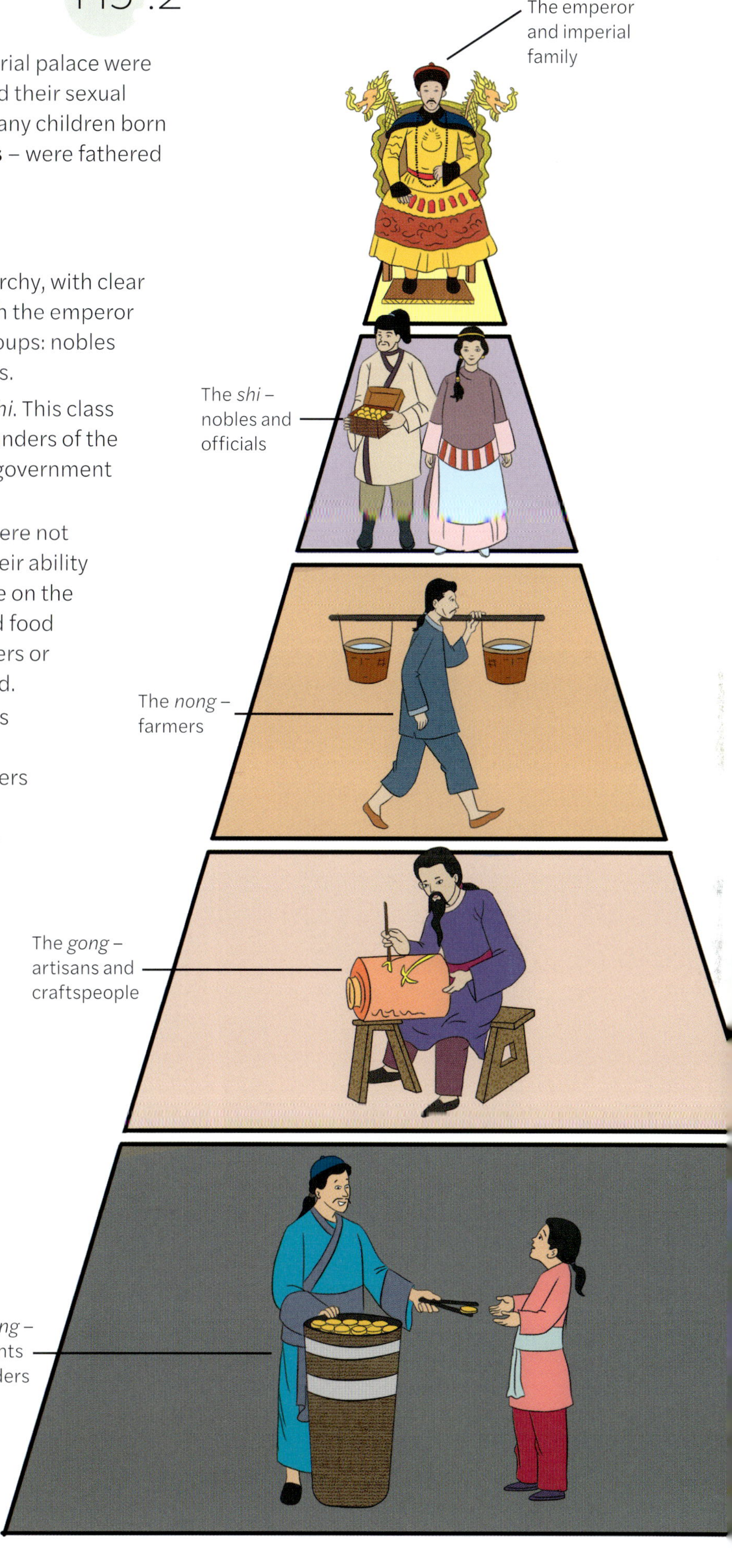

Source 2

Social hierarchy in ancient China

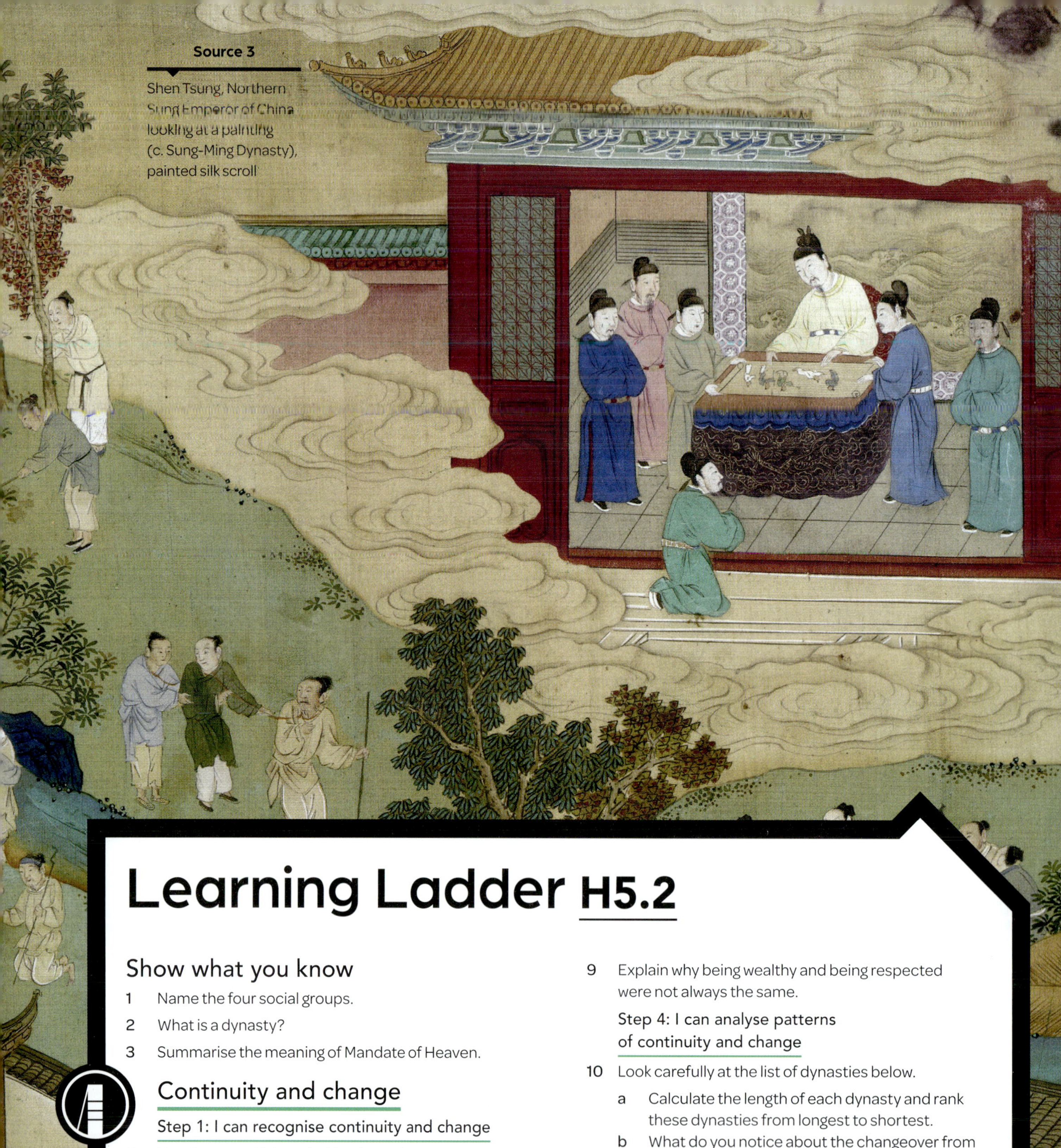
Source 3

Shen Tsung, Northern Sung Emperor of China looking at a painting (c. Sung-Ming Dynasty), painted silk scroll

Learning Ladder H5.2

Show what you know

1 Name the four social groups.

2 What is a dynasty?

3 Summarise the meaning of Mandate of Heaven.

Continuity and change

Step 1: I can recognise continuity and change

4 How long was China ruled by dynasties?

5 How did a new dynasty begin?

Step 2: I can describe continuity and change

6 How did the Mandate of Heaven both ensure continuity and allow change in ancient China?

7 How was respect shown to the emperor to recognise his position in society?

Step 3: I can explain why something did or did not change

8 How did dynasties ensure continuity in China?

9 Explain why being wealthy and being respected were not always the same.

Step 4: I can analyse patterns of continuity and change

10 Look carefully at the list of dynasties below.

a Calculate the length of each dynasty and rank these dynasties from longest to shortest.

b What do you notice about the changeover from the Zhou to the Qin Dynasty? Research what happened during this time and how it was resolved.

Dynasty	Start	End
Xia	2070 BCE	1600 BCE
Shang	1600 BCE	1046 BCE
Zhou	1046 BCE	256 BCE
Qin	221 BCE	206 BCE
Han	206 BCE	220 CE

Continuity and change, page 212

HOW TO

H5.3

key individual

How did Qin Shi Huang rule China?

In 221 BCE, King Zheng of Qin defeated six small kingdoms to unify China for the first time. He renamed himself Qin Shi Huang, meaning 'First Emperor'. The Qin dynasty lasted only 15 years but Emperor Qin Shi Huang introduced great changes in that time.

A strong and organised leader

The Qin dynasty lasted only 15 years – but it is one of the most famous dynasties in Chinese history. Qin Shi Huang needed to unite China and restore peace. To protect against invasion from the north by the Xiongnu tribes, he started building the Great Wall of China.

Qin Shi Huang quickly established control over the conquered states by dividing the country into 36 administrative provinces, each with a governor that reported to him. He developed a single set of laws and a system of roads to link the provinces with the Qin dynasty capital of Xianyang. To make it easier to have centralised control, Qin Shi Huang also established a common currency, common units of measurement and standardised writing.

Qin Shi Huang ran his dynasty with absolute control. He punished those who disagreed with him, and banned the philosophy of Confucianism (see page 192). He ordered that all books be burnt, except those about farming, medicine or divination (predicting the future). Qin Shi Huang wanted history to begin with the rule of the Qin dynasty. Any scholars who did not bring their books to be burnt were killed. People who spoke of old laws or sang old songs were executed.

Qin Shi Huang also ordered that stone tablets inscribed with his achievements be set up around the country. Sima Qian, a Chinese historian of the early Han dynasty (206 BCE–220 CE), recorded the details on one of the tablets:

Source 1

Qin Shi Huang became China's first emperor in 221 BCE. [Portrait of Qin Shi Huang (18th century CE).]

'The Emperor came to the throne and made laws which all the people obeyed.

He inspected the common people in the distant parts and climbed Mount Tai (a sacred mountain).

His obedient people remember his achievements and his goodness.

Under him all things find their right place.

The wise Emperor who controls everything under heaven wakes early, sleeps late and does wise things which have a good effect for a long time.

His power is endless. He is obeyed and his commands will continue forever.'

Source 2

Qin Shi Huang's tablet inscription

Source 3

This painting shows the burning of the books and burying of scholars who disagreed with Chinese Emperor Qin Shi Huang in 213 BCE. [Qin Shi Huangdi, the first Qin Emperor, watercolour on silk.]

A terracotta army

Qin Shi Huang wanted to live forever. He searched for the elixir of life – a magic potion that would give him immortality. To make sure he was equipped in the afterlife, Qin Shi Huang built a tomb complex measuring two kilometres long and one kilometre wide that included a mausoleum where he was buried, an underground palace and buildings for the precious treasures he wanted to have in the afterlife. One and a half kilometres away, he buried an army of life-sized terracotta warriors to command in the afterlife: 8000 soldiers, 130 chariots, 520 horses and 150 cavalry horses. Farmers discovered the site in 1974.

Ancient Chinese historian Sima Qian recorded the construction of the burial site, which took 39 years.

'As soon as the First Emperor became King of Qin, excavations and building were started at Mount Li, while after he won the empire more than 700 000 conscripts from all parts of the empire worked there. They dug through three streams and poured molten copper for the outer coffin, and the tomb was fitted with models of palaces, pavilions and offices, as well as fine vessels, precious stones and rarities.'

Source 5

Description of Qin Shi Huang's burial site by Chinese historian Sima Qian of the early Han Dynasty (206 BCE–220 CE), quoted in Hsien yi and Gladys Young, *Records of the Great Historian (Shi Ji)*, Sima Qian, Peking, 1979

Source 4

Each of the 8000 soldiers in the Terracotta Army is life-sized. The officers are taller than regular soldiers, and the generals are the tallest of them all. There are ten basic face shapes, but each soldier has a different expression. The site is separated into pits, divided by walls of rammed soil. The soldiers are arranged in columns in battle formation. The Emperor wanted the same army that had brought him to power in life to protect him in the afterlife. [Terracotta Army (3rd century BCE), Lintong District, Shaanxi, China.]

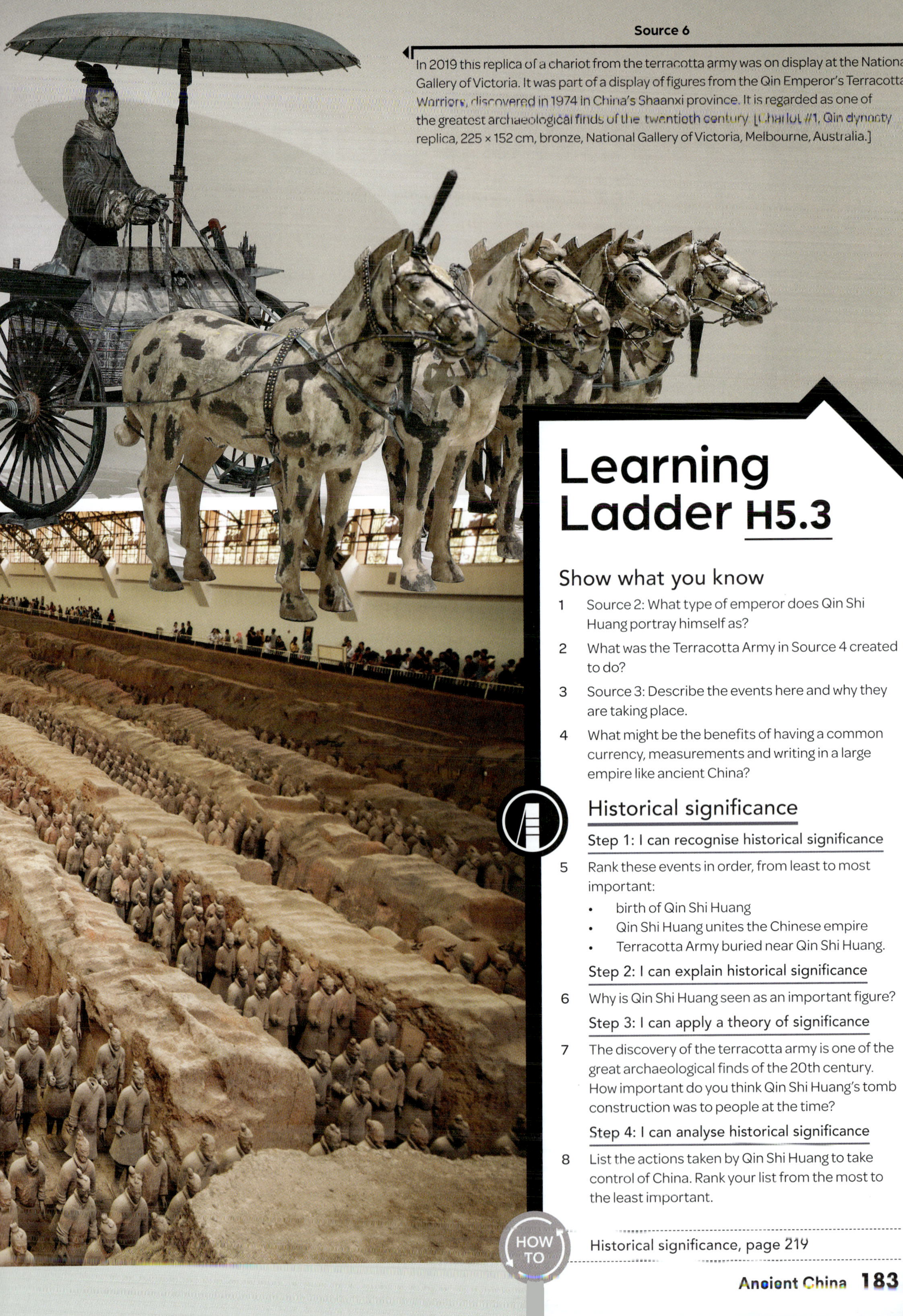

Source 6

In 2019 this replica of a chariot from the terracotta army was on display at the National Gallery of Victoria. It was part of a display of figures from the Qin Emperor's Terracotta Warriors, discovered in 1974 in China's Shaanxi province. It is regarded as one of the greatest archaeological finds of the twentieth century [Chariot #1, Qin dynasty replica, 225 × 152 cm, bronze, National Gallery of Victoria, Melbourne, Australia.]

Learning Ladder H5.3

Show what you know

1. Source 2: What type of emperor does Qin Shi Huang portray himself as?
2. What was the Terracotta Army in Source 4 created to do?
3. Source 3: Describe the events here and why they are taking place.
4. What might be the benefits of having a common currency, measurements and writing in a large empire like ancient China?

Historical significance

Step 1: I can recognise historical significance

5. Rank these events in order, from least to most important:
   - birth of Qin Shi Huang
   - Qin Shi Huang unites the Chinese empire
   - Terracotta Army buried near Qin Shi Huang.

Step 2: I can explain historical significance

6. Why is Qin Shi Huang seen as an important figure?

Step 3: I can apply a theory of significance

7. The discovery of the terracotta army is one of the great archaeological finds of the 20th century. How important do you think Qin Shi Huang's tomb construction was to people at the time?

Step 4: I can analyse historical significance

8. List the actions taken by Qin Shi Huang to take control of China. Rank your list from the most to the least important.

HOW TO

Historical significance, page 219

What was palace life like?

Emperors lived apart from ordinary people. They lived in grand palaces, surrounded by walls and guard towers. Emperors were entertained by secondary wives, known as concubines, and looked after by eunuchs (castrated servants) to make sure that the emperor was the only father of any children.

Grand palaces

Emperor Qin Shi Huang built several elaborate palaces, the largest being the Epang Palace. The palace was 400 metres wide and more than one kilometre long and could hold 10 000 people. More than 700 000 workers laboured to complete it.

Chinese palaces were filled with luxuries, and servants and attendants would meet the emperor's every wish. In 1299 CE, Italian explorer Marco Polo described the imperial palace built by the Yuan dynasty (1279–1368 CE):

'The walls inside are covered with silver and gold and there are paintings of horsemen, dragons and every kind of bird and animal. The vaulted ceiling is also covered with paintings and gold ornamentation. The main reception room can seat more than 6000 people. There is an overwhelming number of rooms; no architect could have designed the palace better. The roof is beautifully painted in many colours – vermillion, green, blue, yellow and so forth – so that it shines like a jewel and can be seen from afar ...'

Source 1

Marco Polo's description of the imperial palace of the Yuan Dynasty. [William Marsden, *The Travels of Marco Polo, the Venetian* (1908 CE).]

Life in the palace

Chinese emperors spent nearly all of their time in their palaces, along with the empress, a harem of concubines and thousands of eunuchs.

The empress was the emperor's official wife, but he also had many concubines whose job it was to keep him entertained and amused. The women in the palace were divided into different grades. The empress was First Grade, and only a son produced by the empress could inherit the throne.

The emperor (and some rich men) also had a harem of concubines. It was an honour to be an emperor's concubine, and wealthy men sent their daughters to the imperial palace, hoping they would catch the attention of the emperor. Concubines entered the palace at a young age – often 13 years old – and were not allowed to leave.

Eunuchs were the male servants of the palace. They did the cooking, cleaning and gardening, as well as teaching the children and guarding the empress and concubines. Eunuchs were usually from poor families, yet sometimes became trusted and powerful members of the imperial court.

'Chambers where the emperor's private property is placed, such as his treasures of gold, silver, gems, pearls, and gold plate, and in which reside the ladies and secondary wives. These rooms are only for him, and no one else has access to them.'

Source 2

Marco Polo's description of the Forbidden City. [Henry Yule, *The Book of Ser Marco Polo*, (1903 CE).]

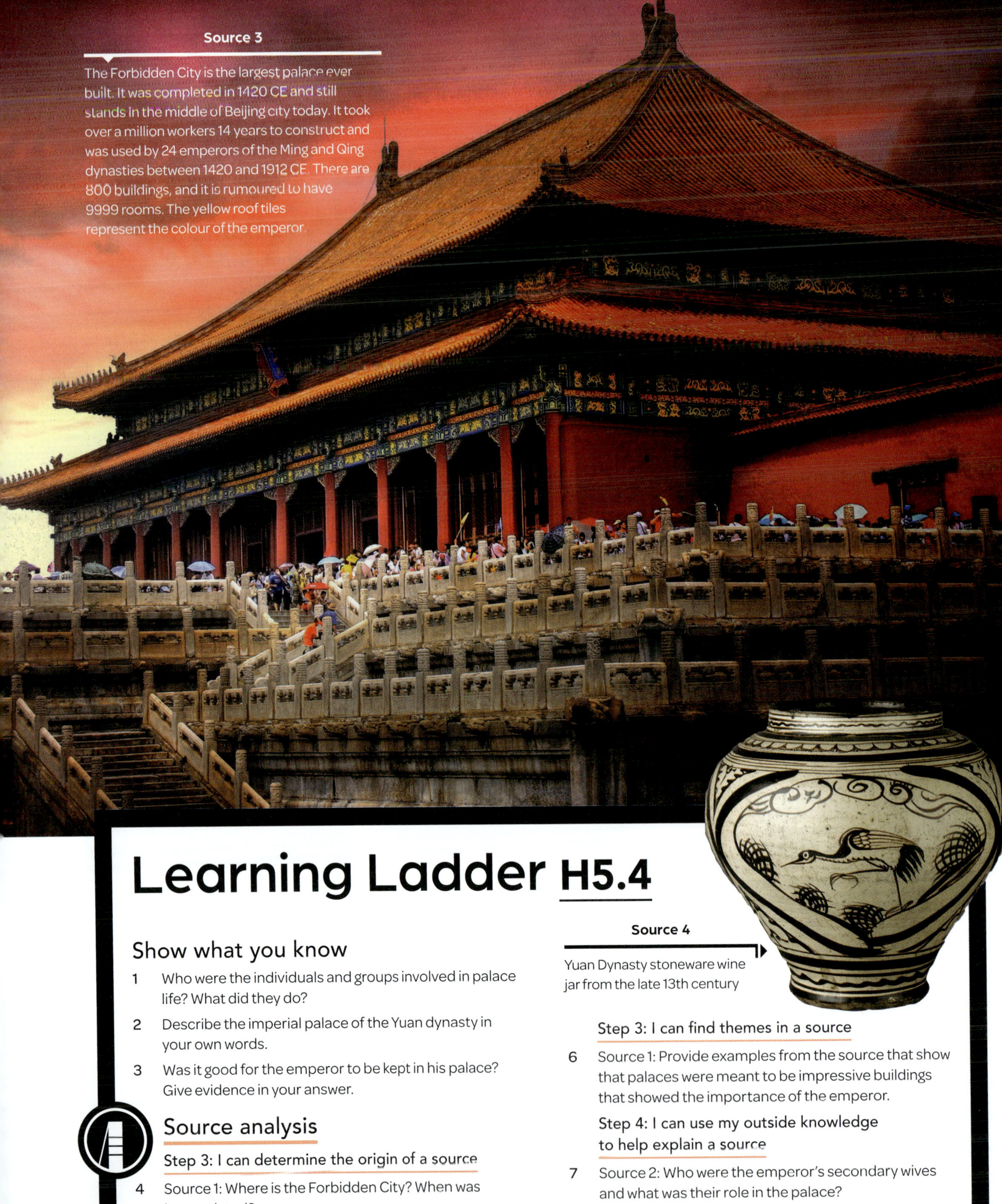

Source 3

The Forbidden City is the largest palace ever built. It was completed in 1420 CE and still stands in the middle of Beijing city today. It took over a million workers 14 years to construct and was used by 24 emperors of the Ming and Qing dynasties between 1420 and 1912 CE. There are 800 buildings, and it is rumoured to have 9999 rooms. The yellow roof tiles represent the colour of the emperor.

Learning Ladder H5.4

Source 4

Yuan Dynasty stoneware wine jar from the late 13th century

Show what you know

1 Who were the individuals and groups involved in palace life? What did they do?

2 Describe the imperial palace of the Yuan dynasty in your own words.

3 Was it good for the emperor to be kept in his palace? Give evidence in your answer.

Source analysis

Step 3: I can determine the origin of a source

4 Source 1: Where is the Forbidden City? When was it completed?

Step 2: I can list specific features of a source

5 Source 3: Give a detailed description of the artistic and architectural style of the Forbidden City palace. Include its colour, shape, patterns, size and structure.

Step 3: I can find themes in a source

6 Source 1: Provide examples from the source that show that palaces were meant to be impressive buildings that showed the importance of the emperor.

Step 4: I can use my outside knowledge to help explain a source

7 Source 2: Who were the emperor's secondary wives and what was their role in the palace?

8 Who were the eunuchs and why were they trusted to be in the palace?

Source analysis, page 209

HOW TO

Why is there a Great Wall in China?

The Great Wall of China is a 21 196-kilometre-long wall that was built to protect China's northern border from invaders. The wall began as a number of smaller walls and fortifications, which were first built around 771 BCE during the Zhou dynasty (1046–256 BCE). Construction on the Great Wall began in 206 BCE, when China's first emperor, Qin Shi Huang, ordered that the existing fortifications be joined so that a single strong wall with lookout towers would protect his empire against invasions by nomadic tribes. The first part of Qin's wall was completed in 206 BCE but the structure we see today was rebuilt during the Ming Dynasty (1368–1644 CE).

Bamboo scaffolding
Bamboo was used to build scaffolding, which helped the workers build the high wall and watchtowers. Bamboo is a very strong grass that is still used for scaffolding today.

Roadway
The top of the Great Wall acted as a road, so troops could quickly move to any point along the wall. The road is six metres wide in places and could fit five horsemen abreast.

Army of workers
Millions of soldiers, farmers, convicted criminals and slaves built the wall, with construction taking more than 1800 years. Guards forced the builders to work hard, and stopped them from escaping.

Building the wall
Local materials were used to build the wall. The outer brick walls were made from local mud. The section between the outer walls was filled with local soil, sand, stones and twigs, transported in baskets. It also contains the bodies of thousands of workers who died working on the wall.

Battlements
Raised sections of brick wall, called battlements, were built during the Ming dynasty. Gaps were left at the very top of the battlement so archers could shoot arrows at enemies.

Great Wall
The completed Great Wall of China was about seven metres tall and stretched for 21 196 kilometres in northern China.

Beacon towers
Watchtowers were built along the wall so guards could watch for invaders. Guards would send signals if they saw invaders, using smoke signals during the day and lights at night.

Deadly work
Accidents were common during construction on the wall. Working conditions were difficult, and at least 100 000 men died.

Source 1
Illustration of the Great Wall of China

Learning Ladder H5.5

Show what you know

1 Under which dynasty did the Great Wall begin?
2 Who built the wall? What dangers did they face while building it?
3 Why was the wall built?
4 Compare how the wall was built to how it might be built today.

Chronology

Step 1: I can read a timeline

5 Use the timeline on page 172:
   a Emperor Qin Shi Huang ordered that a strong wall be built. When was the Qin Dynasty?
   b When was the first part of the wall completed?
   c In which dynasty was the first part of the Great Wall built?

Step 2: I can place events on a timeline

6 Put these events in order, using the same list format as page 207:
   - the Qin dynasty ended in 206 BCE, but it had lasted 15 years
   - the Three Kingdoms Period started in 220 CE and lasted 60 years
   - the Warring States Period started in 476 BCE and ended when the Qin dynasty began.

Step 3: I can create a timeline using historical conventions

7 Using correct conventions, create a timeline with these events on it:

   Parts of the Great Wall may collapse in 2040

   Great Wall named a World Heritage Site by UNESCO in 1987

   The wheelbarrow was invented in China in 118 CE and used to build the Great Wall

   In 2003, the first Chinese man in space said the Great Wall *cannot* be seen from space.

Step 4: I can distinguish causes, effects, continuity and change from looking at timelines

8 Look at the timeline on page 172. Which events added to or broke down China's isolation?

HOW TO
Chronology, page 206

What was life like in a Han dynasty city?

The Han dynasty was a time of peace and prosperity. Chinese cities became the largest and most magnificent in the world. The cities were lively, organised and controlled places where people lived according to their occupation.

The Han dynasty

The Han dynasty was one of China's longest ruling dynasties, lasting for over 400 years. In 202 BCE, the Han army seized control of China following eight years of fighting after the death of Emperor Qin Shi Huang in 210 BCE and the surrender of the last Qin ruler in 206 BCE. Their leader was a peasant named Liu Bang, who renamed himself Gaozu. He was the first peasant to become emperor.

Emperor Gaozu began to restore the people's faith in leadership after they had lived in fear during the Qin dynasty. He did this by:

- disbanding his armies
- excusing men with families from military service
- appointing people to jobs based on exams rather than who their parents were
- reducing taxes
- making laws easier to understand
- promoting Confucianism
- attempting to replace books destroyed by Qin Shi Huang.

Source 1

The Chinese city of Chang'an during the Han dynasty. The city was surrounded by a 12-metre high wall that stretched for 25.7 kilometres.

City life

The walled city of Chang'an was vibrant and crowded. The roads were filled with merchants, shopkeepers, craftsmen, officials, entertainers, soldiers and beggars.

The marketplace was just inside the city gates, which was ideal for travelling merchants. Goods were sourced from far and wide by merchants travelling on the trading route called the **Silk Road** (see page 190). People bought and sold animals and food – including exotic foods such as turtles, snails, panther and owl. Craftsmen sold paintings, pottery, carvings and jewellery. Entertainment was provided by musicians, jugglers and acrobats.

Inside the town walls were further walls that divided the city into sections, called wards. Each ward was identified by the occupation of its residents. For example, all artists' and craft shops were in the same ward. Wards were locked at night and people had a curfew – which meant they had to stay indoors after a set time.

The palace was located in the north of the city. High walls were used to limit access to imperial palace buildings (see page 184), administrative offices, military barracks and storehouses.

Peasants lived in alleyways, in shacks with thatched roofs and walls of mudbrick. They had a single room for living, eating and sleeping, and used shared toilets.

Learning Ladder H5.6

Show what you know

1 What was bought and sold in the market in Chang'an?

2 What did ancient Chinese cities look like?

3 Compare the living conditions of peasants and people living in the palace.

Continuity and change

Step 1: I can recognise continuity and change

4 Describe at least two things that are similar and two things that are different about your nearest capital city and Chang'an.

Step 2: I can describe continuity and change

5 Describe how a typical Chinese home would have had to change to become a palace building.

Step 3: I can explain why something did or did not change

6 What changes did Emperor Gaozu make to restore faith in leadership following the harsh conditions of the Qin dynasty?

Step 4: I can analyse patterns of continuity and change

7 How was the population controlled by the government to ensure stability and continuity?

Continuity and change, page 212

Wooden homes
Unlike other cities in the world, Chinese cities were made of wood. They were quick to build, but easily destroyed by fire.

Courtyards
Houses of wealthy people were built around a courtyard, known as the 'well of heaven'. Around the courtyard were the bedrooms, kitchen, storage room and women's quarters.

Shops
Shops sold goods from all over the world. Merchants used umbrellas – a Chinese invention – to protect their goods from the sun.

How did the Silk Road connect the world?

The Silk Road is the name of the trade route that developed from eastern China to the Mediterranean Sea during the Han dynasty. The journey of Zhang Qian in 138 BCE – when he was searching for an ally – opened up China to the rest of the world, allowing it to begin trading with other lands.

The quest for horses

Trade beyond China's borders began in the Han dynasty. In 138 BCE, a diplomat named Zhang Qian was sent on the long and hazardous journey to Central Asia to find allies who would help China fight against the Xiongnu.

However, Zhang Qian was captured by the Xiongnu, and held captive for 10 years.

Source 1

A bronze sculpture of a Dayuan horse from the second century CE. Dayuan horses had an infection that made them sweat blood. The Chinese considered this a sign that the horses were marked by heaven, making them 'heavenly horses'. [*Flying Horse of Gansu* (25-220 CE), bronze, Gansu Provincial Museum, Lanzhou, China.]

After escaping and arriving back in China, Zhang Qian described to Emperor Wudi the people he called the Dayuan, who lived in the Ferghan Valley with their horses:

> 'The people are settled on the land, plowing the fields and growing rice and wheat. They also make wine out of grapes. The region has many fine horses, which sweat blood; their forebears are supposed to have been foaled from heavenly horses.'

Source 2

Sima Qian, *Records of the Grand Historian*, Han Dynasty, 233 CE, translated from the *Shih chi of Ssu-ma Ch'ie* by Burton Watson, Columbia University Press, 1971

Because of continual warfare among **empires** in Asia, maintaining an empire depended on being able to fight on horseback. The Dayuan horses were a special breed: they were taller, stronger and faster than Chinese horses.

Emperor Wudi ordered Zhang Qian to trade silk for 1000 horses and bring them back along the Silk Road. However, the Dayuan king refused to trade.

Emperor Wudi then sent large armies of up to 60 000 men to fight the Dayuan and seize the horses. After defeating the Dayuan, 3000 horses were brought back along the Silk Road to China, although only 1000 horses survived the journey. Later, with the help of the Dayuan horses, China defeated the Xiongnu.

Trade along the Silk Road

The Silk Road was a 6500-kilometre network of routes over land and sea that connected China to Asia, Europe and the Mediterranean. Camel caravans carried goods along the route through the mountain ranges and deserts of southern Asia.

Ancient China's most important **export** was silk. Silk was highly prized by people in other countries. For a long time, spinning silk from the cocoons of silkworms was a skill known only by the Chinese – and people were executed if they were caught stealing silkworm eggs or cocoons. A thin, strong form of pottery called porcelain was also in demand. Merchants also exported umbrellas, paper, medicines, perfumes, tea, rice, cinnamon, ginger, bronze weapons and mirrors.

Horses and camels were **imported** from central Asia, along with luxury goods such as gold, silver, gemstones and glassware. Exotic items such as grapes, watermelons, peaches, leopards and lions also reached China, along with fabrics, spices, dyes and ivory from India.

The Silk Road also became important for introducing new ideas into China, such as Buddhism (see page 192) and new technologies for making metal. China's secrets for making silk were eventually passed on to others along the Silk Road. The Silk Road also exposed traders to new diseases, such as bubonic plague, which spread along the route.

Source 3

Tea was one of the many items traded on the Silk Road.

Source 4

Map showing the Silk and Spice routes

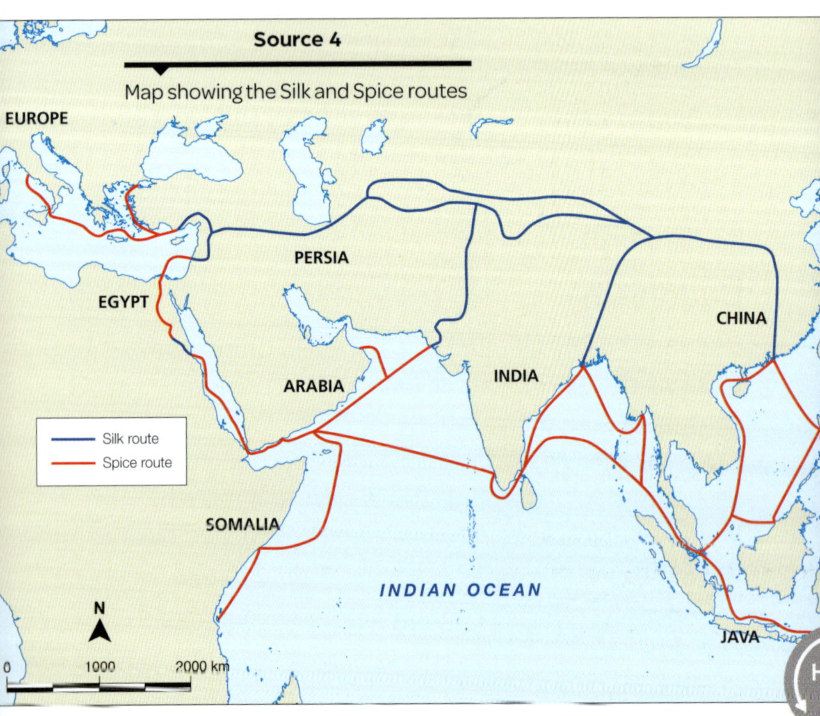

Source: Matilda Education Australia

Learning Ladder H5.7

Show what you know

1 Which diplomat brought information about Dayuan horses back to China?

2 Why did Zhang Qian want horses?

3 Why was silk so precious to China?

4 Compare the exports and imports to and from China via the Silk Road. Which would be more useful to an ancient empire?

Cause and effect

Step 1: I can recognise a cause and an effect

5 Which of these was an effect of trade along the Silk Road?
- Confucianism increased in popularity in China
- European colonisation
- Emperor Wudi went to war
- Islam spread to China
- Sharing of ideas

Step 2: I can determine causes and effects

6 What led to (caused) Zhang Qian being captured?

7 What was an effect of Chinese secrets being shared along the Silk Road?

Step 3: I can explain why something is a cause or an effect

8 Why were camels used to transport goods along the Silk Road and where were they sourced from?

9 Why was silk China's most important export and how was this market protected?

Step 4: I can analyse cause and effect

10 Look carefully at Sources 1 and 2:

a Why do you think the Dayuan King refused to trade for the horses?

b What was the effect of the stalemate in the trade negotiations for the Dayuan horses?

HOW TO

Cause and effect, page 215

What did the ancient Chinese hold sacred?

Chinese people worshipped many different gods. They also followed the teachings of Confucius and Buddha, which encouraged peace and respect for others.

The three ways

Many of the ideas and events in ancient China were shaped by the three major religions or philosophies: **Taoism**, **Confucianism** and **Buddhism**. Taoism was a religion, whereas Confucianism and Buddhism were philosophies about how to live your life.

Closely tied in with the three major beliefs were customs relating to how people were buried and how they arranged their homes – and even ceremonies about how to prepare and share a pot of tea.

Ancestors, gods and goddesses had been worshipped by the Chinese since the Shang dynasty (c. 1600–1046 BCE). The gods were thought to control natural occurrences such as the weather. Natural disasters such as earthquakes or floods were signs that the gods were unhappy. Incense was burned and food offered to keep the gods and their ancestors happy.

Taoism

The founder of Taoism is believed to be Laozi (604–531 BCE). The Tao, or 'the way', is a natural force in the universe that you can live by. Laozi preached that happiness came from living a simple life in harmony with nature. Through a balanced diet and through exercise and breathing techniques such as **kung fu**, you could live forever.

In Taoism all living things are two-sided: the ***yin*** and the ***yang***. The *yin* was the female force: dark, cool and passive. The *yang* was the male force: light, hot and active. When *yin* and *yang* were in balance, there was harmony in the universe. When they were out of balance, famines, disease and natural disasters occurred.

Confucianism

Confucius (551–479 BCE) lived at the same time as Laozi. Confucius taught that people should be honest, brave and knowledgeable and never violent or arrogant. The path to happiness lay in obeying the law, doing one's duty and respecting older people – especially parents and grandparents.

Confucius travelled for many years and had a great influence on Chinese beliefs, because even if people became Taoists or Buddhists, they still followed the teachings of Confucius. Here are some of his teachings:

- Do not do unto others what you would not want others to do to you.
- Choose a job you love, and you will never have to work a day in your life.
- Learning without thinking is useless. Thinking without learning is dangerous.
- Our greatest glory is not in never falling, but in rising every time we fall.

Buddhism

Buddhism began in India and arrived in China from India via the Silk Road (see page 191) in the first century CE. Buddhists believe that after people die, they are reincarnated – reborn into another life. Living a proper life would stop the cycle of rebirth, and this stage, called *nirvana*, could be reached by leading a good life and no longer wanting things.

Buddhist temples contain a statue or painting of Buddha and are places of spiritual reflection and meditation. Buddhist monks give up their possessions to lead a life of poverty.

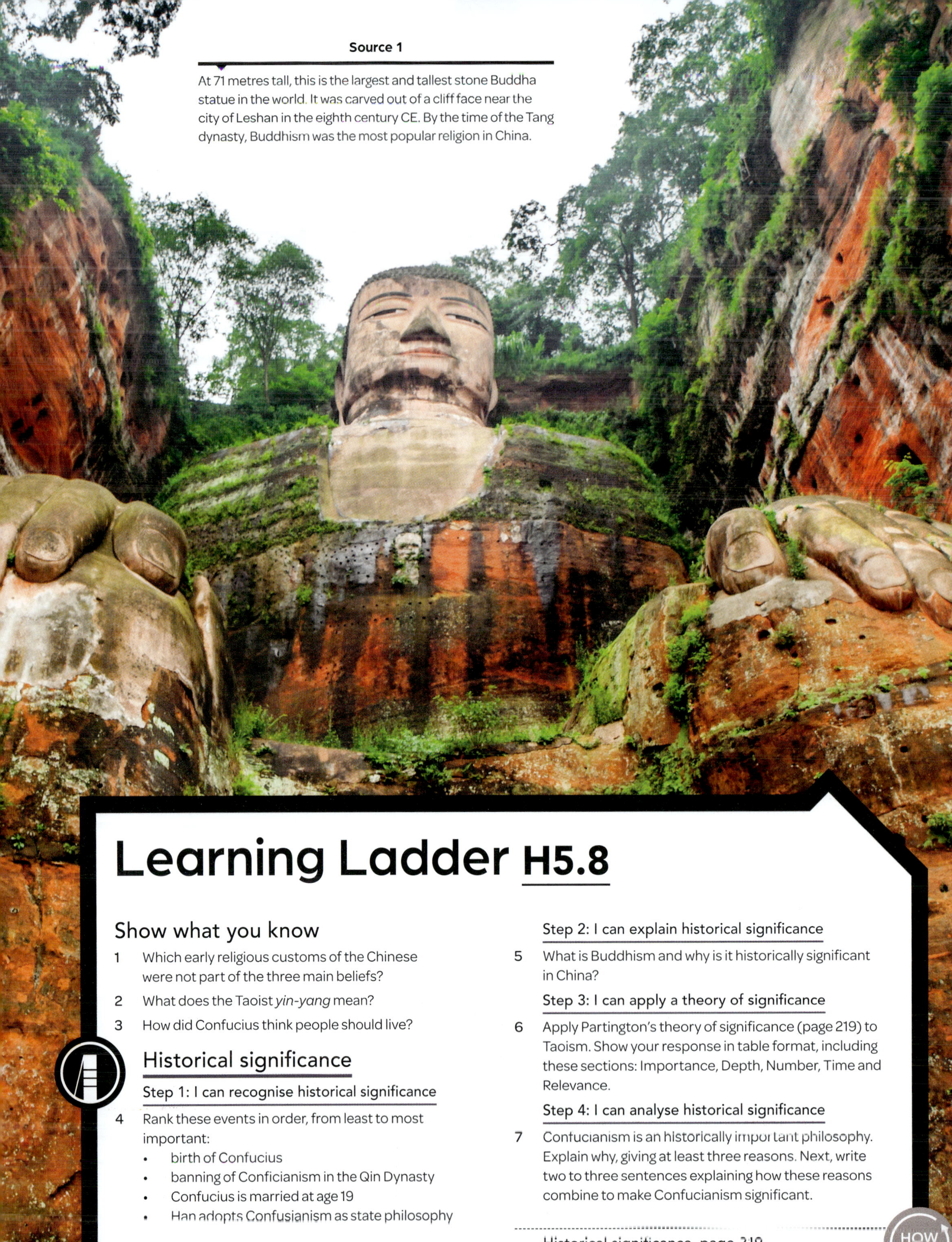

Source 1

At 71 metres tall, this is the largest and tallest stone Buddha statue in the world. It was carved out of a cliff face near the city of Leshan in the eighth century CE. By the time of the Tang dynasty, Buddhism was the most popular religion in China.

Learning Ladder H5.8

Show what you know

1 Which early religious customs of the Chinese were not part of the three main beliefs?

2 What does the Taoist *yin-yang* mean?

3 How did Confucius think people should live?

Historical significance

Step 1: I can recognise historical significance

4 Rank these events in order, from least to most important:

- birth of Confucius
- banning of Conficianism in the Qin Dynasty
- Confucius is married at age 19
- Han adopts Confusianism as state philosophy

Step 2: I can explain historical significance

5 What is Buddhism and why is it historically significant in China?

Step 3: I can apply a theory of significance

6 Apply Partington's theory of significance (page 219) to Taoism. Show your response in table format, including these sections: Importance, Depth, Number, Time and Relevance.

Step 4: I can analyse historical significance

7 Confucianism is an historically important philosophy. Explain why, giving at least three reasons. Next, write two to three sentences explaining how these reasons combine to make Confucianism significant.

Historical significance, page 219

HOW TO

What was family life like in ancient China?

In ancient China, several generations of a family lived under one roof, and everyone worshipped their ancestors. The oldest member of the family had the best room, and they were cared for by their children. Marriages were arranged by the parents and, after the wedding, the bride lived with the husband's family. Women had very few rights outside the home.

Source 1

The tradition of ancestor worship continues today. This family is visiting the grave of an ancestor and making an offering to show their respect for them.

Ancestor worship

The ancient Chinese believed that a person's spirit lived on in the **afterlife**, and would watch over their family. Families had small shrines in their houses dedicated to their dead ancestors, and they would *kowtow* (see page 176), burn incense and make sacrifices to show their respect.

Source 2

This well-preserved mummy was discovered in China in 1971. The woman, Xin Zhui, died from a heart attack 2200 years ago. Her tomb contained combs, vases, musical instruments, containers of food and 162 small models of servants to support her in the afterlife.

When a person died, their grave was filled with possessions they could use in the afterlife, such as food, clothing and weapons. Sometimes wives and concubines were buried alive with the dead emperor. Emperor Qin Shi Huang had an entire terracotta army buried near his tomb, so he could command an army in the afterlife (see pages 182–183).

Family structure

Whether rich or poor, ancient Chinese people lived in large extended families where all generations lived together in the one home. Older people were treated with great respect in the family. The oldest male was the head of the family, and his decisions had to be obeyed.

Men sometimes took a concubine (a secondary wife) to increase the chances of producing more sons. Daughters had little value and no social status, whereas sons were highly valued because they carried on the family name.

Women in China

Women had little social status in ancient China, and were expected to obey their fathers and husbands. Only males received an education – girls stayed home to learn how to be housewives and mothers.

A daughter had to marry the husband chosen by her father. She then had to move into her husband's home and completely obey him and her mother-in-law. Wives had to worship their husband's ancestors and be loyal to their husband's family. The only time a woman gained power was when her son became the head of a household.

Source 3

Tiny feet became a mark of beauty from the Tang dynasty (618–907 CE). Girls as young as five would have their toes broken and folded flat against the sole of their foot. Foot binding was outlawed in 1912.

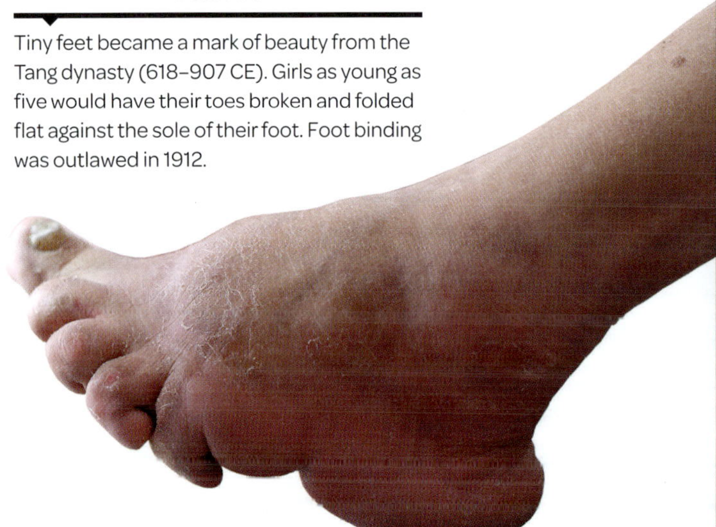

Learning Ladder H5.9

Show what you know

1 How did Chinese people worship their ancestors?

2 Source 3: What was foot binding, and why was it performed?

3 What happened when a woman married a man?

4 Was it a good idea to live with extended family? Support your answer with evidence.

Cause and effect

Step 1: I can recognise a cause and an effect

5 Source 2: If the effect is the practice of burying people with objects, what is the cause?

6 Source 2: What items were buried with this woman and how were they expected to be used?

Step 2: I can determine causes and effects

7 Who was the head of Chinese families and why?

Step 3: I can explain why something is a cause or an effect

8 Explain how men having higher status would have affected the lives of women.

9 Why did men sometimes take a concubine?

Step 4: I can analyse cause and effect

10 What is this article from a Protestant missionary journal (c. 1835) discussing? What was the orginal cause and what is the effect of the practice?

'. . . the practice originated with the infamous Take, the last empress of the Shang dynasty, who perished in its overthrow, B.C. 1123. "Her own feet being very small, she bound them tight with fillets, affecting to make that pass for a beauty which was really a deformity. However, the women all followed her example; and this ridiculous custom is so thoroughly established, that to have feet of the natural size is enough to render them contemptible." Again, the same author remarks, "The Chinese themselves are not certain what gave rise to this odd custom. The story current among us, which attributes the invention to the ancient Chinese, who, to oblige their wives to keep at home, are said to have brought little feet into fashion, is by some looked upon as fabulous. The far greater number think it to be a political design, to keep women in continual subjection. It is certain that they are extremely confined, and seldom stir out of their apartments, which are in the most retired place in the house; having no communication with any but the women-servants."'

Source: 'Small feet of the Chinese females: remarks on the origin of the custom of compressing the feet ...' Chinese Repository 3 (1835): 537–539.

civics+citizenship

How did the ancient Chinese define justice?

Justice is a concept stating that all people should be treated fairly according to the laws and values of society.

Ancient Chinese law

Laws in ancient China were based on the principles of order, obedience and respect for the emperor and elders. Laws were established by orders from the emperor, and a list of rules and punishments issued by each dynasty.

The only code of punishments that has survived from ancient China is the Kaihuang Code, which was developed during the Sui dynasty (581–618 CE) and adopted by later dynasties. This code defines the Ten Abominations and the Five Punishments.

The Ten Abominations

The Ten Abominations was a list of offences that threatened civilised society. The first three offences were punished by death.

1 Plotting a rebellion.
2 Plotting to destroy royal buildings.
3 Plotting to tell national secrets.
4 Harming or murdering your parents, grandparents or elder relatives.
5 Murdering three or more innocent people.
6 Showing disrespect to the emperor or his family.
7 Seeking entertainment during the three-year mourning period for a parent.
8 Harming or suing one's husband or elder relatives.
9 Murdering a superior or local official.
10 Having an affair with your father's or grandfather's concubine.

The Five Punishments

These punishments were for slaves.

1 Bamboo lashes to the buttocks.
2 Stick lashes to the back, buttocks or legs.
3 Compulsory manual labour and beatings with a large stick.
4 Exile to a remote location.
5 Death by strangulation or decapitation.

Ancient Chinese justice

Local magistrates investigated the facts of a case, decided whether the accused was guilty or not, and used the code to determine the sentence. No death sentence could be carried out without approval from the emperor.

The accused person was presumed guilty. However, they could not be convicted of a crime unless they confessed, so torture was often used to gain a confession.

Australian law and justice

In modern Australia laws need to reflect the values of society and be understood by most people. The three main principles of the Australian legal system are:

- *Fairness* to achieve equal outcomes for all
- *Equality* with no discrimination
- *Justice* to provide a fair outcome according to the laws and values of society.

The rule of law

The *rule of law* is a principle that means that all people are subject to the law, including those who make and enforce the laws. The rule of law also means that Australians are governed by laws, not the decisions of individual government officials.

Source 1

Ancient Chinese punishments included beatings and exile, as well as death by beheading or suffocation. [Alexander, William, *The punishment of the Bastinado* (1805 CE).]

To achieve the rule of law, there must be checks or limits on the use of government power. The Australian Constitution states that the power to govern should be distributed between three groups, with no one group having all the power:

1 the **Parliament** passes the laws
2 the **Executive** creates regulations to enforce the laws
3 the **Judiciary** interprets the laws.

The right to a fair trial and representation

Any Australian accused of breaking a law will be given a fair chance to prove their innocence. The accused person must go to court, where they have the right to a fair trial. Each person can present their own case in court – or they can hire a lawyer to represent them. People who cannot afford a lawyer can have one provided free by the legal aid service.

Presumption of innocence

In Australia, any accused person is presumed to be innocent until proven guilty. This means that the **prosecutor** – the name given to the lawyer who conducts the case against the accused person – needs to produce evidence that will prove their claims against the accused person. So it is up to the prosecutor to prove or disprove any disputed facts in court.

Learning Ladder H5.10

Civics and citizenship

Step 1: I can identify topics about society

1 Source 1: Which of the Five Punishments is shown here? Why do you think this punishment was performed in front of other people?

2 Outline the 'rule of law' in your own words.

Step 2: I can describe societal issues

3 Rank the Ten Abominations from most severe to least severe. Give reasons to support your most severe and least severe selections.

Step 3: I can explain issues in society

4 What is presumption of innocence and how does it shape the role of the prosecutor in legal trials?

Step 4: I can explain different points of view

5 Was there a presumption of innocence in ancient Chinese law? Compare this aspect of law between ancient China and modern Australia.

Step 5: I can analyse issues in society

6 Fairness, equality and justice are the three main principles that underpin the Australian legal system. Undertake research to uncover what action is taken to ensure that people with disabilities, who speak languages other than English or who are financially disadvantaged have the right to a fair trial.

What inventions did China give to the world?

The ancient Chinese were remarkable inventors, and many of their inventions are still in use today. Among the countless Chinese inventions are four that changed the world: the magnetic compass, papermaking, printing and gunpowder.

Magnetic compass

Historians believe the magnetic compass was invented about 206 BCE, during the Han dynasty. Early Chinese compasses were made using lodestone – a stone that is naturally magnetic. The lodestone was carved into a spoon-shaped needle and placed on a bronze plate. As the plate was moved, the spoon spun until it stopped with its point facing south. For this reason, early compasses were called 'south-pointers'.

The earliest compasses were used to create buildings that matched the principles of *feng shui*. It wasn't until the Song dynasty in the 11th century CE that compasses were used for navigation.

Source 1

The first magnetic compasses were known as south-pointers.

Papermaking

The invention of paper and printing had a huge impact on communication around the world. Before the invention of paper, people wrote on bones, tortoise shells, wood, bamboo – and even silk. Books were made from wood or bamboo.

The person credited with inventing papermaking is Cai Lu, who was head of the imperial workshops. In 105 CE, Cai Lu came up with a process for making paper from bark, cloth and old fishing nets. These raw materials were cheap and easily available, making it possible to produce large quantities of paper.

The technique for papermaking slowly spread out from China to other parts of the world: first to Korea in 384 CE, then Japan in 610 CE, arriving in India by the 11th century CE. Eventually China's papermaking technique spread around the whole world, because paper provided a lighter, cheaper and better surface for writing on than anything that had existed before.

Source 2

The papermaking process using bamboo

Printing

Block printing was invented by the Chinese in 750 CE, during the Tang dynasty. Blocks were cut from wood and used to print cloth, which was then used to produce scrolls and books. Cheap printed books became widely available in China during the Song dynasty (960–1279 CE).

Gunpowder

The first recorded use of gunpowder was in 142 CE, when Wei Boyang described a mixture of three powders that would 'fly and dance' violently. By 300 CE, a Jin dynasty scientist named Ge Hong described the explosion and wrote down the ingredients of gunpowder: sulphur, charcoal and saltpetre.

At first, gunpowder was rolled up in paper and used to make fireworks, but by 904 CE it was being used in weapons. A step from fireworks led to rockets, which were soon being used by the Chinese military, and cannons. Early cannons were made from bamboo and, after lighting gunpowder at one end, a stone cannonball could be shot out the other end. Bamboo was soon replaced by another Chinese invention – cast iron – and metal cannons were created.

The Chinese tried to keep gunpowder secret, but the 'formula' got out once they started using rockets and cannons against their enemies.

Knowledge of gunpowder spread rapidly through the rest of the world when the Mongols of the Yuan dynasty started fighting against European armies. In 1267 CE, English philosopher Roger Bacon described his introduction to gunpowder in his science book *Opus Majus*:

> 'From the violence of that salt called saltpeter so horrible a sound is made by the bursting of a thing so small, no more than a bit of parchment, that we find exceeding the roar of strong thunder, and a flash brighter than the most brilliant lightning.'

Source 3

Roger Bacon quoted in Joseph Needham, *Science and Civilisation in China: Volume 5*, Cambridge University Press, 1987.

China invention timeline 600 BCE–1030 CE

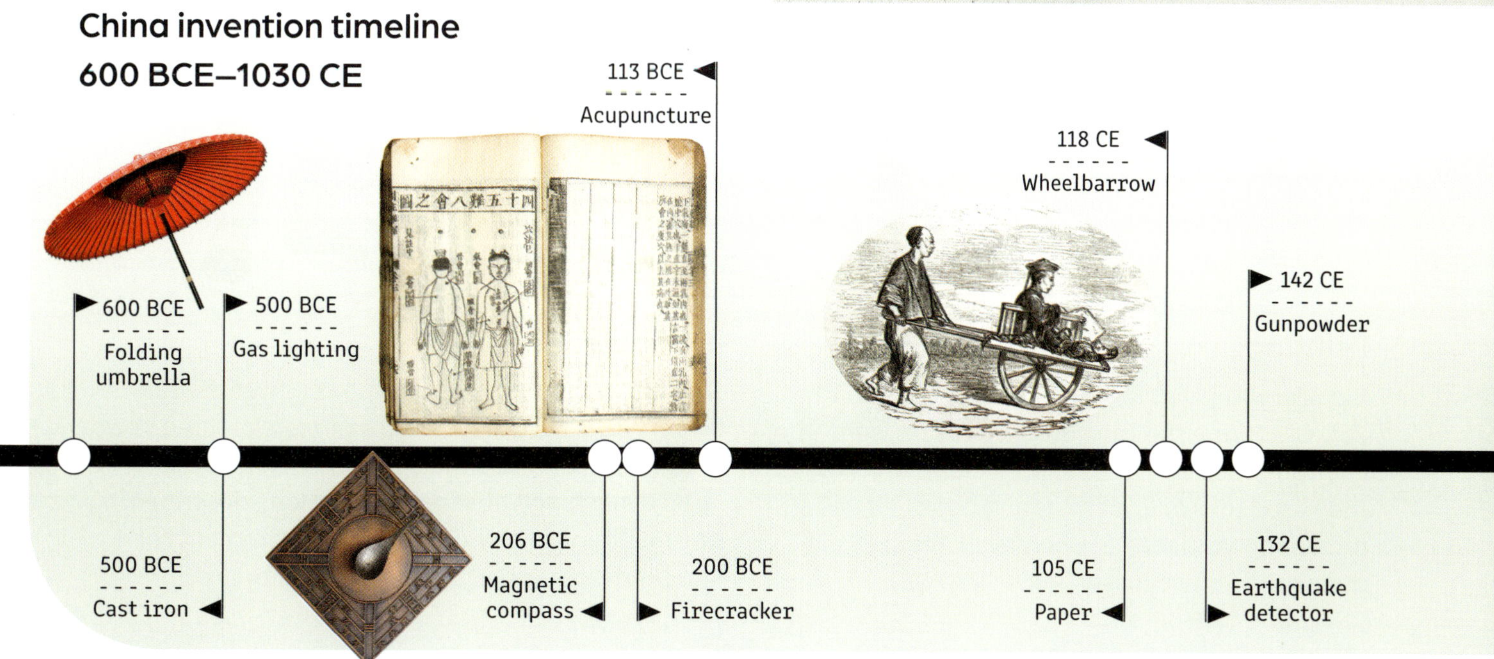

Source 4

Two Chinese soldiers are seen here using gunpowder in an early cannon. From an illustration created in 1847 CE. [Illustration from *L'Illustration, Journal Universel*, No. 231 (1847 CE).]

Learning Ladder H5.11

Show what you know

1 What did Chinese people write on?

2 Source 1: What were compasses used for?

3 Why did use of papermaking and gunpowder spread around the world so quickly?

4 What materials were needed to make these inventions?
- a magnetic compass
- b paper
- c gunpowder

Chronology

Step 1: I can read a timeline

5 Use the timeline below to answer these questions.
- a When was the folding umbrella invented?
- b What major items were invented in a 37-year period from 105 CE?
- c How many years were there between the discovery of the firecracker and gunpowder?

Step 2: I can place events on a timeline

6 Put these Chinese inventions in order from earliest to latest.
- Kite (2800 years ago)
- Toothbrush (1498 CE)
- Paper money (9th century CE)
- Tea production (2737 BCE)

Step 3: I can create a timeline using historical conventions

7 Using correct conventions, create a timeline with these events on it.

Bronze (1700 BCE)

Iron smelting (1050 BCE)

Porcelain (2000 years ago)

Silk (6000 years ago)

Step 4: I can distinguish causes, effects, continuity and change from looking at timelines

8 What other inventions on the timeline to the left relied on the invention of paper?

9 Create a chronological list to show inventions that followed the invention of gunpowder in 142 CE.

HOW TO
Chronology, page 206

700 CE – Playing cards

725 CE – Mechanical clock

c. 750 CE – Woodblock printing

c. 868 CE – Printed book

c. 1030 CE – Moveable type printing

Masterclass

Learning Ladder

Step 1

a I can read a timeline

From the timeline on pages 172–173, write out all the dynasties listed in order, from longest lasting to shortest. Include how long each dynasty lasted.

b I can determine the origin of a source

Source 3: When and where is this from?

c I can recognise continuity and change

Which of these things are still the same, and which have changed? Put them on a scale, with 'most changed' at one end and 'least changed' at the other:

- Chinese religion
- Chinese clothing
- Chinese language
- Chinese political system
- Chinese food.

d I can recognise a cause and an effect

Which of these is a cause of merchants being low on the social hierarchy?

- they did not produce anything
- they rejected religion
- they were poor
- they were too powerful

Which of these is an effect of Confucian principles?

- improvement of the environment
- increase in technological advance
- more warfare
- stable society

e I can recognise historical significance

Rank these in order, from least to most historically important:

- invention of the compass
- invention of gunpowder
- invention of paper
- invention of printing.

Step 2

a I can place events on a timeline

Read these statements and put the events and developments in the correct order, from earliest to most recent. Note there are six items listed.

Xia dynasty began c. 2100 BCE

About 900 years after the Xia dynasty began, silk was first made in China

Shang dynasty fell in 1045 BCE after lasting 721 years

Source 1

The Great Wall

Work at the level that is right for you or level-up for a learning challenge!

Source 2

Forbidden City, Beijing.

Source 3

Spear head from the eastern Zhou dynasty (770–476 BCE). The spear head is bronze with a metallic inlay.

Iron-making began in China about 500 BCE

In 770 BCE, the Western Zhou dynasty ended and the Eastern Zhou dynasty begun. The Western Zhou lasted for 274 years and the Eastern Zhou lasted 549 years.

b I can list specific features of a source

Describe the spear head in Source 3 in detail. Include discussion of its shape, material, size, proportions, colour and markings.

c I can describe continuity and change

Look at the timeline on pages 172–173. What do you notice about the length of dynasties? Explain your answer in detail, referring to the timeline.

d I can determine causes and effects

What caused men to have higher status in society than women?

What was the effect of Qin Shi Huang introducing a common currency?

e I can explain historical significance

Which dynasty affected more people: the Shang dynasty or the Ming dynasty? Explain whether this makes one of them more significant than the other.

Step 3

a I can create a timeline using historical conventions

Take the events from Question 2a and put them on a timeline. Use the correct conventions, including equal spacing for equal years.

b I can find themes in a source

How did the artisan who produced the spear head in Source 3 make it into a beautiful weapon? Discuss its shape, material, size, proportions, colour and markings.

c I can explain why something did or did not change

Why did Chinese move from using bronze to using iron?

d I can explain why something is a cause or an effect

Source 2: Write three to four sentences explaining why palaces were so lavish and finely decorated.

e I can apply a theory of significance

Consider the Silk Road:

- How important was it to traders during ancient times?
- How deeply were the lives of those traders affected?
- Which different groups of people would have been affected by trade on the Silk Road?
- How long has there been trade along this route?
- Is trade in this area still occurring today?

Step 4

a I can distinguish causes, effects, continuity and change from looking at timelines

Look at the timeline on pages 172–173.

- Which of those events are linked by cause and effect?
- Which events show continuity? What makes you say that?

b I can use my outside knowledge to help explain a source

What was happening during the period the spear head in Source 3 was made? What technology were Chinese people using at the time? How does this help you to know more about the spear head?

c I can analyse patterns of continuity and change

The Chinese population grew a lot over time. It used to be one among many countries that had big populations. Today it is the most populated country in the world. Why do you think this has happened?

d I can analyse casue and effect

List four different causes for the building of the Great Wall (Source 1). Rank them in order, from most important to least important. Write them in sentences, using the vocabulary from the History How-To section on pages 226–227.

Masterclass

e I can analyse historical significance

Without using Partington's model of significance, describe five reasons why Qin Shi Huang was an important figure in Chinese history.

Step 5

a I can describe patterns of change

Look at the timeline on pages 200–201. What is the pattern of technological improvement you notice? Focus on technological improvements and inventions, then write a paragraph showing how Chinese skills changed.

b I can use the origin of a source to explain its creator's purpose

Source 3: Why might an artisan of the Zhou period have wanted to create such a beautiful weapon? What might this say about Chinese culture at the time?

c I can evaluate patterns of continuity and change

China's population has grown a lot since ancient times. What do you make of this change? Is it a good thing or a bad thing? Give evidence for your answer.

d I can evaluate cause and effect

Do you think city life during the Han dynasty would have made people healthy or unhealthy? Provide historical evidence in your response.

e I can evaluate historical significance

For centuries, many Chinese women had their feet bound to improve their status and likelihood of marriage. Is this historically important? Does using Partington's model of significance help us to answer this question? Are there other things that should be considered when deciding how important this custom was?

Historical writing

1 Structure

Imagine you are writing an essay: 'Chinese history shows the importance of stability over change. Discuss.'

Create an essay plan, using the TEEL structure.

2 Draft

Using the drafting and vocabulary suggestions on page 226, draft a 300 word essay (at least 15 sentences) responding to the topic.

2 Edit and proofread

Using the editing and proofreading guidelines on page 227, edit and proofread your draft.

Historical research

3 Organise and present information

Imagine you are completing a research project comparing ancient Chinese and modern Australian justice.

Create a graphic organiser that would help you to organise the information you gather. Use pages 223–224 to take at least six dot point notes on this topic, putting them in your graphic organiser.

Capstone

How can I understand ancient China?

In this chapter, you have learnt a lot about ancient China. Now you can put your new knowledge and understanding together for the capstone project to show what you know and what you think.

In the world of building, a capstone is an element that finishes off an arch or tops off a building or wall. That is what the capstone project will offer you, too: a chance to top off and bring together your learning in interesting, critical and creative ways. You can complete this project yourself, or your teacher can make it a class task or a homework task.

Scan this QR code to find the capstone project online.

mea.digital/GHV7_H5

H6

History How-To

History has its own set of skills to help us analyse and understand societies in the past and the key ideas, people and changes that shape the world we live in today. Historical skills are based around interpreting sources of evidence from the past, promoting debate and encouraging investigation.

Chronology

Chronology is the arrangement of events into the order they happened. Historians put events in order, from earliest to most recent. They do this to:

- **identify continuity and change**
- **determine cause and effect (both short-term and long term)**
- **create a narrative of how history unfolded.**

step 1 I can read a timeline

When giving answers to questions, include part of the question in your answer. This way, your answer is a full sentence by itself, and you could use it in a piece of history writing. If a question asks, 'When did the Roman Empire fall?' you should answer with, 'The Roman Empire fell in 476 CE', not just '476 CE'.

Working out the years between dates of BCE and CE is the same as when you work out the difference between positive and negative numbers in maths.

For example, how many years were there between 490 BCE and 150 CE?

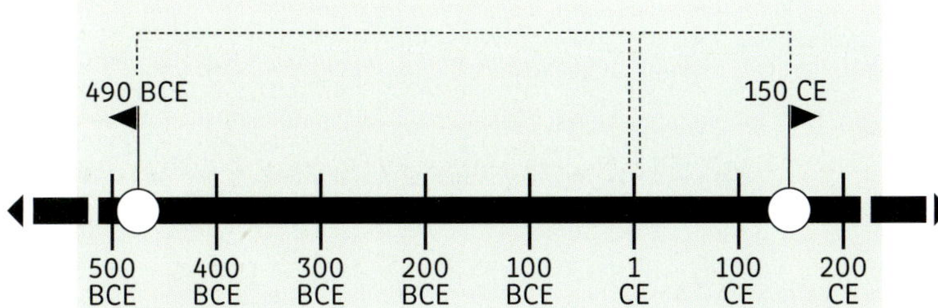

490 BCE is the same as −490

So, the difference in years between 490 BCE and 150 CE would be:

490 plus 150

= 490 + 150 = 640 years

step 2 I can place events on a timeline

Placing events on a timeline involves putting them in order from earliest to most recent. The order goes like this:

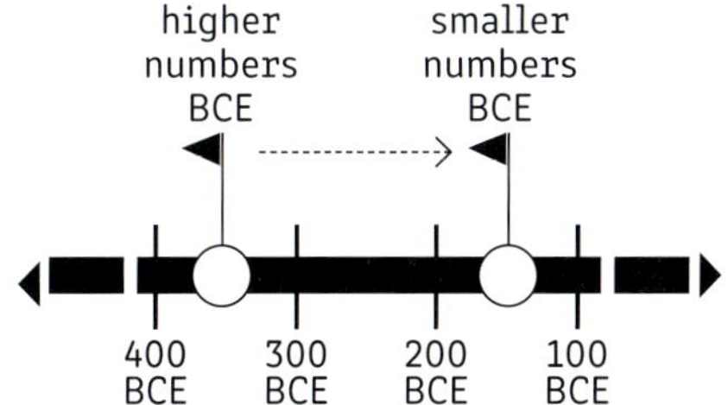

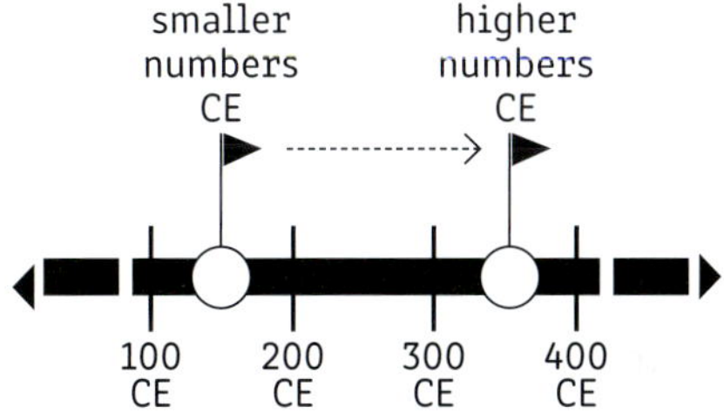

It's just like in maths with positive and negative numbers. An event that lasts for 10 years BCE will start with a higher number and go *down*. An event that lasted for 10 years in CE will start with a lower number and go *up*.

For example:

700–690 BCE = an event taking 10 years

700–710 CE = an event taking 10 years

Source 1

An arch from the Roman Empire

I can create a timeline using historical conventions

Refresh your understanding of time words on page 4.

Timeline creation process

Follow these steps to make a timeline.

1 Select events to include on your timeline.
2 Put them in order, from earliest to latest.
3 Work out the earliest and latest dates you want to include.
4 Pick a span of years that your timeline will cover, so that the earliest and latest dates will fit.
5 Choose which unit of time you want to use: years, decades or centuries.
6 Work out how many segments your timeline will need. Figure this out based on how much space you have on your sheet of paper.
7 Draw a straight line and divide it up into segments.
8 Number the segments into the units you selected in Step 5.
9 Label the events on your timeline.

If an event goes over many years, add it as a span rather than a line.

If there is a large gap between the events on your timeline, add a zigzag line to show there has been a jump in time.

Use a zigzag line to show a break in sequence on your timeline.

Timeline creation process example

1 Choose events to include on your timeline:

776 BCE first Olympics, 399 BCE Death of Socrates, 460–404 BCE Peloponnesian Wars

Source 2

An artist's impression of a battle at Syracuse. [*Peloponnesian War: the naval battle in the harbor of Syracuse* (19th century), wood engraving, 27.4 x 19.3, Granger Historical Picture Archive.]

2 Put them in order from earliest to latest.

776 BCE first Olympics, 460–404 BCE Peloponnesian Wars, 399 BCE Death of Socrates

3 Work out the earliest and latest dates you want to include.

776 BCE is the earliest, 399 BCE is the latest

4 Pick a span of years that your timeline will cover, so that the earliest and latest dates will fit.

800 BCE – 350 BCE (450 years)

5 Choose the unit of time you want to use: years, decades or centuries.

Segments of 50 years

6 Work out how many segments your timeline will need. Figure this out based on how much space you have on your sheet of paper.

Nine blocks of 50 years. I'm using an A4 exercise book. I have about 20 cm to draw in. Nine blocks at 2 cm per block would give me 18 cm.

7 Draw a straight timeline and divide it up into segments.

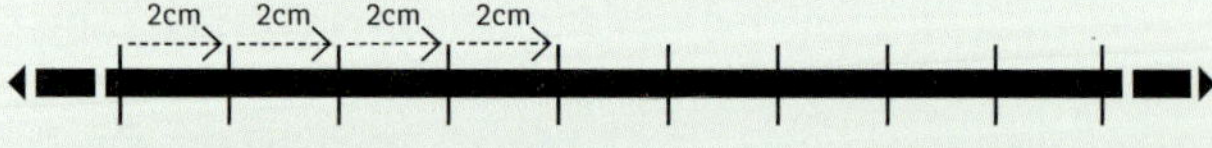

Nine blocks at 2 cm per block

8 Number the segments into the units you selected in Step 5.

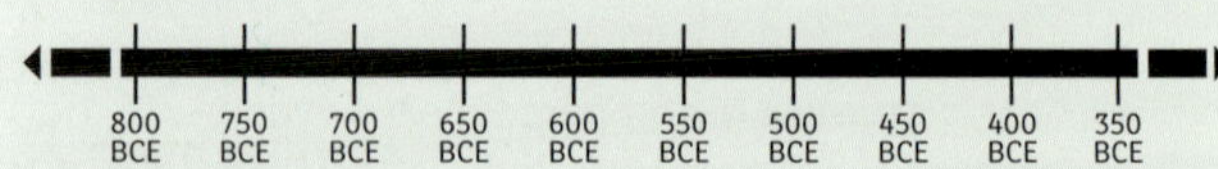

9 Add the events you want to include on your timeline.

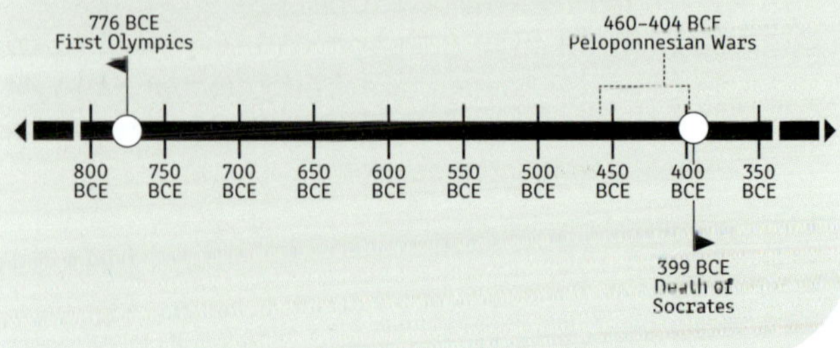

step 4 I can distinguish causes, effects, continuity and change from looking at timelines

Causes and effects

Distinguishing cause and effect on a timeline involves looking at the timeline for events and developments that are linked.

In looking for events that are linked, ask yourself:

- are both events about the same thing? Perhaps two or more events are about religion, war, trade, social upheaval or changes in thinking.
- are both events about different things, but you suspect there is a cause–effect link between them? For example, there is often a boom in artistic output in times of strong government, so perhaps one event causes the other.
- are more than two things linked? For example, does one event have two effects? Or are there two causes of one effect?

If you have found links, try to confirm your belief. See if you can find other information to back up your belief.

Continuity and change

To be able to distinguish continuity and change in a timeline, you need to classify the events or developments in the timeline into things that can continue or change.

For example, a statement such as 'The assassination of Julius Caesar results in change from Roman Republic to Roman Empire' doesn't show continuity or change because it is a one-off event. You need to work out what *type* of event it is. In this case, it could be considered a few different types of events: assassination of a leader, change from one type of political system to another.

You can then look at other parts of the timeline, or compare it to the present, and ask yourself, do we still see assassinations of leaders? Do we still see changes from one type of political system to another?

You can either look for continuity and change within the timeline, or by comparing timeline events with later times, like the present.

It is easiest to compare timeline events with the present, because it is the time we know most about. For example, seeing 'Great Pyramids built' on a timeline, could show continuity with the present, because large monuments and buildings are still being created today. But if we saw 'Romans take control of Egypt', we might say this is a change. In modern times, we do not see empires conquering each other.

step 5 I can describe patterns of change

Describing patterns of change from looking at a timeline involves bringing lots of events together. You should find a *group* of events that are linked.

Things you need to remember about patterns of change include the following.

- You can show a pattern of change from within the timeline, or from the timeline to a later period you know about, such as the present.
- Patterns of change won't necessarily be all about the same thing. A pattern of change in history is like finding a theme in a book.
- Think about what is changing; for example, trade, society, life-expectancy, technology.
- Think about the pattern of change; for example, quick improvement, slow evolutionary improvement, quick decline, slow and steady decline, rises and falls.
- Typical things that could be considered patterns of change in history include:

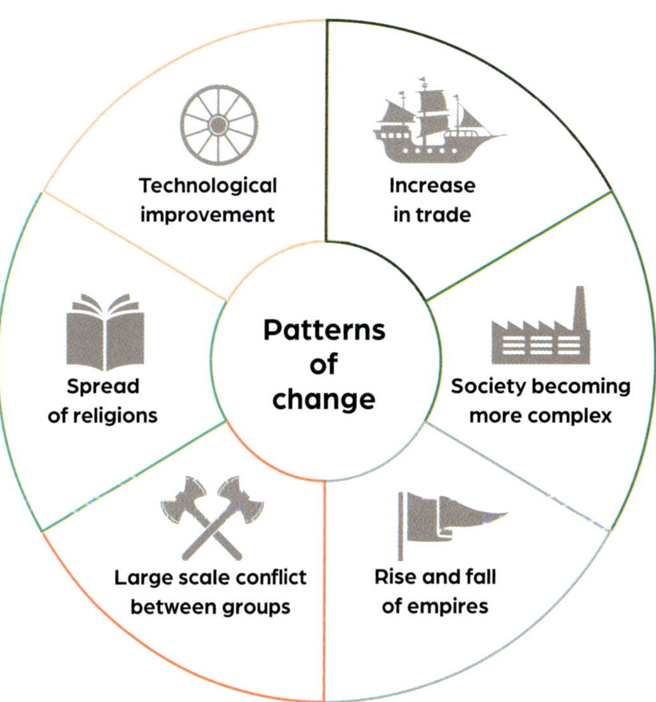

Source analysis

Source analysis asks us to look at evidence and ask, 'How do we know what we know about the past?' A good source analysis interprets and makes meaning of the source.

First, you need to ask good questions of a source. These questions include:

- **who created it?**
- **when was it created?**
- **what was the author or creator's purpose?**
- **what is the historical context of the source?**

You should use multiple sources, where possible.

I can determine the origin of a source

To figure out who produced a source and when it was produced, look at the clues in the source. Often there will be a caption that tells you this information.

I can list specific features of a source

Listing detailed features usually means writing more. Listing general features might sometimes look like a summarised version. Listing detailed features would be the long version, describing as much as you can.

General features	Detailed or specific features
Obvious features	Minor details
The most important things in the source	Everything in the source, whether or not you think it is 'important'
Using vague words such as 'big, small, very, good …'	Using specific words and phrases, such as 'three times bigger than, small in comparison to …, in the background …, useful for …'

I can find themes in a source

There is often a theme in a source. The theme can help you uncover more meaning in a source. A theme is something that you might notice after recognising specific features.

Some examples of themes and how they might be shown are listed in this table.

Theme	How this might be shown in a source
Beauty	A statue of a handsome person with a muscular body and symmetrical facial features
Faith	A decoration on a vase showing people offering sacrifices to a god
Good vs bad	A statue showing two figures fighting – one that looks like an angel and one that looks like a demon
Hierarchy	An image going from top to bottom, showing gods on the top, then people, then animals, then rocks
Man vs nature	A building placed in a natural place, dominating it; e.g. a temple on a mountain
Technology in society – good or bad	Good: a statue of a person using a new farm implement looking happy. Bad: a painting of people happily working in the fields, and another part of the picture shows a machine doing the work and those people unhappy
War or conflict	A decoration on a building depicting a large-scale war or conflict

Source 3

A Greek amphora (or container). What themes are depicted on this amphora? [Neck-amphora jar (c.500 BCE), terracotta, Metropolitan Museum of Art.]

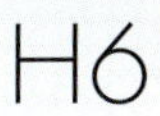

If you think there is an underlying theme, after looking at all the details in a source, always give evidence *from the source* to back up your answer.

Not all sources have themes. If the source is a simple farm tool, it might not have a 'theme' like the amphora shown in Source 3. However, you could still comment on the type of society that produced it. For example, what materials did they use? How advanced were they? What did they farm? How might they have used the tool?

step 4 I can use my outside knowledge to help explain a source

A visual source can tell us a lot about a time in history. However, we can also use information that we know about a period to help understand a visual source.

These are two different skills. They both help us to understand, but they should not be confused.

Evidence from visual source ▷ **helps to understand** ▷ **time in the past**

Outside knowledge about the time ▷ **helps to understand** ▷ **the visual source**

Both evidence from the visual source and outside knowledge about the era can help us gain understanding.

Before using outside knowledge to help understand a source, you first need knowledge about the period. Then, think what knowledge is relevant to the source, as outlined in this table:

What's in the visual source?	Outside knowledge that might help to understand it
Rock painting of people	Knowing how kinship systems work in Indigenous culture
Vase with fine detail	Knowing how craft technology allowed fine detail to be produced
Mummified corpse	Knowing why Egyptians mummified people
Helmet	Knowing how common warfare was in ancient Greece
Engraved stone writing	Knowing who was Roman emperor at the time, and what kind of ruler he was

step 5 I can use the origin of a source to explain its creator's purpose

Here are three ways you could explain a creator's purpose. You could:

1 use *who* created it to explain *why* it was made
2 use *when* it was created to explain *why* it was made
3 use both when it was created *and* who created it to explain why it was made.

The third explanation is the best.

Here are some questions to ask of the source:

- What do you know about who created the source?
  - How old were they?
  - What gender were they?
  - What job did they do?
  - What position did they have in society? For example, were they powerful? Or powerless?
  - What beliefs did they have?
- What was going on at the time the source was made?
  - Were there any important events taking place?
  - What was going on politically?
  - What biases might people have had?
  - What was normal behaviour in society?
- Why was it produced?

The purpose of a source refers to what the source was originally made for. Don't get confused and think about what *we*, as historians, might use it for. Sources aren't usually created to leave records for historians.

Try to get into the head of the creator *at the time they were making the object*. For example, where they trying to:

- influence people?
- sell something?
- tell their version of events?
- make art? If so, who would enjoy it?
- make something practical, such as a tool?

The table below provides two examples of using the origin of a source to explain its purpose, using Plato's dialogue (play) *The Republic* and the Egyptian pyramids.

Source	Origin: Who created the source	Origin: When the source was created	Why you think the source was created (its purpose)	Using the origin of the source to explain its purpose
The Republic (a play with Socrates as main character)	Plato (one of Socrates' students)	Around 380 BCE	For Plato to state his political views, but make it look like they were the views of Socrates	*The Republic* is a play about politics, where Socrates' student Plato describes his views about how the state should be ruled. It was written 20 years after Socrates' death.
Pyramid	Egyptian workers, under orders from the pharaoh	2500 years ago, during the Egyptian dynastic period, a time when religion focused heavily on the afterlife	As a tomb for the pharaoh	The pyramids were created as tombs for important leaders. We know that these monuments were built under orders from the pharaoh. We also know that at the time people believed you needed wealth in the afterlife. So the pyramids were there to store the wealth that leaders would need in the afterlife.

Source 4

Relief block with the names of Amenemhat I and Senwosret I (c. 1962–1952 BCE), limestone, Metropolitan Museum of Art

Continuity and change

One reason we study history is to see how life was different in the past. We can also see how an idea or a piece of technology evolved over time.

Continuity and change are on a scale, between 'No change' and 'Completely different', as shown below.

Continuity and change can exist at the same time: some things stay the same, while others change. Change can be fast or slow, happen gradually or in a burst.

Scale of change from least (on the left) to most (on the right)

No change at all (e.g. human DNA)	A bit different (e.g. food)	Quite different (e.g. attitudes to race, gender and sexuality)	Completely different (e.g. transportation technology)

step 1

I can recognise continuity and change

The easiest kind of continuity or change to notice is from a historical period to the present, because we know a lot about the present. This table looks at continuity and change in ancient Greece, ancient Rome and ancient China.

Civilisation	Continuity from then until now	Change from then until now
Ancient Greece	Democracy	Slavery is outlawed
Ancient Rome	Large powers have permanent armies	There is more equality between men and women
Ancient China	Buddhism and Confucianism are still common	China is more open to ideas from the rest of the world

When you are trying to recognise continuity and change, ask:

1. Did it exist in the past and *still exists* now? That is continuity. You can say: 'This situation still exists today and represents continuity from the past to the present.'
2. Did it exist in the past but *doesn't exist* now? That is change. You can say: 'This situation is different today, and represents a change from the past to the present.'

Source 5

The Leshan Giant Buddha is 71 metres tall. It was carved out of a mountain in Sichuan province, China, in the eighth century CE.

I can describe continuity and change

Describing continuity and change is more difficult because you have to recognise it yourself (without help) and describe it, rather than just noticing it.

Civilisation	Earlier time	Later time	Continuity between these times	Change between these times
Ancient Australia	60 000 years BP: Small number of first peoples in the north	100 CE: First peoples had spread across the entire continent	People lived in tribal groups	People were spread across the whole of Australia
Ancient Egypt	5000 BCE: Early settlers along the Nile	1450 BCE: Egypt at its peak	People relied on the Nile for irrigation	Powerful rulers controlled many people

After recognising continuity or change, you must then write descriptive sentences. These sentences should use historical evidence to back up your claim.

I can explain why something did or did not change

Explanations require you to answer the question *why?* When explaining continuity and change, you need to:

- recognise it
- describe it
- know what caused it, or what effect it had.

I can analyse patterns of continuity and change

Examples of patterns of continuity include:

- close family bonds
- the importance of food
- the impact of disease
- natural disasters.

Examples of patterns of change include:

- the improvement in technology
- the rise in population
- the spread of new ideas.

Source 6

Kom Ombo temple on the Nile River, Egypt

step 5

I can evaluate patterns of continuity and change

Evaluation is a 'higher-order thinking skill'. It involves remembering what you have learned and being able to use that information to make sense of something or make a judgement.

Evaluation can involve:

- assessing whether a theory or belief is true; e.g. Is it true that all empires rise and fall?
- comparing different ideas; e.g. Which is a better life for humans: being a nomad or being a farmer?
- judging between different things; e.g. Some people think Roman civilisation was more influential than Greek civilisation: who is right?
- asking, 'So what?'; e.g. Was it a good or bad thing? Or was it both good and bad? A historian might include both positive and negative aspects to form a balanced view, as shown in this table. (See the key in the next column for an explanation of the colours.)

Each balanced view in the table contains four elements:

1 A statement about whether the situation was a continuity or a change.

2 A statement showing the positive aspect of it.

3 A statement showing the negative aspect of it.

4 A statement summarising the balance.

There is no 'right' answer when evaluating. Having a balanced view is not always the best answer, either. For example, it would be difficult to argue for the benefits of Hitler's policy to exterminate specific groups of people. However, some answers are better or worse than others. Better answers:

- use more historical evidence as examples in their evaluation
- use more logical reasoning – they show directly how beneficial the patterns of change were.

Situation	Positive thing	Negative thing	Balanced view
Continuity: slavery was a major part of Greek civilisation	It allowed a small elite group enough leisure time to produce cultural works and develop philosophy, art and architecture	It was a violation of human rights	Slavery was a common feature of Greek civilisation for the entire time it flourished. Widespread slavery allowed for a small group of elites to engage in cultural pursuits, leaving us with a legacy of art, architecture and philosophy. On the other hand, countless individuals suffered greatly from being enslaved. While we can accept there were some benefits to the use of forced labour, it left a stain on the history of ancient Greece.
Change: Chinese civilisation occupied more and more territory	Shared culture brought about stability that protected the civilisation against invasion from outside forces	Many groups of people had their unique ways of life destroyed	Chinese civilisation was dynamic and changed greatly over the course of thousands of years, gradually occupying more and more territory in the far East. Benefits flowed from this situation, such as when those within the Chinese cultural sphere were safer from war and conquest. However, many cultural groups from territories conquered by China lost their individuality and some of their cultural uniqueness after being incorporated into the greater Chinese civilisation. Therefore, the expansion of Chinese civilisation brought peace, but at the expense of cultural diversity.

Cause and effect

Cause and effect is visible every day. For example when we open the fridge, the little light comes on; when we wave our hand, the bus stops for us.

There are short-term and long-term causes. The short-term cause of the fridge light coming on is opening the door. The long-term cause is because we are hungry. Equally, there are short-term and long-term effects. After I wave my hand, the short-term effect is that the bus stops. The long-term effect is that I arrive at school on time.

Cause and effect requires understanding which events are linked and why. When we say things are linked by cause and effect, we say they have a *causal link*. This means that one thing *caused* the other.

Most things that happen have multiple causes and effects – some of which are more important than others. Two main types of causes are:

- historical actors: the individuals or groups involved; e.g. Alexander the Great, Julius Caesar, Qin Shi Huang, the First Triumvirate
- historical conditions: social, political, economic, cultural and environmental; e.g. mass unemployment, drought, world religions and bad harvests.

However, just because one event happens after another, it doesn't always mean that the first event caused the second event; for example, when the rooster crows in the morning, we don't think it makes the sun rise. You also need to be able to tell a believable story about why something caused its effect.

Events in history are not inevitable. When we study cause and effect, it can seem like things were always going to work out in a certain way. Yet, change a few conditions and things could have happened differently. If Alexander the Great had fallen off his horse as a boy and died, would Greek civilisation have spread so far and seem so significant today? It is easy, with hindsight, to think our cause and effect explanations are perfect. But we need to be cautious when we make claims about cause and effect, as many events happen at random.

Source 7

The Roman Forum

I can recognise a cause and an effect

Recognising cause and effect means correctly choosing from a list of possibilities. For example, which of these is most likely to be the cause of the Peloponnesian Wars between Athens and Sparta?

A Rome conquered Greece centuries later.

B Athens and Sparta both wanted to control all of Greece.

C Persians had better technology than the Greeks.

D Sparta is in Greece.

E Athens was founded before Sparta.

The only one of these options that is linked to the conflict is B. Option A refers to a time *after* the wars, so it can't be the cause. Option C refers to Persians, not Greeks, so it is not relevant. Option D and Option E just tell us facts that wouldn't necessarily lead to conflict.

For events to be causally linked:

- one event must come before the other
- you must be able to tell a believable story about why one event caused the other
- if possible, you should have some historical evidence that one event caused the other.

There is not necessarily a right or a wrong answer, but there are better or worse answers. Better answers use more historical evidence, and have more logical reasoning.

I can determine causes and effects

Determining cause and effect means deciding what the cause or effect of something might be. You need to have knowledge of the period to do this.

Examples of historical causes include:

- conditions:
  - social
  - political
  - economic
  - cultural
  - environmental
- actors:
  - individuals
  - groups.

Once you suspect that two things are linked, see if one of the above items is the cause.

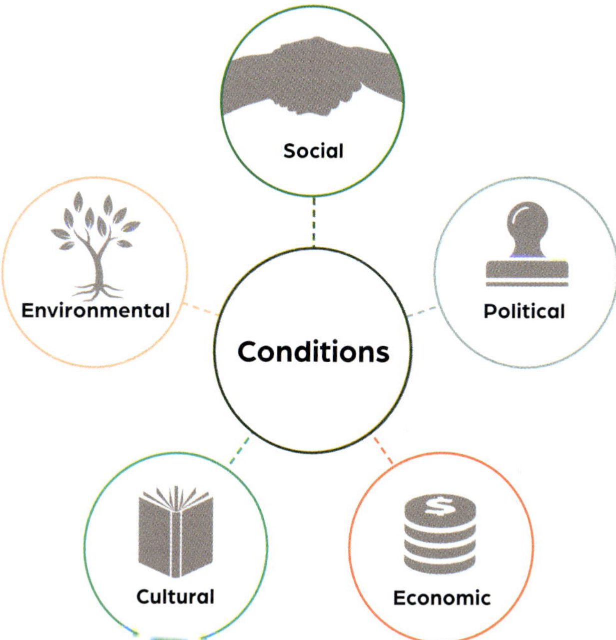

Type of cause	Example cause	Example effect
Social conditions	Widespread slavery in the Roman Empire	Uprising led by Spartacus
Political conditions	Imminent attack on Greece from Persians	Different Greek city–states form alliance
Economic conditions	Very little economic growth in Rome	Rome conquers other territories to increase size of economy
Cultural conditions	Stability is valued in Chinese society	Confucianism, which favours stability, becomes widespread
Environmental conditions	Struggle over resources in Africa	Migration from Africa to the Middle East, India, Asia and Australia
Individual actor	Alexander the Great conquers new territory	Greek culture is spread to the Middle East and West and Central Asia
Group actor	Roman senators worry that Caesar is acting like a king	Senators assassinate Caesar, but Rome becomes an Empire anyway

I can explain why something is a cause or an effect

Explaining cause and effect involves stating *how* or *why* a cause led to an effect.

Cause	Effect	Explaining how the cause led to the effect
During the early Stone Age, east Africa became overpopulated	People migrated to the Middle East, India, Asia and Australia	There weren't enough resources in east Africa after a while. Hunting and gathering requires a lot of land so, as the population grew, there wasn't enough land. To avoid starvation, people were forced to find new land elsewhere. They moved north to the Middle East, then east toward India, Asia and Australia.
Alexander the Great conquered the known world and spread Greek culture	Greek culture, such as building styles, dress, religion and language, spread to the Middle East and West Asia	Alexander the Great was a conqueror who defeated Greece, Egypt, Persia and parts of India. He loved Greek culture. As he conquered territory, his soldiers moved into the new lands. The invaders mixed with the local people. Over time, they began to share the way they did things. This led to the existing culture being mixed with the new Greek culture. It created a new type of culture, known as 'Hellenistic' culture. People from areas conquered by Alexander begun to live differently, with new clothing styles, architecture and religious rituals, and language also changed.

Source 8

Territory conquered by Alexander the Great

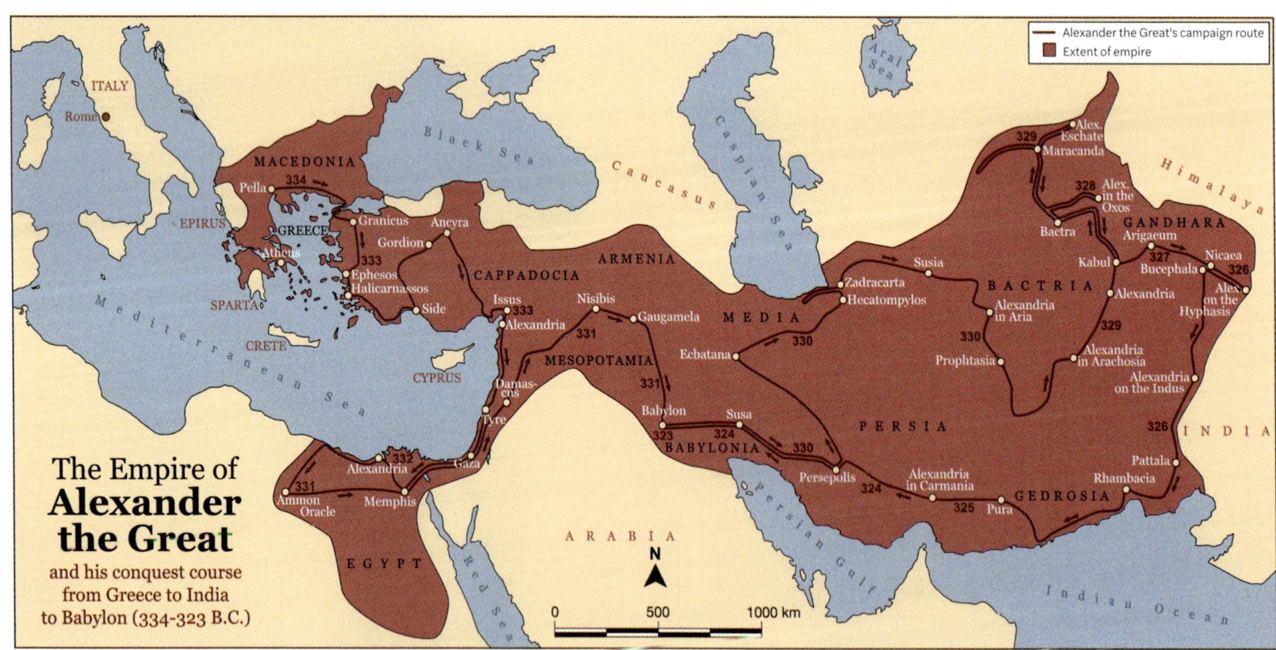

step 4 I can analyse cause and effect

Analysing means the ability to break down something into its parts. If you can identify these different parts, and explain how together they make up the whole, you are analysing. If you can explain the rules or theories that show how these parts are organised, you are analysing.

The first step is being able to break something down into its parts. For example, what caused Egypt to fall to the Roman Empire?

We can break the cause into four parts:

1 Rome had superior military forces
2 Rome needed the food resources that Egypt had
3 Egypt was undergoing a series of revolts at the time
4 Cleopatra's ambition.

Breaking down the cause is the first part of the analysis. Now we can try to explain how these combined causes would have led to Rome conquering Egypt. For example:

> Rome needed the food resources that Egypt had, and had a more powerful army. Because Rome needed something (food) and had the ability get it (the army), it was always likely that Rome would conquer Egypt and bring it into its empire.

Here we have linked Cause 1 and Cause 2 together.

Another way to analyse cause and effect is to look at how strong the causal link is. Causes can have different forces, as shown in this cause and effect scale:

Less	More
Small causal link	**Large causal link**
Gradual, minor, almost no part, short-lived, partly, partial, to some extent, small extent	Radical, powerfully, significant, important, to a great extent, considerable, main, crucial

Once you have decided how strong a cause was, here are some words you can use to describe it.

- If something had only a minor effect you could say:

  Indigenous lifestyles were only affected to a small extent by the introduction of dingos.

- If something had a major effect you could say:

  Indigenous lifestyles were significantly affected by European colonisation.

step 5 I can evaluate cause and effect

Evaluation is a 'higher-order thinking skill'. Evaluation can be:

- assessing whether a theory or belief is true or not; e.g. Some historians think climate change destroyed Harappan civilisation in the Indus River Valley. Is this true?
- comparing different ideas; e.g. Which is a more important effect of the Pyramids being built: improvement in engineering or modern-day tourism?
- judging between different things; e.g. Some historians think Greek civilisation caused Roman civilisation to be so dominant. Others think their dominance came mostly from within. Who is right?
- asking, 'So what?'; e.g. Were the causes or effects good or bad, or perhaps a bit of both? A historian may include both positives and negatives to form a balanced view.

Look at Step 5 in 'Historical significance' (page 222) for more tips on evaluating.

Source 9

Dingos were introduced into Australia about 4000 years ago.

Historical significance

How do we decide what is important to learn about from the past? How do we decide which events or time periods have historical significance?

We can use a model or theory to help us decide. A useful model is Geoffrey Partington's model of significance.

Partington's model states that you can determine historical significance by asking the following questions:

1 Importance: How important was it to people living at the time?
2 Depth: How deeply were people's lives affected?
3 Number: How many people were affected?
4 Time: For how long were they affected?
5 Relevance: How relevant is it to the present?

The table below shows some examples using Partington's model of significance.

I can recognise historical significance

Recognising historical significance means looking at a list of events or developments and deciding how important they are. (Significant means important; something worth noting.)

You should have some way of determining significance. You might ask: Was it important back then? Were people deeply affected? Did it affect a lot of people for a long time? Is it still relevant to modern times? The more you answer 'Yes' to these questions, the more significant the event was.

Historical significance is not a black and white issue, as it can be shown on a significance scale, like this one:

Least significant — Most significant

Name of Plato's mother | The date Caesar died | Invention of the compass | Farming developed

	Importance	Depth	Number	Time	Relevance
Introduction of the dingo	Quite important, as dingos were companions, protectors and hunters	Quite deep. They were more than pets, and had sacred status	A large proportion of Indigenous Australians, so about 400 000–700 000 people	For about the last 4000 years	Relevant to Indigenous Australians but less so to other people. They are more like pets or pests now
Invention of pyramids	Kind of important. A pyramid was the burial site of the pharaoh, who was seen as a god	Not that deeply. They were mostly for dead bodies	About 20 000–30 000 people built them	About 2500 years between being built and the end of Egypt as a major power	Not that relevant. They are mostly just tourist attractions now

I can explain historical significance

Explaining historical significance means asking *how* or *why* something is important.

Below are some examples of significant and less significant events, based on Partington's model.

Good explanations of historical significance will discuss more than one of these elements.

Partington's model of significance is just an aid to your thinking. There are other things that could explain whether a historical event was significant. For example, a person might be important if they changed other people's ideas, or provided a good or bad example of how to live. An event might be important if it reveals underlying themes or patterns in history.

	More significant	Less significant
Importance	The notion of 'citizenship' was very important historically in the Roman Empire. All people were affected by it, either by being granted citizenship privileges or by being outside its protection and status.	The exact nature of Roman religion is not that important in history because there were so many gods, and some were even borrowed from other civilisations.
Depth	The annual flooding of the Nile is important because many people relied on it for food.	The various lovers that the pharaoh had aren't a big deal historically because they didn't affect many people that deeply at the time.
Number	The colonisation of Australia by Europeans is significant historically because it affected all Indigenous people at the time – perhaps 750 000 people.	How many people arrived on the first fleet is less historically important because it was the arrival of Europeans, not necessarily the initial amount of people, which had the bigger impact.
Time	The invention of farming is very important because it has affected human life for the past 11 000 years.	The invention of drones can't be considered historically significant yet because they have only been around for a few years.
Relevance	The spread of Chinese influence across East Asia is significant because China is a powerful country today and has a lot of influence.	Traditional Chinese clothing is not that important historically because hardly anyone dresses like that anymore.

Source 10

A jacket from the Qing Dynasty. [Man or woman's jacket (Wedding or Theatrical) (18th century), silk, The Metropolitan Museum of Art.]

step 3 I can apply a theory of significance

Applying Partington's theory would mean looking at a set of events and ranking them against his categories. When applying his theory, use phrases such as those in the table below.

Importance	Issue A was more significant than Issue B because it was more important to people at the time.
Depth	Issue A is more important in history, because it affected people more deeply at the time than Issue B did.
Number	Issue A deserves the status of historical importance more than Issue B. Put simply, it affected more people.
Time	Issue B has been shown to be less important historically than Issue A, because it didn't last as long.
Relevance	Issue A is a more significant event than Issue B because it is still relevant to the present. It helps us to understand the modern world, whereas Issue B doesn't help as much.

step 4 I can analyse historical significance

Analysing means the ability to break down something into its parts. If you can identify these different parts, and explain how together they make up the whole, you are analysing. If you can explain the rules or theories that show how these parts are organised, you are analysing.

In the next paragraph, the significance of Greek philosophy is analysed. Three main points are made rather than just one, and an example of how each leads to significance is added. (See below the paragraph for a colour key.)

> Greek philosophy is significant because of several factors. First, it is long-lasting. The Ideas of Socrates, Plato and Aristotle were developed over 2000 years ago and are still discussed today. Second, it has affected many lives. Greek logic is the basis for computer programming, which affects billions of people today. Third, it is still very relevant today. Greek philosophy helps us understand morality, justice and which political system is best. Therefore, Greek philosophy is important because it is long-lasting and affects many people, even today.

Key:

Main point

Claim

Evidence backing up claim

Overall statement

Look at Step 4 in 'Cause and effect' (page 218) for more tips on analysing.

Source 11

Famous Greek philosophers [Raphael, detail from *The School of Athens* (1509–11), fresco, 500 × 700 cm, Apostolic Palace Vatican City.]

step 5

I can evaluate historical significance

Evaluating can mean asking, 'So what?' In terms of evaluating historical significance, evaluating could be:

- questioning Partington's model of significance:
  - Perhaps the model suggests that Event A is more important than Event B, but you don't agree. Evaluating would involve you explaining why you think the model of significance doesn't give the right result in this instance.

Questioning the model of significance	Some historians think the more people are affected by an event, the more important it was. There were more people under the Roman Empire than in Athens during its golden age, so we could assume that Roman history is more important. However, the heights that civilisation reached during the Golden Age of Athens were arguably greater than those of Rome. Its art, literature, philosophy and architecture were all superior. Perhaps historical importance needs to take the quality of the effect into account, not just how much of it there is (the quantity).
Judging the worth of an important event	Many people think Greek philosophy is significant, but there is more to it than that. Philosophy did not start with the ancient Greeks. They got their ideas from earlier thinkers, going all the way back to the Sumerians. Also, most of the ancient philosophy that is still talked about today has been changed over the centuries or reinterpreted so it makes more sense to modern readers. So, while it is *somewhat* important, Greek philosophy is just another chapter in the evolution of ideas.

  - Maybe you think some of the questions about significance are more important than others. Perhaps you think that relevance to today is more important than how important it was to people at the time.
- making a judgement call about the worth of important events:
  - Were these events 'good' or 'bad' in some way? When making an evaluation like this, make sure you define what 'good' or 'bad' mean in this context.

Look at Step 5 in 'Cause and effect' (page 218) for more tips on evaluating.

Source 12

A bronze statue of Caesar in Rome. [Statue of Julius Caesar (c.46 BCE), bronze, Roman Forum.]

Historical writing

Step 1: I can identify writing purpose

If you are given a writing task, it will usually involve certain 'task words', such as *analyse, argue* and *compare*. These task words are explained below.

- *Analyse*: look at the features of something, showing the relationships between the parts, how they're related and why they're important
- *Argue*: make a case for or against something
- *Compare*: discuss two things, emphasising what is the same and what is different between them
- *Contrast*: discuss two things, emphasising what is different between them
- *Describe*: write a detailed description of something, showing what something looks like, what it is for and how it works. Don't judge.
- *Discuss*: write about something, talking about the arguments for and against an issue. Provide a balanced description, but make a judgement at the end.
- *Evaluate*: make a judgment about something, but back it up with lots of evidence
- *Explain*: answer the question 'why?' about something. Go into detail about the reasons for it, causes of it and effects of it.
- *Justify*: provide reasons why a decision was or should be made, or why a conclusion was reached
- *Summarise*: briefly state the main points. Leave out the details.

After you know your purpose, figure out:

- *what kind of information you need to gather.* This relates to Stage 2 of the history-writing process: gathering information. Gather the right kind of information – but avoid gathering lots of irrelevant material.
- *how that information should be organised.* You will eventually need to write up any information you have gathered. How you do that – and what structure your writing takes – should be determined by the purpose of the writing.

Here is an example history writing question, and how it can be tackled: 'Discuss how important Julius Caesar was in the creation of the Roman Empire.'

This is a 'discuss' type question, so it is asking us to write about the topic, discussing arguments for and against it. Both sides should be discussed but a judgement should be made.

Information needed to answer the question:

- details about the political career of Julius Caesar (but *not* details about every aspect of his life)
- details about the creation of the Roman Empire. Gather information about the transition from the earlier form of Roman civilisation (the Roman Republic).

How should that information should be organised? A graphic organiser like the one shown below is a great way to structure your note-taking.

	General information	Evidence he helped create the Roman Empire	Evidence he did *not* help create the Roman Empire
Julius Caesar's political career	○ ○ ○	○ ○ ○	○ ○ ○
Information about the creation of the Roman Empire	○ ○ ○	○ ○ ○	○ ○ ○

I can gather information

Good history writing involves providing lots of evidence. The more *relevant* information you use, the better. Relevant means the information is closely connected to what is being studied.

Gathering information will involve taking notes from historical sources. Academic historians look at many primary and secondary sources. For most school projects, you are likely to rely on secondary sources. Secondary sources provide a wide range of information that is easily accessible to young people. Textbooks and reference books are easy to get, relatively cheap, easy to read and contain pictures, facts, explanations and examples.

Follow these steps when taking notes for your history writing:

1 Purpose: *why* am I taking notes?

2 Organise: use a graphic organiser or codes

3 Skim-read the source. This is so:
   - you can look for topics, headings etc.
   - you *don't* have to read the entire source.

4 Find the *most* important information *for your purpose*:
   - rewrite it in your own words
   - write as briefly as possible

   include keywords, and definitions of any words you don't know.

I can organise information

Here are two ways to organise your information: a graphic organiser or codes.

A graphic organiser is best for:

- when you know in advance the kind of information you will be taking notes about
- when the question you are answering has obvious parts to it that you can divide information into; e.g. for and against.

Using codes is a process that involves:

- taking notes
- reading through your notes several times and seeing what patterns, themes or categories emerge
- making up a code for each pattern, theme or category; e.g. 'W' for *war*; 'I' for *individual*; 'SP' for *Sparta*
- going through your notes and writing the code beside each point
- rewriting your notes in the code categories (This is much easier if you have taken notes electronically, because you can change their order without having to rewrite them.)
- using your notes in their coded categories to form the basis of your essay structure.

With either of these methods, don't forget to ask which notes you should *not* use. You will always take notes that you thought were important but later realise don't actually matter. Get rid of them. Remember: the final written piece is what is most important, not your notes. Don't worry that you spent time writing those notes in the first place, because only your final piece of writing matters. Next time, try to take fewer irrelevant notes.

Here is an example of the process, with all the steps from note-taking to organising your notes

Essay question: 'Was trade or conflict more influential in shaping Egyptian society?'

1 Original notes

- copied military technology from others
- exported grain, papyrus, flax, cloth
- had large army of charioteers
- had natural protection: desert and sea
- imported luxury items: silver, spices, jewels
- launched attacks on neighbours
- sometimes behind in military technology
- self-sufficient
- no permanent army

2 Put into a graphic organiser, the notes look like this:

	Influential	Not influential
	Yes, it was quite influential	No, it wasn't that influential
Trade	○ exported grain, papyrus, flax, cloth ○ imported luxury items: silver, spices, jewels	○ quite self-sufficient
Conflict	○ launched attacks on neighbours	○ no permanent army ○ had natural protection: desert and sea